Crop Breeding and Production

Crop Breeding and Production

Edited by Kiara Woods

SYRAWOOD
PUBLISHING HOUSE

New York

Published by Syrawood Publishing House,
750 Third Avenue, 9th Floor,
New York, NY 10017, USA
www.syrawoodpublishinghouse.com

Crop Breeding and Production
Edited by Kiara Woods

© 2018 Syrawood Publishing House

International Standard Book Number: 978-1-68286-568-2 (Hardback)

Cataloging-in-Publication Data

Crop breeding and production / edited by Kiara Woods.
    p. cm.
Includes bibliographical references and index.
ISBN 978-1-68286-568-2
1. Plant breeding. 2. Crops. 3. Agriculture. 4. Agronomy. I. Woods, Kiara.
SB123 .C76 2018
631.53--dc23

TABLE OF CONTENTS

PREFACE

This book has been a concerted effort by a group of academicians, researchers and scientists, who have contributed their research works for the realization of the book. This book has materialized in the wake of emerging advancements and innovations in this field. Therefore, the need of the hour was to compile all the required researches and disseminate the knowledge to a broad spectrum of people comprising of students, researchers and specialists of the field.

Modern crop breeding includes techniques to produce crop hybrids that exhibit better growth and disease resistance. This field takes the help of molecular biology or plant biotechnology to genetically modify crop traits. Crop breeding increases yield as well as improves nutritional value of a crop. Adaptability to harsh climatic conditions as well as tolerance against natural calamities such as droughts are another aspect of this field of study. Most of the topics introduced in this book cover new techniques and the applications of crop breeding. Coherent flow of topics, student-friendly language and extensive use of examples make this book an invaluable source of knowledge.

At the end of the preface, I would like to thank the authors for their brilliant chapters and the publisher for guiding us all-through the making of the book till its final stage. Also, I would like to thank my family for providing the support and encouragement throughout my academic career and research projects.

Editor

Comparison of Removal Efficiencies of Different Household Cleaning Methods in Reducing Imidacloprid and Triabendazole Residues on Cherry Tomatoes

Zhi-Yuan Meng[1,2], Yue-Yi Song[1,2], Xiao-Jun Chen[1,2], Ya-Jun Ren[1,2], Chun-Liang Lu[2], Li Ren[1,2], Hua-Chen Gen[1,2], Jia-Xin Zhu[1,2], Quan Yuan[1,2], Teng-Fei Li [1,2] & Zhi-Ying Xu[3]

[1] School of Horticulture and Plant Protection, Yangzhou University, Yangzhou, Jiangsu, P.R. China

[2] Joint International Research Laboratory of Agriculture & Agri-Product Safety, Yangzhou University, Yangzhou, Jiangsu, P.R. China

[3] Yangzhou Polytechnic College, Yangzhou, Jiangsu, P.R. China

Correspondence: Xiao-Jun Chen, School of Horticulture and Plant Protection, Yangzhou University, Yangzhou, Jiangsu, P.R. China. E-mail: cxj@yzu.edu.cn

Zhi-Ying Xu, Yangzhou Polytechnic College, Yangzhou, Jiangsu 225000, P.R. China. E-mail: yzxuzhiying@163.com

Abstract

People have paid much attention on the pesticide residues in agricultural products at present. However there are less concern on processed agricultural products or on final consumption of processed foods, even though most processed foods are being finally consumed. In this paper, pesticide residues such as imidacloprid and triabendazole on the cherry tomatoes were cleaned by conducting different household cleaning methods in the way of normal vegetable cleaning. Results showed that it can effectively remove the residual imidacloprid on cherry tomatoes by soaking first them in water and then rinsing them with running tap water, wherein the removal rates were 30.59%-65.24% and processing factor were 0.3476-0.6941. While to remove the residual triabendazole on cherry tomatoes, we first soaked them in 0.1% edible vinegar solution and then rinsed them with running tap water, which can also effectively remove the triabendazole residues on the cherry tomatoes, with removal rates reaching 29.31%-74.01% and processing factor reaching 0.2599-0.7069. Our research provides an inherent relationship between pesticide residues and cleaning approaches as well as important theoretical basis for risk assessments of agricultural food.

Keywords: cherry tomatoes, imidacloprid, triabendazole, removal efficiency

1. Introduction

Compared with ordinary tomato or vegetable, cherry tomato is rich in nutrients such as organic acids, sugar, vitamins and minerals vitamin C (Li et al., 2012). Cherry tomato is a type of leguminous vegetable, which is being daily consumed in China. Cherry tomato is the host of thrips, pod borer, cutworm, aphids, leaf miner and other major pests, therefore chemical pesticides will be inevitably applied during its growth process. As a kind of neonicotinoid insecticide, imidacloprid enjoys advantages including high efficiency, broad spectrum against insects such as *Thysanoptera Homoptera, Coleoptera, Diptera* and *Lepidoptera* and other insects, and good uptake characteristics in protecting plants. Moreover, imidacloprid is of excellent effect in controlling sucking pests (Zhang et al., 2009; Liu, 2012). Belonging to benzimidazole fungicides thiabendazole has high efficiency, low toxicity, broad-spectrum and good uptake characteristics in protecting plants and plays as fresh-keeping sterilizing agent in the process of fruit storage (Zhao et al., 2009; Li et al., 2011). The application of imidacloprid and thiabendazole in cherry tomatoes production will inevitably cause pesticide residues, which may threat consumers' health.

Currently, a strict regulation has been implemented regarding pesticide residues on agricultural products such as fruits, crops and vegetable. Although some pesticide residues still remain at harvest, the residues tend to decline with the decomposition of pesticide. Great attention has been paid on the monitoring of pesticide residues, however insufficient emphasis has been placed on its cleaning process or on the processed agricultural products today. Most agricultural products are eaten after being cleaned. Therefore, it is very necessary to monitor these novel pesticides (imidacloprid and triabendazole) before and after cleaning. The two pesticides may degrade during food processing or storage processes, and some degradation products may be even more toxic than their parent compounds (St-Amand et al., 2004; Luís et al., 2005; Muhammad et al., 2006; Chai et al., 2009). Therefore, it is in great need to seek an appropriate cleaning method to remove the residual pesticides on agricultural products for both household and factory processing, so as to reduce the potential threat to human health. A systematic study of removal efficiencies of different cleaning methods is helpful for establishing guidelines for food cleaning and handling, with the purpose of minimizing human exposure to pesticide residues on food. Furthermore, it is a critical step for people to understand the removal efficiencies of different household cleaning methods in reducing imidacloprid and triabendazole residues on cherry tomatoes, because this is the basis for characterizing degradation products in the cherry tomatoes during processing, and for revealing the inherent relationship between pesticide residues and processing approaches, so that the theoretical basis for risk assessments of food can be reached.

2. Materials and Methods

2.1 Chemicals and Materials

Dr. Ehrenstorfer GmbH provided the technical grade analytical standard of imidacloprid (purity 98.0%) and triabendazole (purity 98.5%). 70% imidacloprid water dispersible granule (70% imidacloprid WDG) was obtained from Shanghai Dupont Agricultural Chemicals Co., Ltd, and 15% triabendazole suspension concentrate (15% triabendazole SC) was obtained from Jiangsu Bailing Agrochemical Co., Ltd. N-propyl ethylene diamine (PSA) and C18 were obtained from DIKMA Technology Co., Ltd. (USA). Edible vinegar, edible iodine salt, edible sodium bicarbonate, fruits and vegetables cleaning agent were obtained from Zhenjiang Hengguan Industry Co., Ltd., Jiangsu Guhuai Salt Co., Ltd. and Golden Bridge Salt Group, Nanjing Honeydew Park Sugar Co., Ltd., and Xi'an Kaimi Corporation, respectively.

2.2 Preparation of Imidacloprid and Triabendazole Residues on Cherry Tomatoes

The crisp, full pod, non-damage and non-mildewed cherry tomatoes, which were grown in Yangzhou University Experimental Farms, were selected as experimental materials. The upper and lower ends of the cherry tomatoes were removed. A 100 mg kg^{-1} soaking solutions of imidacloprid or triabendazole (70% imidacloprid WDG or 15% triabendazole SC) were prepared in deionized water, respectively. Totally 10 kg of cherry tomatoes were fully immersed in 60 L of the two soaking solutions respectively for 30 min at room temperature, and then dried in open air for 1 h.

2.3 Cleaning Methods

In treatment one (T_1), all cherry tomatoes were cleaned using a sieve with running tap water for 2 min, so that it can guarantee the cherry tomatoes were uniformly cleaned. After that the cherry tomatoes samples were air-dried in a fume hood at room temperature. In the second treatment (T_2), all cherry tomatoes remained uncleaned. Three replicates were applied for each treatment.

We investigated the soaking effects when using different solutions such water, 0.1% edible vinegar, 0.1% edible salt solution, 0.1% edible sodium bicarbonate solution and 0.1% fruit and vegetable cleaning solution. By soaking cherry tomatoes (1 kg) in 4 L of water, 0.1% water, 0.1% edible vinegar solution, 0.1% edible salt solution, 0.1% edible sodium bicarbonate solution, and 0.1% fruit and vegetable cleaning solution, for 5, 10, 15, 30, 45 and 60 min, respectively. Cherry tomatoes in control group remained uncleaned. After that the cherry tomatoes were rinsed with running tap water for 2 min for three times, and then air-dried in a fume hood. Imidacloprid or triabendazole residues on cherry tomatoes were extracted, cleaned-up and analyzed using HPLC. Removal efficiency and processing factor of imidacloprid and triabendazole were calculated respectively according to the contents of their residues on cherry tomatoes.

2.4 Extraction of Imidacloprid and Triabendazole Residues from Cherry Tomatoes

Cherry tomatoes (5 g) were homogenized in 30 mL of acetonitrile for 3 min using high-speed homogenizer. For recovery experiment, imidacloprid and triabendazole were added to the cherry tomatoes samples, respectively, and the final concentration of imidacloprid was 5.00, 1.00, and 0.20 $\mu g\ g^{-1}$, so was that of triabendazole. In addition, a blank control group was set up and all experiments were repeated three times.

Imidacloprid and triabendazole were extracted from cherry tomatoes using QuEChERS method after applied with different cleaning treatments. Sodium chloride (1.5 g) and anhydrous magnesium sulfate (6 g) were added to the samples before conducting centrifuging treatment at 4000 rpm for 5 min. After that, it can collect about 2 mL of supernatant, which was then added into the centrifuge tubes containing MgSO$_4$ (150 mg), PSA (25 mg), and C$_{18}$ (25 mg) for 5 min of centrifuging. Finally, the sample solution was filtered using a 0.22 μm membrane and analyzed by HPLC.

2.5 Instrumental Analysis

Treatments of separation and detection were carried out using L-2000 Series HPLC (Hitachi Co., Japan) which was equipped with UV detector. The imidacloprid residues were quantitatively determinded by HPLC as conducted by Lu and Wang (Wang et al., 2012; Lu et al., 2013). It is worth noting that the L-2000 HPLC system used in the experiment was equipped with a binary pump, auto plate-sampler, column oven, and UV detector. The treatment of separation was performed using ODS chromatographic columns C$_{18}$ (250 mm× 4.6 mm (i.d.), 5 μm) under the conditions of 25 °C, wavelength 270 nm, mobile solvents CH$_3$CN:H$_2$O = 30:70 (V:V) and isocratic at 1.0 mL min^{-1}. 10 μL of aliquot was injected directly into the HPLC system to test imidacloprid and quantified with external standard peak area. The triabendazole residues were quantitatively determined using HPLC by Wang and Zhao (Wang et al., 2008; Zhao et al., 2009). The treatment of separation was conducted using ODS chromatographic columns C$_{18}$ (250 mm× 4.6 mm (i.d.), 5 μm) under the conditions of 25 °C, wavelength 298 nm, mobile solvents CH$_3$OH:H$_2$O = 60:40 (V:V) and isocratic at 1.0 mL min^{-1}.

2.6 Calculation Methods for Removal Rate and Processing Factor

Removal rate and processing factor of each treatment were calculated. The removal rate and processing factor can be calculated by formulas as below, respectively.

$$\text{Removal rate(\%)} = \frac{\text{Initial concentration - Residues concentration}}{\text{Initial concentration}} \times 100 \tag{1}$$

$$\text{Processing factor} = \frac{\text{Residues concentration}}{\text{Initial concentration}} \tag{2}$$

All data were analyzed through analysis of Duncan multiple comparison. All the experiments were repeated three as means±standard error of mean. Different lowercase letters after the number was of 5% significant difference, while different uppercase letters after the number was of 1% significant difference.

3. Results

3.1 Fortified Recoveries of Imidacloprid and Triabendazole from the Cherry Tomatoes

To figure out the extrication efficiency and clean-up procedures, recovery experiments were conducted at different fortification levels to verify the reliability and validity of analytical method. The control samples of cherry tomatoes were spiked at 5, 1.0, and 0.2 mg kg^{-1} respectively followed by methodology as described above. According to Table 1, it can be seen that the average recovery rate was over 80%. In addition, the coefficients of variation were changed from 1.89% to 4.08% and changed from 1.89% to 4.35%, respectively.

Table 1. The recoveries of imidacloprid and triabendazole from the cherry tomatoes

Pesticides	Fortification levels (mg kg^{-1})	Average recoveries (%)	Standard deviation	Coefficient of variance (%)
imidacloprid	5.0	92.72	0.0878	1.89
	1.0	89.52	0.0358	4.00
	0.2	82.03	0.0067	4.08
triabendazole	5.0	93.71	0.0884	1.89
	1.0	88.67	0.0226	2.55
	0.2	84.20	0.0073	4.35

Note. The remaining amount was the average of three replicates.

3.2 Removal Efficiencies of Different Cleaning Methods in Removing the Residual Pesticide on Cherry Tomatoes

3.2.1 Removal Efficiency of the Cleaning Method Using Running Tap Water

According to Table 2 that the removal rate and the processing factor were 49.37% and 0.5043, respectively, after the cherry tomatoes with imidacloprid residue were cleaned with running tap water for 5 min. In constrast, the removal rate and the processing factor for cherry tomatoes with triabendazole residues were 33.93% and 0.6607, respectively after cleaning with simply running tap water for 5 min.

Table 2. Removal efficiencies of residual imidacloprid and triabendazole on cherry tomatoes by cleaning with running tap water

Samples	Treatment method	Initial concentration (mg kg^{-1})	Residues concentration (mg/kg)	Removal rate (%)	Processing factor
Imidacloprid	Tap water 5 min	4.473±0.4435	2.2648±0.0892	49.37±1.9953	0.5063±0.0199
Triabendazole	Tap water 5 min	4.3640±0.2564	2.8833±0.4289	33.93±9.8271	0.6607±0.0983

Note. The experiments were set up in a completely randomized design. All the experiments were repeated thrice as means ± standard error of mean.

3.2.2 Removal Effects of Residual Imidacloprid and Triabendazole on Cherry Tomatoes by First Soaking in Tap Water and Then Cleaned with Running Tap Water

According to Figure 1, it can be seen that the removal rates were 30.59%-65.24% and the processing factors were 0.3476-0.6941 after the cherry tomatoes with imidacloprid residues were first soaked in the water for several times and then cleaned with running tap water for 2 min. In contrast, the removal rates reach 19.13%-60.73% and the processing factors reach 0.3927-0.8087 after the cherry tomatoes with triabendazole residue were first soaked in the water for several times and then cleaned with running tap water for 2 min.

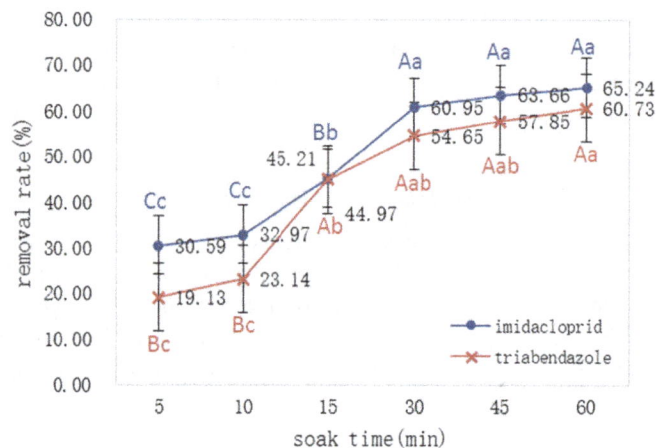

Figure 1. Removal efficiencies of residual imidacloprid and triabendazole on the cherry tomatoes by first soaking in tap water and then cleaned with running tap water

3.2.3 Removal Efficiency of Residual Imidacloprid and Triabendazole on Cherry Tomatoes by First Soaking in Edible Salt Solution and Then Cleaned with Running Tap Water

According to Figure 2, it can be seen that the removal rate were 23.61%-50.76% and the processing factors were 0.4924-0.7639 after the cherry tomatoes with imidacloprid residues were first soaked in the 0.1% edible salt solution for different times and then cleaned with running tap water for 2 min. In constrat the removal rates were 22.62%-63.95% and the processing factors were 0.3605-0.7738 after cherry tomatoes with triabendazole residues were first soaked in the 0.1% edible salt solution for different times and then cleaned with running tap water for 2 min.

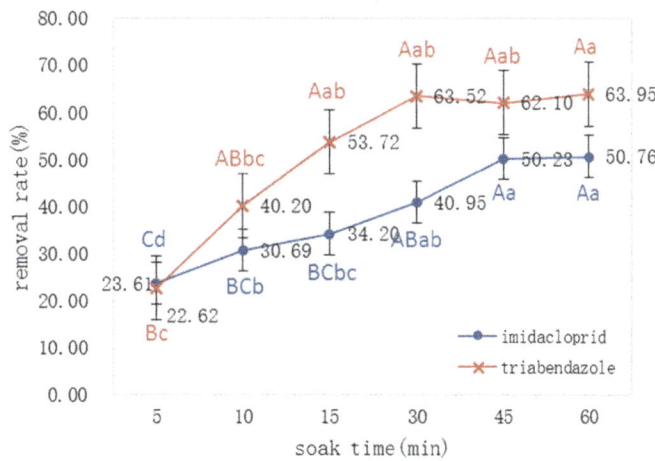

Figure 2. Removal efficiencies of residual imidacloprid or triabendazole on cherry tomatoes by first soaking in the salt solution for different times and then cleaned with running tap water

3.2.4 Removal Efficiency of Residual Imidacloprid or Triabendazole on the Cherry Tomatoes by First Soaking in the Edible Vinegar Solution and Then Cleaning with Running Tap Water

According to Figure 3, it can be seen that the removal rate were 19.97%-64.79% and the processing factors were 0.3521-0.8003 after the cherry tomatoes with imidacloprid residues were first soaked in the 0.1% edible vinegar solution for different times and then cleaned with running tap water for 2 min. In contrast, the removal rates were 29.21%-74.01% and the processing factors were 0.2599-0.7069 after the cherry tomatoes with triabendazole residues were first soaked in the 0.1% edible vinegar solution for different times and then cleaned with running tap water for 2 min.

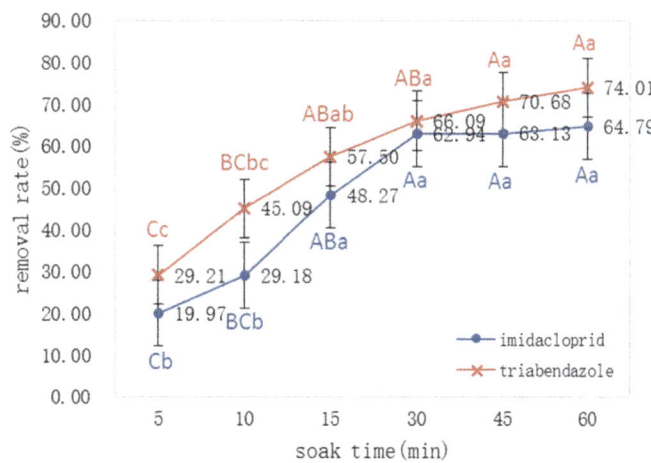

Figure 3. Removal efficiencies of residual imidacloprid and triabendazole on cherry tomatoes by first soaking in the 0.1% edible vinegar solutions for different times and then cleaned with running tap water

3.2.5 Removal Efficiency of Residual Imidacloprid and Triabendazole on the Cherry Tomatoes by First Soaking in the Edible Sodium Bicarbonate Solution and Then Cleaned with Running Tap Water

As shown in Figure 4, the removal rate were 15.42%-61.04% and the processing factors were 0.3896-0.8458 after the cherry tomatoes with imidacloprid residues were first soaked in the 0.1% edible sodium bicarbonate solution for different times and then cleaned with running tap water for 2 min. In contrast, the removal rates were 21.81%-59.55% and the processing factors were 0.4045-0.7819 after the cherry tomatoes with triabendazole residue were first soaked in the 0.1% edible sodium bicarbonate solution for different times and then cleaned with running tap water for 2 min.

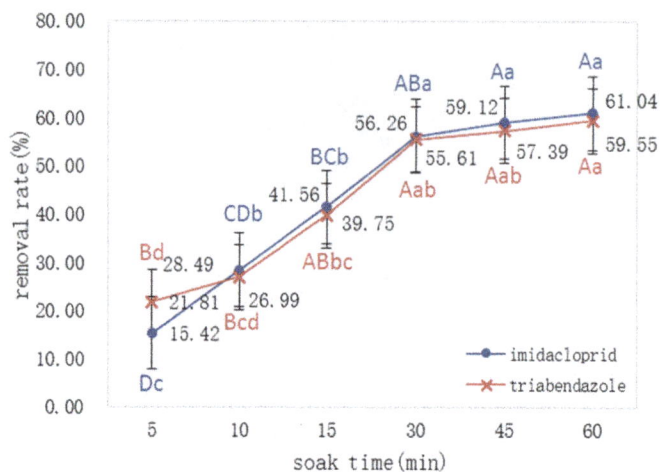

Figure 4. Removal efficiencies of residual imidacloprid or triabendazole on cherry tomatoes by first soaking in the 0.1% sodium bicarbonate solution for different times and then cleaned with running tap water

3.2.6 Removal Efficiencies of Residual Imidacloprid or Triabendazole on Cherry Tomatoes by First Soaking in the Fruit and Vegetable Cleaning Solution and Then Cleaned with Running Tap Water

As shown in Figure 5, the removal rate were 17.67%-56.26% and the processing factors were 0.4376-0.8232 after the cherry tomatoes with imidacloprid residues were first soaked in the 0.1% fruit and vegetable cleaning solution for different times and then cleaned with running tap water for 2 min. In contrast, the removal rates were 22.83%-57.81% and the processing factors were 0.4219-0.7717 after the cherry tomatoes with triabendazole residues were first soaked in the 0.1% fruit and vegetable cleaning solution for different times and then cleaned with running tap water for 2 min.

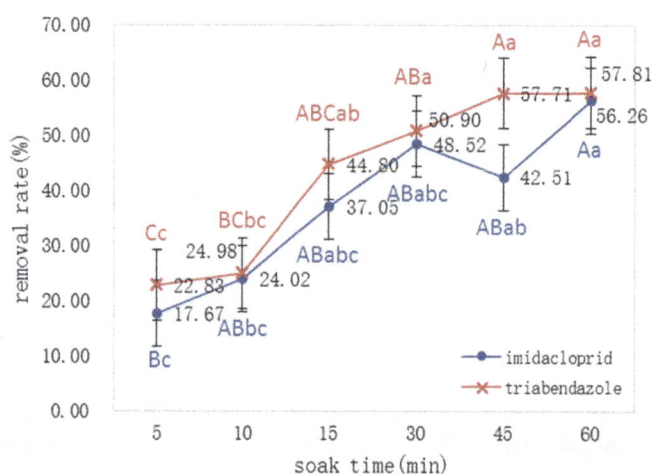

Figure 5. Removal efficiencies of residual imidacloprid or triabendazole on cherry tomatoes by first soaking in the 0.1% fruit and vegetable cleaning solution for different times and then cleaned with running tap water

As shown in Figure 6, it can conclude that cleaning method of first soaking in the water and cleaning with running tap water is the preferred method. After being soaked in water for 30 min, the repeated times for soaking will not affect the removal efficiency of residual imidacloprid. Imidacloprid enjoys good solubility in water which is 0.61 g/L and KowlogP = 0.57 with low lipid water distribution coefficient (Jeschke et al., 2011; Liu, 2012). The residual imidacloprid may easily enter into the water when the cherry tomatoes are soaked in the water.

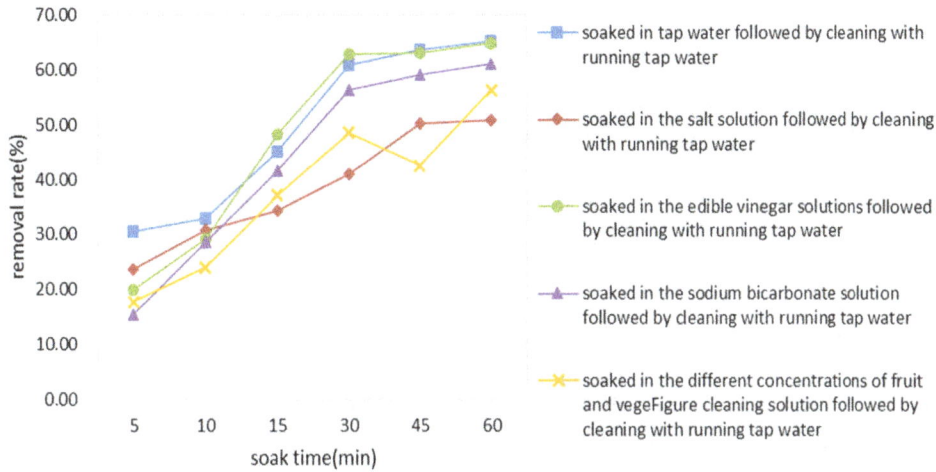

Figure 6. Removal efficiency of residual imidacloprid on cherry tomatoes by soaking in different solutions for different times

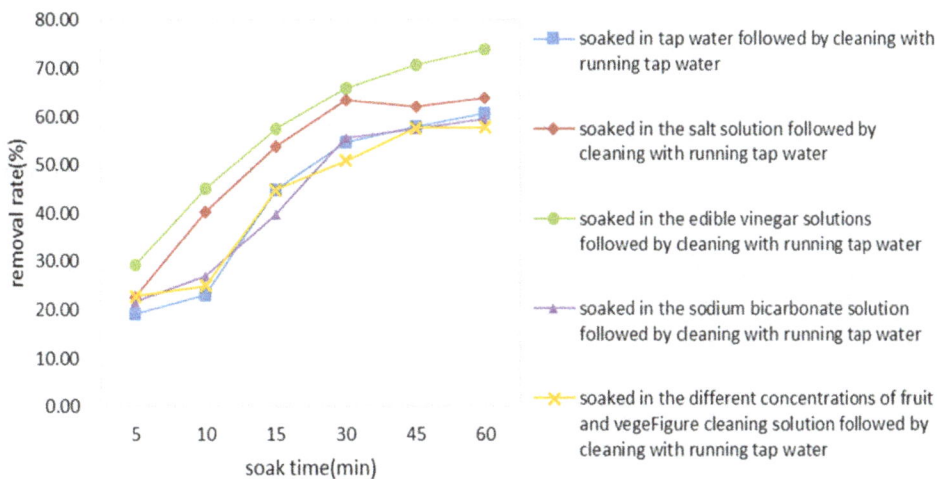

Figure 7. Removal efficiency of residual triabendazole on cherry tomatoes by soaking in different solutions for different times

As shown in Figure 7, it can conclude that the cleaning method of first soaking in the 0.1% edible vinegar and then cleaned with running tap water was the preferred method. After the residue triabendazole cherry tomatoes being soaked in the 0.1% edible vinegar for 30 min, the repeated times for soaking will not affect the removal efficiency of residue triabendazole on cherry tomatoes. Triabendazole enjoys good solubility in the water which is 0.05 g/L and KowlogP = 2.39 (pH = 7) with high lipid water distribution coefficient (Liu, 2012). The residual triabendazole on cherry tomatoes may easily enter into 0.1% edible vinegar when being soaked in 0.1% edible vinegar.

4. Conclusions

In this study, pesticide residues such as imidacloprid and triabendazole on the cherry tomatoes were cleaned using different cleaning techniques in the way of vegetable cleaning habits. The cleaning methods included rinsing with running water, cleaning after soaking in water, edible vinegar, edible salt, sodium bicarbonate solution or fruit and vegetable cleaning solution, respectively. It concluded that using the method of first soaking in water and then rinsing with running tap water, it could remove the imidacloprid residues on the cherry tomatoes most effectively, with removal rates reaching 30.59%-65.41% and processing factor reaching 0.3476-0.6941. In contrast, based on the method of first soaking in 0.1% edible vinegar solution and then rinsing with running tap water, it can remove the triabendazole residues on the cherry tomatoes most effectively, with removal rates reaching 29.31%-74.01% and processing factor reaching 0.3521-0.7069. Our research provides the inherent relationship between pesticide residues and cleaning approaches, as well as the important theoretical basis for risk assessments of agricultural food.

Acknowledgements

We gratefully acknowledge the financial support received from the Science and Technology Project of Jiangsu Province (BK20130443), Key Research and Development Program of Jiangsu Province (BE2015354), Key Research and Development Program of Yangzhou City (YZ2015029), Practicality and Innovation Training Project for Graduate Students in Jiangsu Province (KYLX15_1373) and Innovative Practice Programs of Yangzhou University for Undergraduates (x2015656, x2015662).

References

Chai, L. K., Mohd-Tahir, N., & Hansen, H. C. B. (2008). Determination of chlorpyrifos and acephate in tropical soils and application in dissipation studies. *International Journal of Environmental and Analytical Chemistry, 88*, 549-560. http://dx.doi.org/10.1080/03067310802002508

Jeschke, P., Nauen, R., Schindler, M., & Elbert, A. (2011). Overview of the status and global strategy for neonicotinoids. *Journal of Agricultural and Food Chemistry, 59*(7), 2897-2908. http://dx.doi.org/10.1021/jf101303g

Li, H. F., Nie, J. Y., Li, J., Xu, G. F., & Wu, Y. L. (2011). Study on the degradation trends of thiabendazole in orange,banana and mango during storage. *Journal of Yunnan Agricultural University, 26*(3), 340-344.

Li, L. J., & Wu, R.Y. (2012). Optimization of extraction process for lycopene from cherry tomatoes. *Food Science, 33*(22), 158-161.

Liu, C. L. (2012). *World pesticide: Pesticide Volume*. Beijing: Chemical Industry Press.

Lu, H. Y., Li, B., Chen, Z. Z., Gao, X., Lin, X. R., Wei, C. T., & Zhang, Y. Y. (2013). Determination of imidacloprid and acetamiprid in green tea infusion by high-performance liquid chromatography with dispersive solid phase extraction. *Food Science, 34*(20), 203-206.

Luís, P. D. M. P., Lourival, C. P., Luiz, L. F., Luiz, R., & Pimentel, T. (2005). Kinetics of carbosulfan hydrolysis to carbofuran and the subsequent degradation of this last compound in irrigated rice fields. *Chemosphere, 60*, 149-156. http://dx.doi.org/10.1016/j.chemosphere.2005.02.049

Muhammad, I. T., Shahzad, A., & Ishtiaq, H. (2006). Degradation and persistence of cotton pesticides in sandy loam soils from Punjab, Pakistan. *Environment Research, 100*, 184-196. http://dx.doi.org/10.1016/j.envres.2005.05.002

St-Amand, A. D., & Girard, L. (2004). Determination of acephate and its degradation product methamidophos in soil and water by solid-phase extraction (SPE) and GC-MS. *International Journal of Environmental and Analytical Chemistry, 84*, 739-748. http://dx.doi.org/10.1080/03067310410001729600

Wang, D. F., Chen, L. H., You, J., Huang, Z. H., Zheng, J. C., & Wang, R. L. (2008). Simultaneous determination of thiabendazole and carbendazim residues in fruits and vegetables by solid phase extraction and HPLC. *Agrochemicals, 47*(6), 443-447.

Wang, X. H., Wan, X. C., & Hou, R. Y. (2012). A method for determination of imidacloprid residue in tea with HPLC-UV. *Journal of Tea Science, 32*(3), 203-209.

Zhang, M. F., Fan, J. Y., Zhang, H. W., Zhang, X. Z., & Ma, X. G. (2009). Research development of studies on neonicotinoid insecticides.*World Pesticide, 31*(1), 22-26.

Zhao, L. J., Yue, H., Guo, D. L., Cui, J., & Zhao, S. C. (2009). Ultra-performance liquid chromatography-tandem mass spectrometry for the determination of benzimidazoles residues in apples. *Food Science, 30*(12), 209-212.

Response of Maize (*Zea mays* L.) Secondary Growth Parameters to Conservation Agriculture and Conventional Tillage Systems in Zimbabwe

Regina Hlatywayo[1], Blessing Mhlanga[1], Upenyu Mazarura[1], Walter Mupangwa[2] & Christian Thierfelder[2]

[1] Department of Crop Science, Faculty of Agriculture, University of Zimbabwe, Mount Pleasant, Harare, Zimbabwe

[2] International Maize & Wheat Improvement Centre (CIMMYT), Mount Pleasant, Harare, Zimbabwe

Correspondence: Regina Hlatywayo, Department of Crop Science, Faculty of Agriculture, University of Zimbabwe, P.O. Box MP 167, Mount Pleasant, Harare, Zimbabwe. E-mail: regina.hlatywayo@gmail.com

Abstract

Previous, research focused mainly on the effects of conservation agriculture (CA) and conventional practices (CP) on crop yield mostly. A study was conducted at five sites in Zimbabwe from 2012 to 2014 to investigate effects of CA and CP practices on emergence, chlorophyll content, early vigour and grain yield of different maize varieties using 12 hybrids and 4 open pollinated varieties (OPVs). The experiment was laid as a 4×4 alpha lattice design with three replications. Emergence was higher under CA (75%) at University of Zimbabwe (UZ) in 2012/13 and Domboshawa Training Centre (DTC) (67%) compared to CP (71% and 39% respectively). Lower early plant vigour was observed under CA compared to CP at most sites. CA had lower leaf chlorophyll content during the early crop growth stages compared to CP. However, at some instances, CA had higher leaf chlorophyll content (45 units) than CP (35 units) at 78 days after sowing in Zimuto 2012/13. For maize yield, CA outperformed CP on a sandy loamy soil (3050 kg ha^{-1} vs 2656 kg ha^{-1}) and clay soil (4937 kg ha^{-1} vs 4274 kg ha^{-1}). However, on a sandy soil, CP outperformed CA (1764 vs 1313 kg ha^{-1}). Our results suggest that tillage effects on early maize plant vigor, leaf chlorophyll content and the final yield can be site and season specific. Furthermore, a delay of nutrient release for plant uptake under CA systems was found and potentially implies investigations of new fertilization strategies for such cropping systems.

Keywords: chlorophyll, cropping systems, early vigour, grain yield, nitrogen content

1. Introduction

Agriculture systems based on tillage and removal/or burning of crop residues pose a threat to food security through accelerated soil degradation (Thierfelder, Mutenje, Mujeyi, & Mupangwa, 2014). Maize yields are reportedly declining to the highest average yield levels of 1.1 tons ha^{-1} in the past decade (Thierfelder & Wall, 2009). The need to implement more sustainable ways of farming in the smallholder sector is therefore increasingly important. Several ways present potential solutions in addressing food insecurity and soil degradation, and this range from new and adapted maize varieties to good agronomic practices (Cairns, Sanchez, Vargas, Ordoñez, & Araus, 2012). Conservation agriculture (CA), based on minimum soil disturbance, permanent soil organic cover, and the use of diverse crop rotations/associations, has the potential for addressing the current food insecurity and soil degradation on smallholder farming systems (Thierfelder et al., 2014). Compared with conventional plough- or hoe-based cultivation practices (CP), CA practices have been shown to increase and stabilize maize yields in studies conducted in different parts of the world (Govaerts, 2009; Thierfelder et al., 2014; Mhlanga, Cheesman, Maasdorp, Mupangwa, & Thierfelder, 2015). This increase in crop yields under CA could be attributed to increased soil moisture induced by residue retention, increased faunal abundance achieved by reduced disturbance of the soil and residue retention, and increased nitrogen (N) fixation from crop rotations with legumes (Mhlanga et al., 2015; Mutema, Mafongoya, Nyagumbo, & Chikukura, 2013; Thierfelder & Wall, 2009). However, the smallholder farmers still regard CP as an easier farming method based on how easily weeds are controlled at the beginning of the season through ploughing which buries the weeds

(Derksen, Lafond, Thomas, Loeppky, & Swanton, 1993). Furthermore, CP improves soil aeration but this benefit is outweighed by the detrimental effects it has on the environment, leading to yield decreases in the longer term (Reicosky, Sauer, & Hatfield, 2011). Moreover, CA in some instances reduces yield through waterlogging, reduced soil temperatures and N lockup (Chikowo, Mapfumo, Nyamugafata, Nyamadzawo, & Giller, 2003).

The growth and development of maize is affected by management practices such as soil fertility and weed management. Differences in crop performance are likely to occur between CA and CP as they involve different management strategies. Altuntas and Dede (2009) observed an increased maize crop emergence under CA compared to CP. On the contrary, Hayhoe, Dwyer, Stewart, White, and Culley (1996) found a slow and uneven maize crop emergence under CA compared to CP and this was directly linked to low temperatures under CA systems which are generally triggered by residue retention. Burgess, Mehuys, and Madramootoo (1996) pointed out possible effects of allelopathy of the maize residues usually retained under CA systems to the emerging maize seed. Cairns et al. (2012) investigated *in-vivo* chlorophyll content in maize and found a strong and positive relationship between leaf chlorophyll content and grain yield. The relationship between the two parameters was stronger especially towards and during maize flowering in the same study. However, because of different soil conditions under CA compared to CP *i.e.* the temporal unavailability of N through immobilization in maize stover mulched soils, photosynthesis levels are affected especially during the early stages of crop growth (Verhulst et al., 2011a). The temporal unavailability of N is caused by the increased levels of immobilization which is triggered by the high C: N ratio of mulching material such as maize stover (Wortman, 2006). Furthermore, the temporal unavailability of N has an effect on early vigour of the maize crop as growth and vigour are dependent on photosynthesis (Lundy, Pittelkow, Linquist, Liang, Van Groenigen, Lee, & Van Kessel, 2015). Low soil temperatures under CA systems affect germination, vigour and growth of crops (Duiker & Haldeman, 2006). In a study carried out by Verhulst et al. (2011b), the initial maize growth was slower under CA compared with CP although this did not translate into significant yield differences at harvest. The slower initial maize growth under CA may be due to the slower mineralization of N into the soil that could be delayed by up to three weeks (Chikowo et al., 2003).

On the other hand, soil tillage increases mineralization of soil organic matter (SOM) because the soil environment becomes well aerated (Barbera, Poma, Gristina, Novara, & Egli, 2012). However, increased mineralization leads into high potential loss of carbon (C) and nitrogen (N) from the soil through erosion, leaching and mining by plants (Barbera et al., 2012). As a result, maize under CP can have increased access to soil nutrients leading to high early plant vigour and a higher rate of growth compared with maize under CA (Chikowo et al., 2010). This creates differences in growth of plants under CA and CP cropping systems. Information on effects of tillage systems on secondary growth parameters in maize is still limited. The aim of this study was to investigate the effects of different management systems (*i.e.* CA and CP) on maize secondary traits such as early vigour, leaf chlorophyll content, and grain yield. The hypothesis of the study was that different management systems affect growth and yield of maize (on above mentioned parameters). This was tested on a set of locally produced maize varieties under different environments in Zimbabwe.

2. Materials and Methods

2.1 Study Description

The experiment was established in the 2009-10 season with maize varieties evaluated under CA tillage system only. For comparison of tillage systems, CP tillage system was introduced in 2011-12 with maize varieties grown in both systems. This current study however, presents data from the 2012-13 and 2013-14 cropping seasons. The study was carried out at five experimental locations: University of Zimbabwe (UZ) (17.73°S; 31.04°E and 1483 meters above sea level (m.a.s.l), Domboshawa Training Centre (DTC) (17.62°S; 31.17°E and 1500 m.a.s.l), Madziva (17.00°S; 31.43°E and 1169 m.a.s.l), Hereford (17.42°S; 31.44°E and 1054 m.a.s.l), and Zimuto (19.85°S; 30.88°E and 1223 m.a.s.l). The soils at UZ are characterized by a high clay content of more than 40% and are classified as *Chromic luvisols* (Nyamapfene, 1991). Domboshawa Training Centre has soils classified as *Gleyic luvisols* and has a 5% clay content generated from granitic parent material (Thierfelder & Wall, 2009). Hereford soils are heavy red clays with up to 40% clay content and classified as *Chromic luvisols* (Nyamapfene, 1991). Madziva is characterized by sandy soils which are classified as *Gleyic luvisols* generated from granite parent material (Thierfelder et al., 2012). Dominant soils at Zimuto are *Arenosols* generated from granitic sands of low inherent fertility and less than 5% clay content (Thierfelder & Wall, 2012).

2.2 Maize Varieties

Twelve commonly grown maize hybrids and four open pollinated varieties (OPVs) from different sources were used in this study. The choice of maize genotypes was based on commercial availability in Zimbabwe and wide adoption in southern Africa (Kassie et al., 2012). The genotypes can be subdivided into early maturing (*i.e.* SC403, PAN 413, PHB 3253 and PHB P2859W), medium maturing (*i.e.* PAN 53, Pristine 1, SC 513, SC 533, SC 635, ZAP 51, ZAP 61, ZS 261, PGS 63) and late maturing (*i.e.* SC 637) hybrids; and early maturing (*i.e.* ZM 401 and ZM 309) and medium maturing (*i.e.* ZM 523 and ZM 525) OPVs.

2.3 Experimental Design

The experiment was set as an alpha lattice design at all sites with the CA and CP plots laid adjacent to each other, replicated three times and blocked four times against slope. At all sites, gross plot size measured 4.8 m × 3.6 m (*i.e.* 17.28 m^2) with a net plot size of 3.8 m × 1.8 m (*i.e.* 6.84 m^2).

2.4 Agronomic Management

Residues (maize stover) were uniformly spread over the CA plots at a rate of 3 t ha^{-1} while residues were removed in the CP plots before disc ploughing to a depth of 25 cm prior to the beginning of the season. Maize varieties under CA were planted in rip lines created by a Magoye ripper (a ripper tine attached to an ox-drawn mouldboard plough) at all sites except for UZ were basins were created with hand hoes during the dry winter period. Basins under CA were only used at UZ due to the unavailability of animal draft power. Under the practice of CP, planting was done using basins at all sites. Maize was planted at 90 cm between lines and 50 cm between plants with 3 seeds per station thinned to 2 to achieve a population of 44, 444 plants ha^{-1} at all sites. Maize plants in both CA and CP received basal fertilizer [Compound D (7 N:14 P$_2$O$_5$:7 K$_2$O)] at the rate of 14 kg N ha^{-1}, 12.2 kg P ha^{-1}, and 11.6 kg K ha^{-1} at sowing.Top dressing was applied to both CA and CP in two splits, at four and seven weeks after crop emergence to a total of 69 kg N ha^{-1} in the form of ammonium nitrate (34.5% N). Weeds were controlled by spraying a tank-mixture of glyphosate [*N-(phosphono-methyl)glycine*], atrazine [*1-Chloro-3-ethylamino-5-isopropylamino-2,4,6-triazine*] and metolachlor [(*RS*)-*2-Chloro-N-(2-ethyl-6-methyl-phenyl)-N-(1-methoxypropan-2-l)acetamide*] at a rates of 2.5 l ha^{-1}, 3.5 l ha^{-1} and 1.0 l ha^{-1} active ingredient (ai), respectively, immediately after sowing the seed in the CA plots. In CP plots, weeds were controlled through tillage at the beginning of the season. This was mainly to control grasses and broadleaved weeds. In both CA and CP plots, manual weeding was conducted after crop emergence during the growing period each time weeds reached approximately 10 cm in height or radius for weeds with a stoloniferous growth habit. Maize stalk borer (*Busseola fusca* L.) was controlled using Dipterex at a rate of 1.6 kg ai ha^{-1} applied in granular form into the maize funnel at early signs of attack.

2.5 Measurements

2.5.1 Rainfall

Rainfall was measured at each site using a standard rain gauge located in an open space next to the experimental location. Readings were taken in the morning after each rainfall event.

2.5.2 Soil Sampling and Chemical Analysis

Soil sampling for fertility analyses was done at each experimental location using the stratified random sampling method in each replicate of each cropping system. Three samples were taken from each replicate using a soil auger at three soil depth levels *i.e.* 0-10 cm, 10-30 cm and 30-90 cm. A composite sample was then created representing each depth and each cropping system. Soil samples were submitted to the Soil Chemistry and Soils Institute of the Department of Research and Specialist Services of Zimbabwe (DR & SS) to analyze for chemical and physical properties of the soils.

2.5.3 Maize Crop Emergence

Days to 50% emergence were determined by counting the number of seedlings emerging in the two central rows of each plot daily for 10 days. Emergence percentage was recorded as the number of plants which emerged out of plants which were expected to emerge (60 plants) within the two central rows.

2.5.4 Maize Plant Height, Leaves per Plant and Dry Matter (Early Vigour)

At 6 weeks after sowing, 5 plants were selected from each plot and height measured from the root crown to the top of the innermost leaf. The number of leaves (omitting the bottom dry leaves) per plant were counted from the same sample and the average number of leaves calculated. To determine the above ground biomass, a destructive sampling method from the border rows was done and 5 plants were selected randomly, oven dried to a constant weight at 65 °C for 72 hours and the dry weight recorded.

2.5.5 Leaf Chlorophyll Content

In vivo chlorophyll content of leaves was estimated using a portable chlorophyll meter (SPAD-502®, Minolta, Tokyo, Japan). Measurements were taken weekly starting from 6 weeks after sowing on the upper most extended leaf on 5 randomly selected plants per plot. However, prior to flowering (*i.e.* about 9 weeks after sowing) onwards, the ear leaf was used instead as the sampling leaf reached senescence stage as this had the greatest contribution of assimilates to the sink (ear).

2.5.6 Yield Measurements

Grain and stover were harvested for yield determination from the net plot measuring 2 rows by 4.8 m. The number of plants and cobs per net plot were recorded. The fresh cobs and stover were weighed and their weight recorded. A sub-sample comprising five cobs was extracted from each plot and the fresh weight of cobs recorded. A stover sub-sample of approximately 500 g was taken and its fresh weight measured. The stover and cob sub-samples were dried in an oven and the dry weight recorded. The grain weight was determined after shelling for each plot. Grain moisture was determined using a Dickey-john mini GAC® moisture tester and yield was expressed at 12.5% moisture content.

2.6 Data Analysis

Data were subjected to analysis of variance (ANOVA), generalized linear models, and multiple linear regressions. This was done to identify significant effects of tillage systems on emergence, early vigour and leaf chlorophyll and the contribution of the primary traits to grain yield using GenStat version 12 (VSN International, 2002) statistical packages. Where the effects of tillage systems were significantly different, separation was done using the Tukey's test (HSD) at $P < 0.05$ probability level.

3. Results

3.1 Rainfall Data across Sites and Seasons

In the 2012-13 season, all sites received above 600 mm of rainfall and had an effective growing season of about 80 days except for Zimuto which had rainfall below 400 mm and an effective growing period of 60 days (Figure 1). There was a short dry spell of about 15 days at Zimuto immediately after planting and a sharp increase in rainfall immediately after planting at DTC and Hereford of about 100 mm in 10 days. In the 2013-14 season, there was a longer growing period of over 80 days compared with the 2012-13 season. However, Zimuto and UZ sites experienced a short dry spell of 10 days after planting in 2013-14 season (Figure 1).

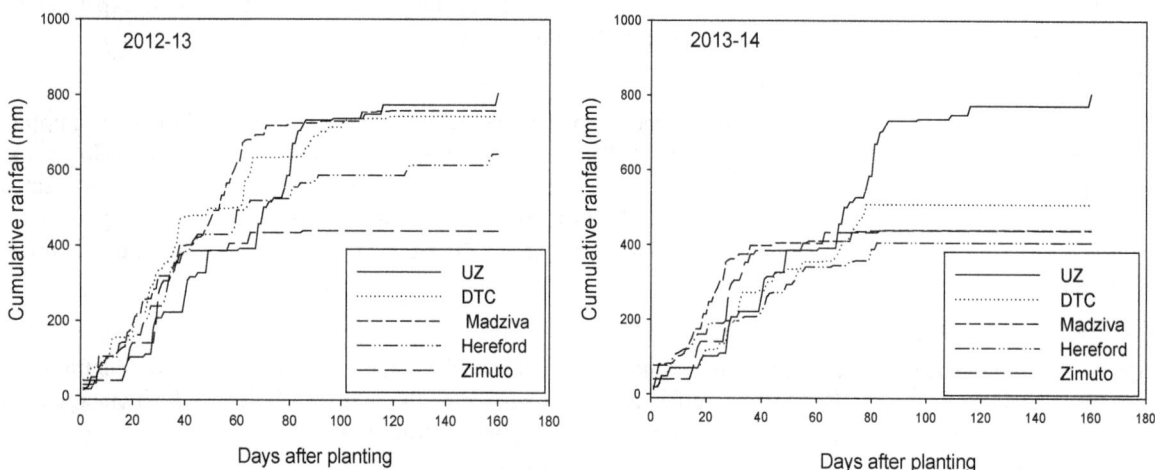

Figure 1. Rainfall patterns at UZ (University of Zimbabwe), DTC (Domboshawa Training Centre), Madziva, Hereford and Zimuto in the 2012-13 and 2013-14 seasons

3.2 Soil Chemical and Physical Properties at the Five Experimental Sites

Soil chemical and physical properties were different at the five study sites (Table 1). Hereford and UZ sites had higher clay content (< 40%) while Madziva and Zimuto had soils with a sandy texture (< 14% clay). Generally the light textured soils had less amounts of phosphorus (P) (< 18 mg/100 g) and potassium (K) (< 0.013 mg/100

g) and were acidic (< 5.9) compared to the heavy textured soils. Heavy textured soils had a pH range of 6-7. Madziva had soil with the lowest clay content (8% in the top layer) and was the most acidic (pH < 6). The highest total N levels (> 50 ppm) were recorded in the heavy textured soils compared with levels (< 34 ppm) in light textured soils. Higher levels of available N under CA compared with CP were found at Domboshawa, Hereford and UZ whilst CP soils had a higher available N content at Zimuto and Madziva. Furthermore, Zimuto and Madziva had more acidic soils under CA compared with CP. However, at the other sites, pH was similar in both cropping systems (Table 1).

Table 1. Soil properties at five experimental sites at UZ farm (University of Zimbabwe), DTC (Domboshawa Training Centre), Madziva, Hereford and Zimuto on the onset of 2012-13 season

Location	Depth (cm)	CA					CP				
		Clay (%)	Available N (ppm)	Total N (ppm)	P (mg/100 g)	pH	Clay (%)	Available N (ppm)	Total N (ppm)	P (mg/100 g)	pH
UZ	0-30	50	35	80	16	6.1	48	30	63	29	6.1
	30-60	60	27	50	4	6.4	52	21	44	12	6.2
	60-90	41	76	104	5	6.5	42	26	45	10	6.3
DTC	0-30	23	34	43	28	5.1	24	19	28	9	5
	30-60	37	28	27	5	5.8	30	10	15	5	6.2
	60-90	31	27	43	7	6.2	34	16	36	3	5.9
Madziva	0-30	8	16	31	12	5.3	5	22	30	10	5.3
	30-60	8	1	9	3	5.2	6	10	11	3	5.3
	60-90	13	22	31	7	5.9	9	14	22	6	5.6
Hereford	0-30	46	16	66	11	6	48	7	52	6	6.1
	30-60	47	12	21	1	6.3	51	10	16	5	6.3
	60-90	53	7	12	1	6.4	53	11	15	11	6.5
Zimuto	0-30	14	21	34	18	5.9	11	1	27	4	5.2
	30-60	14	16	22	1	5.1	12	2	10	1	5.1
	60-90	13	5	13	2	5.7	15	1	10	4	5.7

3.3 Emergence under CA and CP

In 2012-13 season, emergence was significantly ($P < 0.05$) higher under CA at UZ (75%) compared to CP (71%). Emergence was significantly higher in CP (63%) at Zimuto in the same year compared with CA (43%) (Figure 2). In the 2013-14 season, CA (67%) had significantly ($P < 0.05$) higher emergence than CP (39%) at DTC. Conventional practice had significantly ($P < 0.05$) higher emergence at UZ (63%) and Zimuto (80%) compared with CA (58% and 62% respectively) (Figure 2). High emergence percentage showed a positive relationship with grain yield *i.e.* one unit increase in emergence resulted in 0.19 units increase in grain yield under CA and 0.007 units under CP (Table 2). Grain yield decreased by 0.03 and 0.04 units for every one unit increase in days to 50% emergence under CA and CP respectively (Table 2). The multiple linear regression accounted for over 50% variation on the measured parameters both under CA and CP systems.

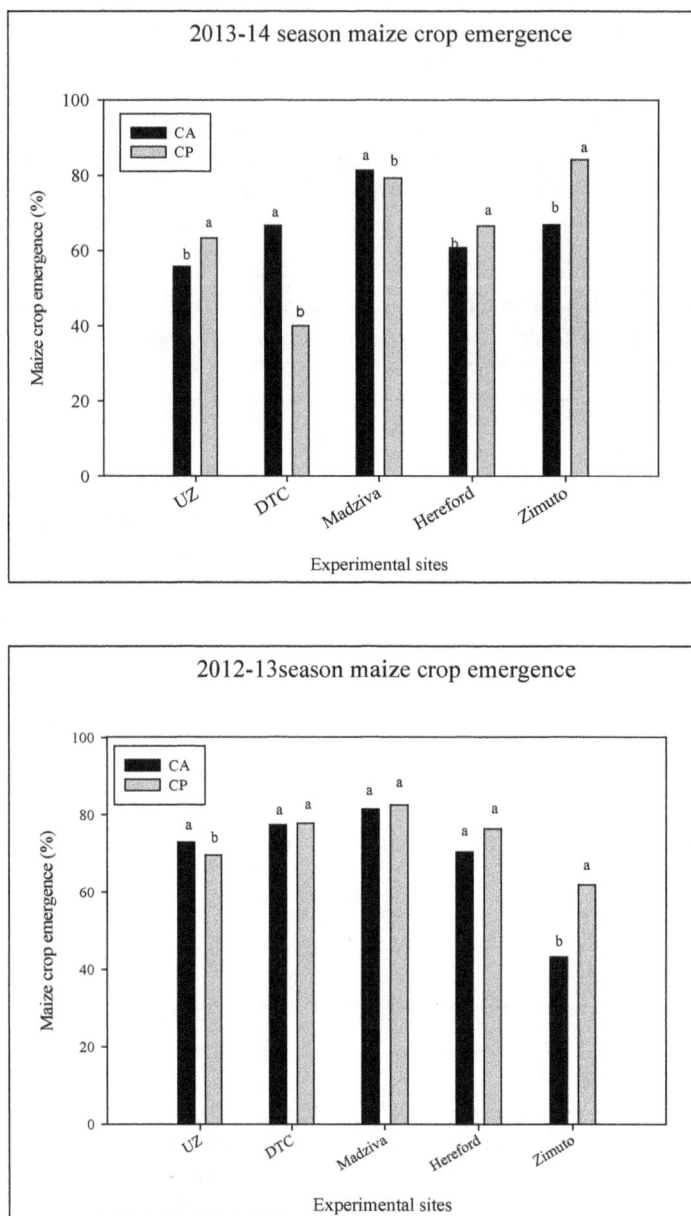

Figure 2. Maize plant emergence percentage under conservation (CA) and conventional practices (CP) at UZ (University of Zimbabwe), DTC (Domboshawa Training Centre), Madziva, Hereford and Zimuto in 2012-13 season

Table 2. The effect of emergence and early vigour parameters, dry matter, crop height, number of leaves and leaf chlorophyll at 6 weeks after sowing (6 WAS on grain yield under CA (conservation agriculture)

	CA		CP	
	Estimate	Probability	Estimate	Probability
Constant (μ)	2.06 ± 0.59	P < .001	1.074 ± 0.354	ns
Emergence (%)	0.019 ± 0.003	ns	0.007 ± 0.0027	P = 0.02
Days to 50% emergence	-0.03± 0.008	ns	0.044 ± 0.01	P < .001
Dry matter (g/5 plants) (6 WAS)	-0.005 ± 0.001	P < .001	-0.008 ± 0.002	P = 0.004
Crop height (cm) (6 WAS)	0.008 ± 0.003	ns	0.03 ± 0.004	P < .001
Number of leaves (6 WAS)	0.4 ± 0.02	P < .001	0.15 ± 0.03	P < .001
Leaf chlorophyll SPAD units (6 WAS)	0.004 ± 0.002a	P < .001	0.0006 ± 0.002	ns
Coefficient of determination	0.516		0.55	
se	1.02		1.04	

3.4 Early Vigour under CA and CP

In the 2012-13 season, CA and CP had significantly (P < 0.05) different total dry matter weight of the maize plants at all sites. Furthermore, the number of leaves per plant at all sites except Hereford, and plant height at DTC and Hereford, were significantly (P < 0.05) different (Table 3). At four of the five sites (DTC, Madziva, Zimuto and Hereford), CP had a significantly (P < 0.05) higher dry matter compared with CA in 2012-13 season. It was only at UZ where CA outperformed the CP treatment. There was no clear trend of CA and CP effects on plant height at most sites. Significant differences in height were only observed at DTC and Hereford where CP out-performed CA (Table 3). During the 2013-14 season all sites had significantly (P < 0.05) more dry mass under CP compared with CA except at Hereford. More leaves were found in the CP system at UZ, DTC, and Zimuto whereas CA outperformed CP at DTC only. A generally lower total dry matter weight was observed at the sandy soil sites, Madziva (9-39 g/5 plants) and Zimuto (< 36 g/5 plants) compared with the clay soil sites.

Table 3. Early vigour (E.V) parameters, dry matter, leaves per plant and plant height at UZ farm (University of Zimbabwe), DTC (Domboshawa Training Centre), Madziva, Hereford and Zimuto under CA (Conservation Agriculture) and CP (Conventional Practices) in 2012-13 and 2013-14 seasons at 6 weeks after crop sowing

		Early vigour parameter								
		Total dry matter weight (g)			Average leaves per plant			Average plant height (cm)		
		CA	CP	P value	CA	CP	P value	CA	CP	P value
2012-13	UZ	80 [a]	55[b]	0.003	6[b]	7[a]	0.001	57	47	NS
	DTC	21.2[b]	48 [a]	0.001	7[b]	8[a]	0.03	35[b]	44[a]	0.004
	Madziva	10[b]	31[a]	0.002	6[b]	7[a]	0.002	40	40	NS
	Hereford	100[b]	207[a]	0.002	9	9	NS	108[b]	126[a]	0.001
	Zimuto	22[b]	30[a]	0.001	5[b]	6[a]	0.003	21	22	NS
2013-14	UZ	66[b]	76[a]	0.003	8[b]	9[a]	0.001	71	77	NS
	DTC	42.52	50	NS	9[a]	8[b]	0.03	48	45	NS
	Madziva	9[b]	39[a]	0.002	4[b]	5[a]	0.002	27	27	NS
	Hereford	250[a]	108[b]	0.001	8	8	NS	82	82	NS
	Zimuto	25[b]	35[a]	0.001	8[b]	9[a]	0.003	35[b]	50[a]	0.001

Note. Means followed by different letters within the same row are significantly different from each other at P < 0.05 probability level.

3.5 Leaf Chlorophyll Content in CA and CP Systems

Maize leaf chlorophyll content consistently fluctuated in both cropping systems throughout the seasons (Figures 3 and 4). There was a significant (P < 0.05) difference between CA and CP systems which was more pronounced in 2012-13 season compared with 2013-14. In the CA system, chlorophyll content was lower compared with CP at initial growth stages of maize at Hereford in the 2012-13 season and at Zimuto in both seasons (Figures 3 and

4). At Zimuto, greater chlorophyll content was observed in CA treatments (43 units) compared with CP plots (37 units) towards the end of the season (Figure 3). In 2013-14 season, no differences ($P > 0.05$) in chlorophyll content were observed at all sites except for Zimuto (Figure 4).

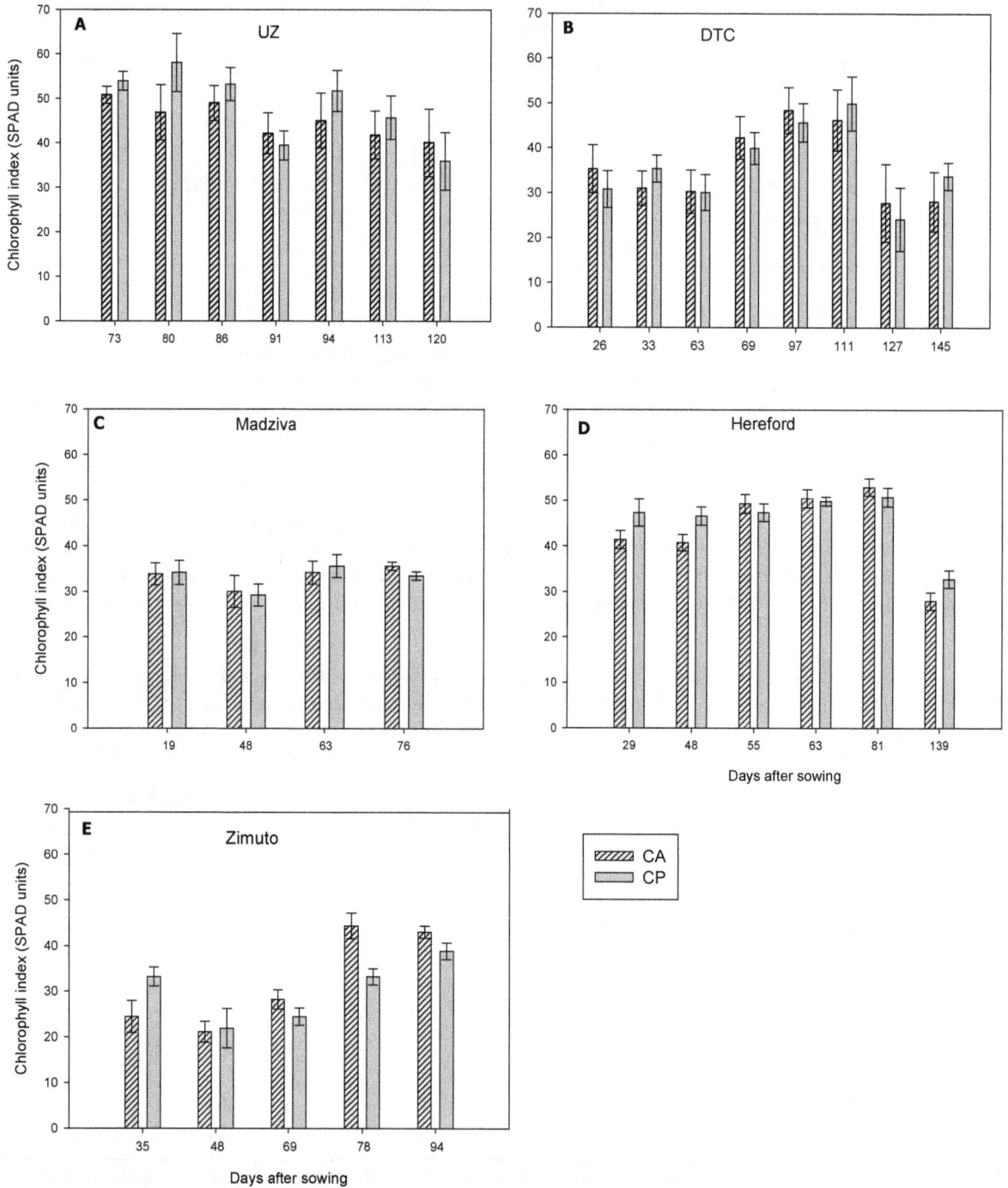

Figure 3. Leaf chlorophyll content at UZ (University of Zimbabwe), DTC (Domboshawa training centre), Madziva, Hereford and Zimuto under CA (conservation agriculture) and CP (conventional practices) in 2012-13 season

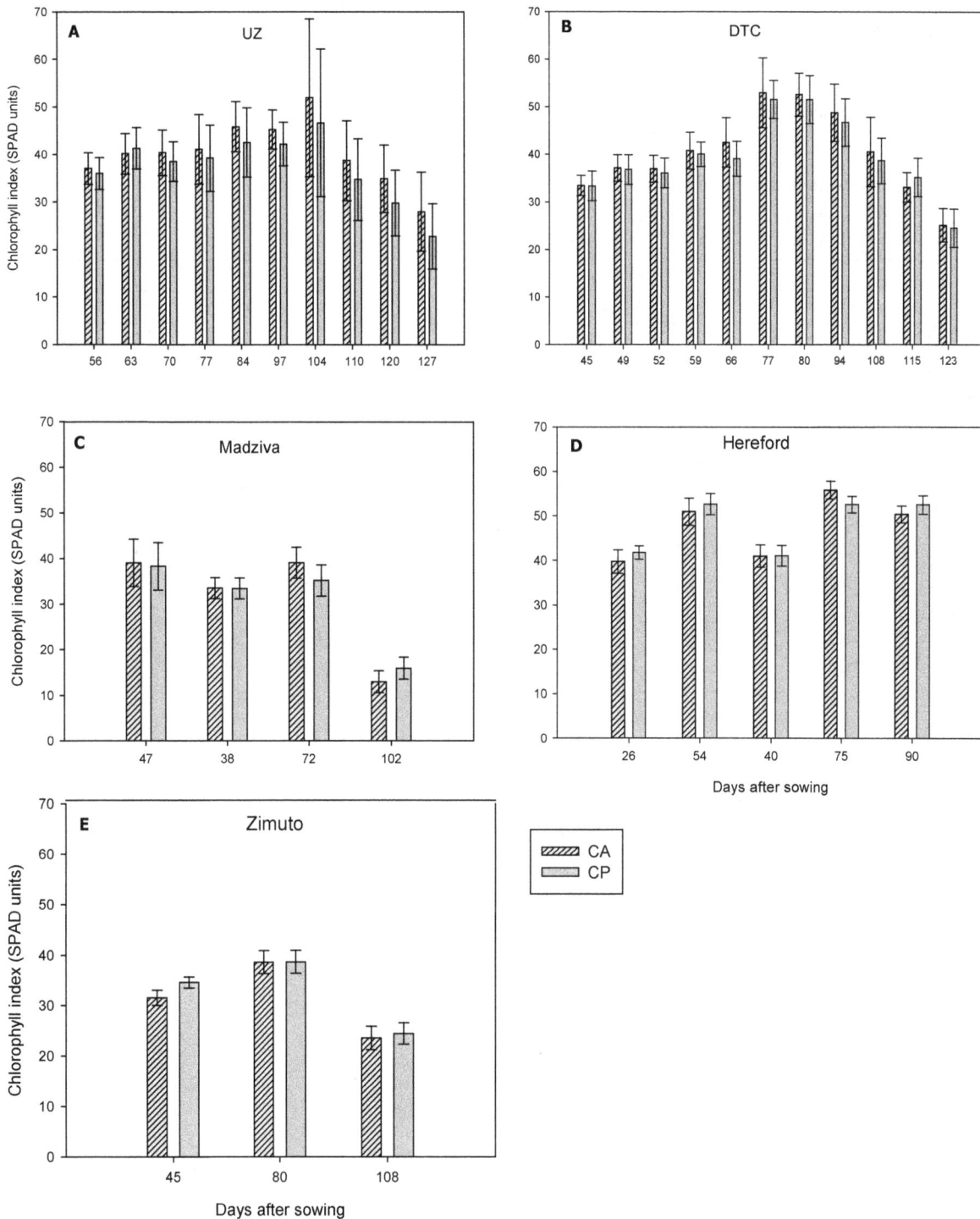

Figure 4. SPAD units (leaf chlorophyll estimates) at UZ (University of Zimbabwe), DTC (Domboshawa Training centre), Madziva, Hereford and Zimuto under conservation agriculture (CA) and conventional practices (CP) in the 2013-14 season

3.6 Grain Yield under CA and CP and Its Relationship with Early Vigour and Chlorophyll Trends

Tillage systems significantly (P < 0.05) affected grain yield at Madziva and DTC in the 2012-13 season and, DTC and Hereford in the 2013-14 season (Table 4). At Madziva CP and CA had a maize yield of 1764 and 1313 kg ha^{-1}, respectively (Table 4). DTC site had 3050 kg ha^{-1} in CA and 2656 kg ha^{-1} in CP in the same season. Hereford site had 4937 kg ha^{-1} in CA and 4274 kg ha^{-1} in CP. Seasons, sites, and varieties showed significant (P

< 0.05) difference (Table 5). Greater yields were attained in the 2013-14 season compared with the 2012-13 season. Hereford had significantly ($P < 0.05$) higher grain yield of 5813 kg ha^{-1} compared with the rest of the sites with Madziva having lowest yield of 1822 kg ha^{-1}. All the sites had a higher grain yield in 2013-14 season compared to 2012-13 season except for Madziva. There was no significant genotype × cropping system interaction and varietal performance between the two tillage systems observed (Table 5).

Table 4. Conservation agriculture (CA) and conventional practice (CP) effects on grain yield (kg ha^{-1}) of different varieties at UZ (University of Zimbabwe), DTC (Domboshawa Training centre), Madziva, Hereford (H/ford) and Zimuto in 2012-13 and 2013-14 seasons

	2012-13					2013-14				
	UZ	DTC	Madziva	H/ford	Zimuto	UZ	DTC	Madziva	H/ford	Zimuto
SC533	3947	4176	1610	3495	682	3127	2729	702	5056	2089
Pristine601	3407	4021	1413	3961	511	4158	3192	599	5238	3021
Pannar53	3283	2125	1336	3484	663	3353	3893	592	4785	2213
Pannar413	2374	2153	1408	3318	835	2301	2063	429	5475	2266
ZM309	2449	2037	799	1811	414	3469	2134	376	4351	1474
Zap51	2760	4030	937	3215	611	2668	2197	682	4434	1258
Zap61	2991	2730	1299	4053	672	3190	2874	729	4777	1463
PHB3253	2785	2912	1025	3518	595	3215	2831	248	5212	3364
ZM525	3614	2619	1213	3233	842	3860	2687	349	4898	2004
ZM401	3237	2517	1343	2677	737	3444	3397	798	4678	2451
PGS 63	3160	4282	1360	4817	640	4906	4826	559	5526	2970
SC637	3896	2649	2056	3390	654	3192	3232	673	4981	1953
ZS265	2982	3564	1625	3053	438	3732	5169	596	5285	3137
SC301	2921	2606	1120	3152	468	3378	2253	815	5609	1766
SC513	2835	4019	1553	2925	447	3657	1897	568	5172	2005
ZS261	3378	2354	915	3636	884	2776	3020	250	3513	1527
CA mean	3126a	3050a	1313b	3359a	631a	3402a	3025b	560a	4937a	2185a
SC533	2948	2265	1733	4781	1075	3693	3745	570	4425	1973
Pristine601	3845	2694	2273	3659	923	3995	4414	664	4097	3353
Pannar53	4471	2858	1748	2825	850	3191	4567	596	4524	2475
Pannar413	3382	2461	1941	3923	716	2979	2040	265	4287	2582
ZM309	3170	2307	2108	4419	641	3448	3251	759	3173	2005
Zap51	3709	2112	1409	3085	889	2900	2848	558	3256	1551
Zap61	4431	3173	1271	1782	586	3269	4641	314	5229	2094
PHB3253	3408	2576	1534	3982	1056	3116	2323	495	4663	2335

Table 4. Continued

| | 2012-13 | | | | | 2013-14 | | | | |
	UZ	DTC	Madziva	H/ford	Zimuto	UZ	DTC	Madziva	H/ford	Zimuto
ZM525	3680	2335	1610	4163	1009	3825	3054	573	3975	3024
ZM401	3488	2260	1580	4459	989	3618	3523	323	4223	2999
PGS 63	2741	3011	2144	4167	892	3633	5104	730	4467	4192
SC637	3668	3233	2145	3424	998	3102	4885	801	4176	2887
ZS265	2769	2652	1369	4012	969	4106	5012	549	4803	3290
SC301	3147	2740	1880	3873	557	3531	3211	770	4770	2194
SC513	3418	2756	1427	2476	883	3650	3723	608	3794	2448
ZS261	2708	3063	2054	3570	547	3092	3375	575	4528	1763
CP mean	**3436a**	**2656b**	**1764a**	**3663a**	**849a**	**3447a**	**3732a**	**572a**	**4274b**	**2573a**
Trial mean	3281	2665	1539	3511	740	3424	3378	566	4606	2379
Site mean	4993b	4355c	1822e	5813a	1929d					
Year mean	2347b					2871a				
SE	720.1									

Table 5. Accumulated analysis of variance output for grain yield (kg ha^{-1}) under conservation agriculture (CA) and conventional practices (CP) at all experimental sites in both seasons.

Source	DF	Sum of squares	Mean of square	Variance	P value
Season	1	85243969	85243969	72.09	<0.001
Site	4	1332132265	333033066	281.63	<0.001
System	1	4229524	4229524	3.58	0.059
Treatment/maize variety	15	79650665	5310044	4.49	<0.001
Season*Site	4	144592773	36148193	30.57	<0.001
Season*System	1	673042	673042	0.57	0.045
Site*System	4	18689359	4672340	3.95	0.004
Season*Treat	15	24820744	1654716	1.4	0.141
Site*Treat	60	84151999	1402533	1.19	0.166
System*Treat	15	28174934	1878329	1.59	0.071
Season*Site*System	4	8445811	2111453	1.79	0.13
Season*Site*Treat	60	93285015	1554750	1.31	0.061
Season*System*Treat	15	15869216	1057948	0.89	0.57
Site*System*Treat	60	59908449	998474	0.84	0.792
Residual	700	827768207	1182526		
Total	959	2807635972	2927670		

Note. Season (2012-13 and 2013-14), Site (UZ (University of Zimbabwe), DTC (Domboshawa Training Center), Madziva, Hereford, Zimuto), System (CA and CP), Treatments (maize varieties).

4. Discussion

4.1 Emergence under CA and CP

Higher emergence under CA compared with CP was found at DTC. This concurred results by Altuntaş & Dede (2009) who reported high emergence rates under CA compared to CP in clay loamy soils. In their study, favorable soil physical properties for maize seedling emergence, which include lower bulk density, were found under CA compared to CP (Altuntas & Dede, 2009). During our study high soil erosion was observed in the CP treatments at DTC because of the heavy rains received during emergence although erosion rates were not measured. This resulted in poor maize crop emergence under CP compared with the CA system. However, CP outperformed CA at UZ, Hereford and Zimuto. This is in line with observations made by Hayhoe et al. (1996), Griffith, Kladivko, Mannering, West, and Parsons (1988) and Dam et al. (2005) which were mainly attributed to low soil temperatures under CA systems compared with CP. Duiker et al. (2006) reported a delayed and uneven emergence under CA, but this delay did not affect vegetative growth or the final crop yield. Hillel (1998) pointed out that soil with more moisture, a scenario under CA compared to CP, heats up more slowly than a dry exposed soil, and this might explain the low temperature that potentially affected emergence. Moreover, since clay soil has ability to conserve more moisture than sandy soils, heating up of a mulched clay soil is even slower than an exposed bare clay soil. This could have affected emergence under CA compared to CP at UZ and Hereford in 2013-14. Zimuto experienced poor maize emergence under CA in both seasons and this could be attributed to delayed weeding and the little rains received at the time of planting. This could have led to a competition for the available moisture between the emerging maize seedlings and the weeds. Delayed weeding was as a result of the unsuitability of chemical weed control particularly glyphosate under very sandy soils due to its slow degradation nature (Locke et al., 2008). Hence manual weeding competed for labor with the seeding operations at the start of the cropping season. This shows that the success of CA also relies on the management practices being followed by farmers and the type of soils (Giller, Witter, Corbeels, & Tittonell, 2009). These inconsistent results indicate the need to further investigate the effect of CA on crop emergence under different soil types. Furthermore, the effects of planting methods under CA must be explored. However, planting basins and direct seeding have shown high emergence percentage compared to ripping in difference parts of southern Africa (Ngwira, Thierfelder, & Lambert, 2012b; Thierfelder & Wall, 2009).

4.2 Maize Early Vigour under CA and CP

Conservation agriculture has shown the potential to improve soil physical and chemical properties (Altundus & Dede, 2009). This has the potential of increasing crop vigour through an improved nutrient cycling. Although plant vigour is determined by the genetics of a plant species, the environment plays a vital role in the exhibition of plant vigour (Negi, Baskheti, & Bhatt, 2015). Furthermore, an improved emergence gives a better chance of high crop vigour. However, high emergence does not always translate into high plant vigour and final yield (Negi et al., 2015). Vigorous plants are associated with resistance to diseases and pests, and good root development which potentially translate to higher yield (Namuco, Cairns, & Johnson, 2009). Lower dry matter and fewer leaves were generally found under CA compared with CP at the experimental sites during the two seasons. Furthermore lower dry matter and fewer leaves were observed under CA in light textured soils compared with heavy textured soils. This is because light textured soils have lower soil fertility and nutrient supply to growing plants is therefore limited compared to heavy textured soils (Chikowo et al., 2010). In addition, the soil chemical and physical analysis in this study revealed a high nutrient base in Hereford, DTC and UZ compared to Zimuto and Madziva. Moreover, at the light textured soil sites, soils from the CA treatment had less available N compared to CP plots. However, CA treatments at all sites had more total N compared with CP. Hence crops with access to high nutrient supply have better chances of having high vigour. The number of leaves per plant and *in-vivo* leaf chlorophyll at 6 weeks after sowing were higher under CP compared with CA. Verhulst et al. (2011a) found a slow initial growth in maize under CA compared with CP. This relates to the availability of nutrients especially N during the early stages of crop growth. High C:N ratio in residue types such as maize stover used in this study potentially caused immobilization of N making its uptake by plants slow (Chikowo et al., 2008). However, at DTC and Hereford more leaves and higher dry matter were found under CA compared with CP.

4.3 Leaf Chlorophyll Trends through the Seasons under CA and CP

High chlorophyll concentration early in the season is a very vital component in a plant's life. It shows how vigorous a crop is performing, and, although it is not the only indicator, gives an indication of the potential yield (Namuco et al., 2009). The chlorophyll content determines the level of photosynthesis and productivity (Egli & Rucker, 2012). Lower chlorophyll content under CA can be explained by a number of reasons, but besides lower temperatures under CA it is mainly due to nitrogen lock-up (Mapfumo et al., 2007; Rasmussen, 1999). Low chlorophyll content in maize leaves under CA has also been observed where different residues are used as soil cover on light textured soil (Mupangwa, Nyagumbo, & Mutsamba, 2016). Carbon and N turnover was found to be 1.5 times slower under CA than CP (Chivenge, Murwira, Giller, & Mapfumo, 2007), which can explain the differences of chlorophyll content between the two systems. Moreover, the extend of differences also depend on clay content of the soil (Chivenge, 2003).

4.4 Grain Yield under CA and CP

Generally there was no significant difference in yield between CA and CP over the sites and seasons. However, higher grain yield was observed under CA compared with CP at DTC in 2012-13 and Hereford in 2013-14. Even though Verhulst et al. (2011b) ascertained a slower growth under CA compared with CP which matches results of current study, this did not translate to yield differences under the two tillage systems. Although some differences were found in the maize growth patterns during the season between CA and CP such as emergence, chlorophyll and some early vigour parameters, this did not translate into yield differences at some sites. However, if these differences are addressed such as poor emergence, lower chlorophyll content at early stages and low vigour, and the potential persistence of foliar diseases under CA systems, yield is highly likely to increase. This study also suggests the need to identify sites which best suits the CA practice. In some sites in this study such as Zimuto, practicing CA showed no yield benefit compared with CP. However, Thierfelder and Wall (2009) found that CA in Zimuto increased yield both on maize and cowpea compared with CP although adoption remained poor. Furthermore, according to this work it is not all seasons that CA results in good yields or CP result in bad yields. Besides, yield performances under CA, other benefits must however, be considered such as reduced erosion and energy use, improvement of soil biological, chemical and physical properties which will bring benefits in the long term (Baudron, Thierfelder, Nyagumbo, & Gérard, 2015).

5. Conclusion

The performance of maize varieties was tested on different soil types and rainfall regimes under two cropping systems (CA and CP) at five locations of Zimbabwe. Based on the results of this study the following conclusions can be made: there is a delay in nutrient accessibility by plants under CA which might be due to the nitrogen immobilization effect of residue mulch in CA systems. This highlights the need for a different fertilization strategy in terms of timing and formulations under CA systems. Chlorophyll content and early vigour characteristics were different under the two cropping systems which indicate the need to select maize varieties that are nutrient-efficient and suitable under CA systems. The study showed that maize yields are affected by a number of traits which emergence and chlorophyll content alone cannot explain. Breeding programs need to select for highly N use efficient varieties to suit CA. Although all the measured establishment and growth parameters showed differences under CA and CP, no grain yield differences were observed at some sites suggesting that if problems under CA such as N lock up are addressed, increases in crop yield are expected. Higher maize yield under CA compared with CP were observed at DTC and Hereford suggesting the site selective effects of CA practices. However, there is need for further investigations as the results of the present study are based on short-term data.

Acknkowledgements

This work was financially supported by the MAIZE Consultative Group on International Agriculture Research (CGIAR) research program and by the work of numerous farmers, field extension officers and researchers in Zimbabwe. Their contribution is gratefully acknowledged. Special thanks to Sign Phiri and Herbert Chipara for logistical and technical support during the course of this work.

References

Altuntaş, E., & Dede, S. (2009). Emergence of silage maize as affected by conservational tillage, ridge and direct planting systems. *Agricultural Engineering International: CIGR Journal.*

Barbera, V. I. T. O., Poma, I., Gristina, L., Novara, A., & Egli, M. (2012). Long-term cropping systems and tillage management effects on soil organic carbon stock and steady state level of C sequestration rates in a semiarid environment. *Land Degradation & Development, 23*(1), 82-91. http://dx.doi.org/10.1002/ldr.1055

Baudron, F., Thierfelder, C., Nyagumbo, I., & Gérard, B. (2015). Where to target conservation agriculture for African smallholders? How to overcome challenges associated with its implementation? Experience from Eastern and Southern Africa. *Environments, 2*, 338. http://dx.doi.org/10.3390/environments2030338

Burgess, M. S., Mehuys, G. R., & Madramootoo, C. A. (1996). Tillage and crop residue effects on corn production in Quebec. *Agronomy Journal, 88*(5), 792-797. http://dx.doi.org/10.2134/agronj1996.000219 62008800050017x

Cairns, J. E., Sanchez, C., Vargas, M., Ordoñez, R., & Araus, J. L. (2012). Dissecting maize productivity: Ideotypes associated with grain yield under drought stress and well-watered conditions. *Journal of Integrative Plant Biology, 54*(12), 1007-1020. http://dx.doi.org/10.1111/j.1744-7909.2012.01156.x

Chikowo, R., Corbeels, M., Mapfumo, P., Tittonell, P., Vanlauwe, B., & Giller, K. E. (2010). Nitrogen and phosphorus capture and recovery efficiencies and crop responses to arrange of soil fertility management strategies in sub-Saharan Africa. *Nutrient Cycling in Agro ecosystems, 88*(1), 59-77. http://dx.doi.org/10.1007/s10705-009-9303-6

Chikowo, R., Corbeels, M., Tittonell, P., Vanlauwe, B., Whitbread, A., & Giller, K. E. (2008). Aggregating field-scale knowledge into farm-scale models of African smallholder systems: Summary functions to simulate crop production using APSIM. *Agricultural Systems, 97*(3), 151-166. http://dx.doi.org/10.1016/j.agsy.2008.02.008

Chikowo, R., Mapfumo, P., Nyamugafata, P., Nyamadzawo, G., & Giller, K. E. (2003). Nitrate-N dynamics following improved fallows and maize root development in a Zimbabwean sandy clay loam. *Agroforestry Systems, 59*(3), 187-195. http://dx.doi.org/10.1023/B:AGFO.0000005219.07409.a0

Chivenge, P. (2003). *Tillage effects on soil organic matter fractions in long term maize trials in Zimbabwe* (MPhil Thesis). University of Zimbabwe, Harare Zimbabwe.

Chivenge, P., Murwira, H., Giller, K., Mapfumo, P., & Six, J. (2007). Long-term impact of reduced tillage and residue management on soil carbon stabilization: Implications for conservation agriculture on contrasting soils. *Soil and Tillage Research, 94*(2), 328-337. http://dx.doi.org/10.1016/j.still.2006.08.006

Dam, R. F., Mehdi, B. B., Burgess, M. S. E., Madramootoo, C. A., Mehuys, G. R., & Callum, I. R. (2005). Soil bulk density and crop yield under eleven consecutive years of corn with different tillage and residue practices in a sandy loam soil in central Canada. *Soil and Tillage Research, 84*(1), 41-53. http://dx.doi.org/10.1016/j.still.2004.08.006

Derksen, D. A., Lafond, G. P., Thomas, G. A., Loeppky, H. A., & Swanton, C. J. (1993). Impact of agronomic practices on weed communities: Tillage systems. *Weed Sci., 41*, 409-417.

Duiker, S. W., Haldeman J. F., & Johnson, D. H. (2006). Tillage × maize hybrid interactions. *Agronomy Journal, 98*, 436-442. http://dx.doi.org/10.2134/agronj2005.0063

Egli, D. B., & Rucker, M. (2012). Seed vigor and the uniformity of emergence of corn seedlings. *Crop Science, 52*(6), 2774-2782. http://dx.doi.org/10.2135/cropsci2012.01.0064

Giller, K. E., Witter, E., Corbeels, M., & Tittonell, P. (2009). Conservation agriculture and smallholder farming in Africa: The heretics' view. *Field Crops Research, 114*(1), 23-34. http://dx.doi.org/10.1016/j.fcr.2009.06.017

Govaerts, B., Verhulst, N., Castellanos-Navarrete, A., Sayre, K. D., Dixon, J., & Dendooven, L. (2009). Conservation agriculture and soil carbon sequestration: between myth and farmer reality. *Crit. Rev. Plant Sci., 28*, 97-122. http://dx.doi.org/10.1080/07352680902776358

Griffith, D. R., Kladivko, E. J., Mannering, J. V., West, T. D., & Parsons, S. D. (1988). Long-term tillage age and rotation effects on corn growth and yield on high and low organic matter, poorly drained soils. *Agronomy Journal, 80*(4), 599-605. http://dx.doi.org/10.2134/agronj1988.00021962008000040011x

Hayhoe, H., Dwyer, L., Stewart, D., White, R., & Culley, J. (1996). Tillage, hybrid and thermal factors in corn establishment in cool soils. *Soil and Tillage Research, 40*(1), 39-54. http://dx.doi.org/10.1016/S0167-1987(96)80005-5

Hillel, D. (1998). *Environmental soil physics: Fundamentals, applications, and environmental considerations.* Academic Press.

Kassie, G. T., Erenstein, O., Mwangi, W. M., La Rovere, R., Setimela, P., & Langyintuo, A. (2012). *Characterization of maize production in Southern Africa: Synthesis of CIMMYT/DTMA household level farming system surveys in Angola, Malawi, Mozambique, Zambia and Zimbabwe.* Socio-Economics Program Working Paper 4, Mexico, D.F.: CIMMYT.

Locke, M. A., Zablotowicz, R. M., & Reddy, K. N. (2008). Integrating soil conservation practice and glyphosate-resistant crops: Impacts on soil. *Pest Management Science, 64*(4), 457-469. http://dx.doi.org/10.1002/ps.1549

Lundy, M. E., Pittelkow, C. M., Linquist, B. A., Liang, X., Van Groenigen, K. J., Lee, J., & Van Kessel, C. (2015). Nitrogen fertilization reduces yield declines following no-till adoption. *Field Crops Research, 183*, 204-210. http://dx.doi.org/10.1016/j.fcr.2015.07.023

Mapfumo, P., Mtambanengwe, F., & Vanlauwe, B. (2007). Organic matter quality and management effects on enrichment of soil organic matter fractions in contrasting soils in Zimbabwe. *Plant and Soil, 296*, 137-150. http://dx.doi.org/10.1007/s11104-007-9304-7

Mhlanga, B., Cheesman, S., Maasdorp, B., Mupangwa, W., & Thierfelder, C. (2015). Contribution of cover crops to the productivity of maize-based conservation agriculture systems in Zimbabwe. *Crop Science, 55*, 1791-1805. http://dx.doi.org/10.2135/cropsci2014.11.0796

Mupangwa, W., Nyagumbo, I., & Mutsamba, E. (2016). Effect of different mulching materials on maize growth and yield in conservation agriculture systems of sub-humid Zimbabwe. *AIMS Agriculture and Food, 1*(2), 239-253. http://dx.doi.org/10.3934/agrfood.2016.2.239

Mutema, M., Mafongoya, P. L., Nyagumbo, I., & Chikukura, L. (2013). Effects of crop residue and reduced tillage on macro fauna abundance. *Journal of Organic Systems, 1.*

Namuco, O. S., Cairns, J. E., & Johnson, D. E. (2009). Investigating early vigour in upland rice (Oryza sativa L.): Part I. Seedling growth and grain yield in competition with weeds. *Field Crops Research, 113*(3), 197-206. http://dx.doi.org/10.1016/j.fcr.2009.05.008

Negi, M., Baskheti, D. C., & Bhatt, S. S. (2015). Evaluation of seed germination and seedling vigour for single cross maize hybrids along with parental lines. *Journal of Hill Agriculture, 6*(2), 171-175. http://dx.doi.org/10.5958/2230-7338.2015.00038.5

Ngwira, A. R., Thierfelder, C., & Lambert, D. M. (2012b). Conservation agriculture systems for Malawian smallholder farmers: Long-term effects on crop productivity, profitability and soil quality. *Renewable Agriculture and Food Systems, First View*, 1-14.

Nyamapfene, K. (1991). *Soils of Zimbabwe* (pp. 75-79). Nehanda Publishers (Pvt) Ltd, Harare, Zimbabwe.

Rasmussen, K. (1999). Impact of ploughless soil tillage on yield and soil quality: A Scandinavian review. *Soil and Tillage Research, 53*(1), 3-14. http://dx.doi.org/10.1016/S0167-1987(99)00072-0

Reicosky, D. C., Sauer, T. J., & Hatfield, J. L. (2011). Challenging balance between productivity and environmental quality. *Tillage impacts.*

Thierfelder, C., & Wall, P. C. (2012). Effects of conservation agriculture on soil quality and productivity in contrasting agro-ecological environments of Zimbabwe. *Soil Use and Management, 28*(2), 209-220. http://dx.doi.org/10.1111/j.1475-2743.2012.00406.x

Thierfelder, C., Cheesman, S., & Rusinamhodzi, L. (2012). A comparative analysis of conservation agriculture systems: Benefits and challenges of rotations and intercropping in Zimbabwe. *Field Crops Research, 137*, 237-250. http://dx.doi.org/10.1016/j.fcr.2012.08.017

Thierfelder, C., Mutenje, M., Mujeyi, A., & Mupangwa, W. (2014). Where is the limit? Lessons learned from long-term conservation agriculture research in Zimuto Communal Area, Zimbabwe. *Food Security*, 15-31.

Thierfelder, C., Rusinamhodzi, L., Ngwira, A. R., Mupangwa, W., Nyagumbo, I., Kassie, G. T., & Cairns, J. E. (2014). Conservation agriculture in Southern Africa: Advances in knowledge. *Renewable Agriculture and Food Systems*, 1-21.

Thierfelder, C., & Wall, P. C. (2009). Effects of conservation agriculture techniques on infiltration and soil water content in Zambia and Zimbabwe. *Soil and Tillage Research, 105*, 217-227. http://dx.doi.org/10.1016/j.still.2009.07.007

Verhulst, N., Kienle, F., Sayre, K. D., Deckers, J., Raes, D., Limon-Ortega, A., ... Govaerts, B. (2011a). Soil quality as affected by tillage-residue management in wheat-maize irrigated bed planting system. *Plant Soil, 340*, 453-466. http://dx.doi.org/10.1007/s11104-010-0618-5

Verhulst, N., Nelissen, V., Jespers, N., Haven, H., Sayre, K., Raes, D., ... Govaerts, B. (2011b). Soil water content, maize yield and its stability as affected by tillage and crop residue management in rain fed semi-arid highlands. *Plant and Soil, 344*(1), 73-85. http://dx.doi.org/10.1007/s11104-011-0728-8

VSN International. (2002). *GenStat for Windows* (12th ed.). VSN International, Hemel Hempstead, UK. Retrieved from http://genstat.co.uk

Wortman, C., Shapiro, C., & Tarkalson, D. (2006). *Composting manure and other organic residues.* University of Nebraska-Lincoln Extension, Institute of Agriculture and Natural Resources, Lincoln, Nebraska, USA.

3

Co-Granulated and Blended Zinc Fertilizer Comparison for Corn and Soybean

Matthew Caldwell[1], Kelly A. Nelson[2] & Manjula Nathan[1]

[1] College of Agriculture, Food and Natural Resources, Division of Plant Sciences, University of Missouri, Columbia, Missouri, USA

[2] Greenley Research Center, Division of Plant Sciences, University of Missouri, Novelty, Missouri, USA

Correspondence: Kelly A. Nelson, University of Missouri Greenley Research Center, 64399 Greenley Place, Novelty, MO 63460, USA. E-mail: nelsonke@missouri.edu

Abstract

A new co-granulated formulation of monoammonium phosphate (MAP) including S and Zn could allow for more uniform nutrient distribution. A six site-year study evaluated the effects of blended phosphorus (P) sources [MAP and diammonium phosphate (DAP)] and zinc amounts (0, 2.2, and 5.6 kg Zn ha^{-1}) compared to co-granulated fertilizer, MicroEssentials® Sulfur-10 (MES10™) (12-40-0-10S) and MicroEssentials Sulfur and Zinc (MESZ™) (12-40-0-10S-1Zn), on corn and soybean response. Fertilizers were broadcast applied for corn and the carry-over effect on soybean was determined. Ear leaf P, S, and Zn concentrations at Novelty in 2013 and 2014 were within the sufficiency range regardless of treatment, even though initial soil test values were low-medium. Yields were similar to the N only control for all site-years except at Novelty in 2013, where MAP+ZnSO$_4$ at 2.2 kg Zn ha^{-1}, MAP+Super Zn at 5.5 kg Zn ha^{-1}, and DAP+AMS were 540 to 570 kg/ha greater. The amount of Zn fertilizer (2.2 vs. 5.6 kg Zn ha^{-1}) also showed no significant effect on yield. Applications of P or Zn generally increased their concentrations in post-harvest soil samples. Fertilizer applied for corn indicated some differences in soybean plant nutrient concentrations, but it had no effect on total plant nutrient uptake, grain yield or quality. At Novelty, soybean plant Zn concentration was greater at 5.6 kg Zn ha^{-1} compared to 2.2 kg Zn ha^{-1}, while Albany showed an increase in whole soybean plant Zn concentration with SuperZn compared to ZnSO$_4$. Carry-over fertilizer from corn showed limited effects on soybean response the following year.

Keywords: corn, fertilizer, phosphorus, soybean, sulfur, and zinc

1. Introduction

Zinc (Zn) is essential to plant survival, with the average plant containing 20 ppm of the micronutrient based on dry weight (Mahler, 2004). Typical soils can contain 0.3 to 2.0 ppm (Mahler, 2004) of plant-available Zn, which is the most common deficient micronutrient in high pH soils (Graham, Asher, & Hynes, 1992). Zn is found in N metabolism pathways that can affect protein synthesis (Fageria, 2004). Deficiencies can cause interveinal chlorosis, bronzing, internode shortening, and epinasty. In severe deficiencies, the root apex can become necrotic. Although Zn is mobile in the plant, its mobility is poor and deficiency symptoms appear first in the upper, young plant leaves. Since Zn is a micronutrient, Zn toxicity is possible, but unlikely (Broadley, White, Hammond, Zelko, & Lux, 2007). Zn fertilizers are available in three major forms: Zn chelate, ZnO, and ZnSO$_4$ (Schulte, 2004). Water solubility greatly influences the availability and effectiveness of Zn fertilizer. In Zn chelate, commonly sold as ZnEDTA, a large organic molecule surrounds Zn and keeps it from leaching, oxidizing, and precipitating (Schulte, 2004). Zinc sulfate is the most common form, due to its low cost and greater solubility (Schulte, 2004), and has traditionally been a steadfast source in Zn fertilizer (Olsen, 1982).

Although soil may contain enough Zn to support a crop through the season, 90% of the Zn is in forms that make it unavailable (fixed, insoluble, or unexchangeable) (Broadley et al., 2007). In most soils, only 0.1 to 2 μg of Zn per gram are exchangeable (Broadley et al., 2007). Soils with large phosphate levels can cause an imbalance in a crops' physiology including a reduction in Zn uptake (Olsen, 1982). This phenomenon is known as P-induced Zn deficiency (Singh, Karamanos, & Stewart, 1986). Zinc-phosphorous interactions are well documented (Halim, Wassom, & Ellis, 1968; Keefer, Singh, Horvath, & Henderlong, 1972; Rehm, Sorensen, & Wiese, 1981, 1983; Robson & Pitman, 1983; Singh, Karamanos, & Stewart, 1988). Phosphorus fertilizer applied in large amounts can induce zinc deficiency in soils with low plant-available zinc (Robson & Pitman, 1983). In soil, P can decrease zinc's solubility (Huang, Barker, Langridge, Smith, & Graham, 2000). When P requirements are met in the plant, root growth is reduced and mycorrhizae infection less common (Amijee, Stribley, & Tinker, 1990). Deficiencies in plants could also be induced by a small concentration of Zn due to rapid growth response to P. Alternatively, large P-to-Zn ratios could cause a metabolic imbalance in cells and lead to P-induced Zn deficiencies (Singh et al., 1988). Zinc deficiency may increase in response to an expression of high-affinity phosphate transporters when P is deficient, likely because the plant utilizes resources from Zn for phosphate transporters (Huang et al., 2000). Increasing applications of both Zn fertilizer and P could help optimize yield (Schnappinger, Martens, & Hawkins, 1969), even though crops take up little Zn during the growing season.

More than 30% of the world's arable land has P-limiting yield potential (Vance, Uhde-Stone, & Allen, 2003). Phosphorus is an essential plant macronutrient that accounts for 3 to 5 g kg^{-1} of a plant's dry weight (Schalchtman, Reid, & Ayling, 1998). Phosphorus, a structural component in nucleic acids (DNA, RNA), transfers energy as adenosine triphosphate (ATP) and maintains cell structure with phospholipids. Though abundant in soil, P occurs primarily in a fixed form or outside of the rhizosphere and so is unavailable for plant uptake. When P is not available in adequate amounts, at least 0.2 mg L^{-1} in soil solution (Pierzynski, McDowell, Sims, & Sharpley, 2005), plants can become deficient. Visual signs of P deficiency include overall stunting of the plant, a purple tint from anthocyanin accumulation, and small necrotic leaf spots. Deficiency typically appears in the lower more mature leaves because P is mobile and translocates to new developing tissue (Briskin, Bloom, Taiz, & Zeiger, 2010). To overcome limited P availability in the soil and maintain soil test P levels, todays growers use P fertilizers commonly available as monoammonium phosphate (MAP) and diammmonium phosphate (DAP).

It is important to optimize sulfur (S) in crops to achieve high yields and grain quality (Tabatabai, 1984). As current environmental laws reduce sulfur emissions from power plants, crop sulfur deficiencies may become more common (Camberato, Maloney, Casteel, & Johnson, 2012). Correspondingly, the need to apply S fertilizers likely will increase in coming years. Sandy soils with small amounts of organic matter and no-till or heavy residue can increase the likelihood of sulfur deficiencies (Camberato & Casteel, 2010). Although S is considered a secondary plant nutrient (primarily because of amount needed), deficiency seriously affects plant growth and yields (Sawyer & Barker, 2012). As organic matter decomposes, it releases sulfate (SO_4^{-2}) into the soil through the process of mineralization (Hergert, 2000). For every one g kg^{-1} of organic matter, 2.25 to 3.36 kg ha^{-1} of sulfate are released annually into the soil, while 10 Mg ha^{-1} of corn (*Zea mays* L.) removes approximately 6.11 kg ha^{-1} of sulfur in grain alone (Schulte & Kelling, 1992).

A patented technology employed in co-granulated fertilizers combine nitrogen, P, S, and Zn into a single prill (MicroEssentials, Mosaic, Plymouth, MN). This allows for uniform distribution and possibly increased uptake of nutrients across a range of crops. MES10 (MicroEssentials Sulfur) contains MAP plus equal amounts of AMS (ammonium sulfate) and elemental sulfur (S) as 100 g kg^{-1} in the co-granulated material. The sulfate is immediately available for plant uptake, though the elemental sulfur must be oxidized by soil bacteria, which allows for season-long sulfur availability (Schulte & Kelling, 1992). MESZ is the same formulation as MES plus one percent ZnO. Microessential Sulfur and Zinc (MESZ) utilizes ZnO as the primary Zn source (Mosaic, 2007). Zinc oxide has the greatest percent of Zn at 72-80% compared to other Zn sources, but it is less water soluble than $ZnSO_4$.

Researchers have studied the effects of co-granulated fertilizers in Iowa with corn (Sawyer & Barker, 2009), and in Arkansas with rice (*Oryza sativa* L.) (Slaton et al., 2010), winter wheat (*Triticum aestivum* L.) (Freeman, Ruffo, & Mann, 2014), and canola (*Brassica napus* L.) (Woolfork, Olson, Mann, & Perez, 2014); however, results have been mixed. Some studies show limited yield differences (Sawyer & Barker, 2009), while others indicate an advantage of MES and MESZ compared to a blend of the same nutrients (Slaton et al., 2010). MESZ increased yields 5.7% compared to MAP and 3.4% compared to MAP+AMS+$ZnSO_4$ (Freeman et al., 2014). In canola, Woolfolk et al. (2014) reported yield increases of 4% at 19 kg P_2O_5 ha^{-1} and 7.1% at 56 kg P_2O_5 ha^{-1}, which were related to less injury to germinating seedlings. Few studies report on the effects of the new co-granulated fertilizers on corn response in the Midwestern U.S., as well as their carry-over impact on soybean [*Glycine max* (L.) Merr.]. The objective of this research was to evaluate corn response to MES10 and MESZ formulations to equivalent blends of DAP or MAP, S, and Zn at two amounts of Zn (2.2 and 5.6 kg Zn ha^{-1}) for the impact on corn (ear leaf nutrient concentration at VT, grain yield, grain quality, and changes to soil test nutrient levels post-harvest) and soybean response (population, plant nutrient uptake, yield, and grain quality).

Table 1. Initial soil characteristics (average ± 1 standard deviation of the mean) 0-15 cm deep at Albany (2013-2014) and Novelty (2011-2014)

Soil characteristics	2011 Novelty	2012 Novelty	2013		2014	
			Novelty	Albany	Novelty	Albany
pH (0.01 M CaCl$_2$)	6.0±0.1	6.2±0.2	5.1±0.6	5.1±0.2	5.7±0.2	5.9±0.3
Neutralizable acidity (cmol$_c$ kg^{-1})	1.9±0.2	1.1±0.4	5.4±5.5	4.5±1.1	2.5±0.1	2.0±0.9
Organic matter (g kg^{-1})	23±1	29±2	20±2	26±3	22±4	29±1
Bray 1P (kg ha^{-1})	15.7±2.4 (VL)[†]	15.7±2.1 (VL)	21.9±8.9 (L)	24.6±7.5 (L)	25.7±2.5 (L)	35.3±4.4 (M)
Exchangeable (1 M NH$_4$OA$_C$)						
Ca (kg ha^{-1})	4547±235	4805±314	3674±381	3618±426	4797±415	5728±235
Mg (kg ha^{-1})	392±37	347±34	328±49	459±64	412±52	716±48
K (kg ha^{-1})	161±11	160±20	128±38	234±44	228±30	206±16
SO$_4$-S (mg kg^{-1})	5.8±1.1 (M)	6.4±0.7 (M)	1.6±0.3 (M)	5.7±0.4 (M)	4.5±0.2 (M)	5.6±0.6 (M)
Zn (mg kg^{-1})	0.2±0.1 (L)	0.5±0.1 (L)	0.3±0.1 (L)	1.0±0.3 (M)	0.8±0.6 (M)	1.0±0.6 (M)
Mn (mg kg^{-1})	16.7±0.8	49.3±7.4	17.2±1.7	-	19.3±3.2	9.8±1.7
Fe (mg kg^{-1})	38±1.0	49.3±7.4	48.3±12.4	-	40.3±3.9	43.3±11.8
Cu (mg kg^{-1})	0.6±0.1	0.6±0.1	0.4±0.1	-	0.6±0.1	0.8±0.1
CEC (cmol$_c$ kg^{-1})	13.7±0.8	13.3±0.7	14.2±3.2	14.6±1.1	14.7±1.2	17.7±0.7

Note. [†] Abbreviations: L, low; M, medium; VL, very low (Buchholz et al., 2004).

2. Materials and Methods

2.1 Corn

In 2011 and 2012, field research was conducted at the Greenley Memorial Research Center (39°56′N, 92°3′W) near Novelty, Missouri. In 2013 and 2014, field research was also conducted at the Hundley-Whaley Center (40°14′N, 94°20′W) near Albany, Missouri. Soil test Bray 1P at Novelty and Albany was very low (15 kg ha^{-1}) to medium (36 kg ha^{-1}) (Table 1). The Novelty sites were a Putnam silt loam (fine, smectitic, mesic Vertic Albaqualfs), while Albany sites were a Grundy silt loam (fine, smectitic, mesic Aquertic Argiudolls). Treatments were arranged in a randomized complete block design with five replications at Novelty and four replications at Albany. Initial (15 cm) soil samples from each replication were collected and analyzed by the University of Missouri Soil and Plant Testing Laboratory using the recommended soil test procedures for Missouri (Nathan, Stecker, & Sun, 2012).

Fertilizer treatments were applied pre-plant for corn in the corn-soybean rotation. Treatments included P source (MAP or DAP), Zn rate (2.2 and 5.6 kg Zn ha^{-1}), and multiple fertilizer technologies (traditional blends or co-granulated fertilizers). Zinc rates were in line with other research showing corn yield increases (Schnappinger et al., 1969). In 2013 and 2014, SuperZn (liquid Zn oxide) (1-0-0-0-40Zn) (Helena, Collierville, TN) was impregnated on the dry fertilizer prills and added at both Novelty and Albany. Co-granulated fertilizers included MicroEssentials Sulfur and Zn (MESZ) (12-40-0-10S-1Zn) and MicroEssentials Sulfur (MES10) (12-40-0-10S-1Zn). In 2013 and 2014, Novelty and Albany had sixteen treatments, including: non-treated control (no fertilizer), nitrogen (N) only, DAP, DAP+ ammonium sulfate (AMS), DAP+ZnSO$_4$ at 2.2 kg Zn ha^{-1}, DAP+ZnSO$_4$ at 5.6 kg Zn ha^{-1}, DAP+SuperZn at 2.2 kg Zn ha^{-1}, DAP+SuperZn at 5.6 kg Zn ha^{-1}, MAP, MAP+AMS, MAP+ZnSO$_4$ at 2.2 kg Zn ha^{-1}, MAP+ZnSO$_4$ at 5.6 kg Zn ha^{-1}, MAP+SuperZn at 2.2 kg Zn ha^{-1}, MAP+SuperZn at 5.6 kg Zn ha^{-1}, MES10, and MESZ. In 2011 and 2012 at Novelty, SuperZn or DAP+AMS were not included, resulting in a total of eleven treatments.

Corn was planted in April or May, depending on yearly weather conditions using no-till (Novelty) or minimum tillage (Albany) into 3 by 9 to 15 m plots, with 76 cm row spacing. Corn followed soybean at all sites, except in 2013 at Albany, a continuous corn site. Management information is available in Table 2.

In 2013 and 2014 at Novelty, ten ear leaves were randomly selected from the middle two rows of each plot. Ear leaf samples were analyzed for P, SO$_4$-S, and Zn concentrations (Bryson, Mills, Sasseville, Jones, & Barker, 2014). The two middle rows of the four-row corn plots were harvested using a plot combine (Wintersteiger Delta, Salt Lake City, UT or Massey 8, Haven, KS) measuring corn grain yields and moisture content. Corn grain yields were adjusted to 150 g kg^{-1} prior to analysis. Individual plot grain samples were collected during harvest and evaluated for oil, protein, and starch using a near infrared (NIR) spectroscopy (Foss Infratec, Eden Prairie, MN).

2.2 Soybean

Soybean was planted into the same plots as corn in April or May, depending on yearly weather conditions. The fields were no-till (Novelty) or minimum tillage (Albany), and plots were 3 by 9 to 15 m. In 2015, soybean was planted in July due to an extremely wet spring. Soybean was planted in 76 cm wide rows at Albany and in 19 cm rows at Novelty. Soybean followed corn in all years. Management information is available in Table 3. Whole plant tissue samples were taken in 2014 and 2015 at Novelty and Albany at R6 (Fehr & Caviness, 1971). Quadrats (0.23 m^2) were randomly selected from the middle two rows of each plot. The plant samples were ground, dried, and analyzed with standard extraction methods for P, S, and Zn concentrations (Bryson et al., 2014).

To determine soybean grain yields and moisture content, the two middle rows of soybean were harvested at Albany (Massey 8, Haven, KS), and a 1.5 m wide section of the plot (four, 38 cm wide rows) was harvested at Novelty (Wintersteiger Delta, Salt Lake City, UT). During harvest, individual plot grain samples were collected and evaluated for oil and protein concentration using near infrared (NIR) spectroscopy (Foss Infratec, Eden Prairie, MN). Soybean yields were adjusted to 130 g kg^{-1} prior to analysis.

2.3 Statistical Protocol

All corn data were analyzed with the Statistical Analysis System (SAS Institute, Cary, NC) using PROC GLIMMIX, and means were separated using Fisher's Protected LSD ($P = 0.05$). Corn data for Novelty in 2011 and 2012 were analyzed separately from 2013 and 2014 data due to the addition of SuperZn in the latter years. Data were combined by year and location when appropriate. Soybean data were combined over years for individual sites for all measurements. Planned contrasts were used to compare Zn sources (SuperZn vs. ZnSO$_4$) and Zn amounts (2.2 vs. 5.6 kg Zn ha^{-1}).

3. Results and Discussion

3.1 Growing Conditions

Research at Novelty from 2011 to 2014 and at Albany in 2013 and 2014 experienced a wide range of precipitation during the growing seasons (March 31 to September 29) (Figure 1). At Novelty in 2011, corn experienced an abnormally dry spring (USDM, 2015) followed by average summer precipitation (532 mm) throughout the growing season. In 2012, the Midwestern U.S. experienced an extreme drought (USDM, 2015), with Novelty receiving only 273 mm of precipitation during the growing season. In 2013, precipitation was average at Novelty (453 mm), and above average at Albany (607 mm).

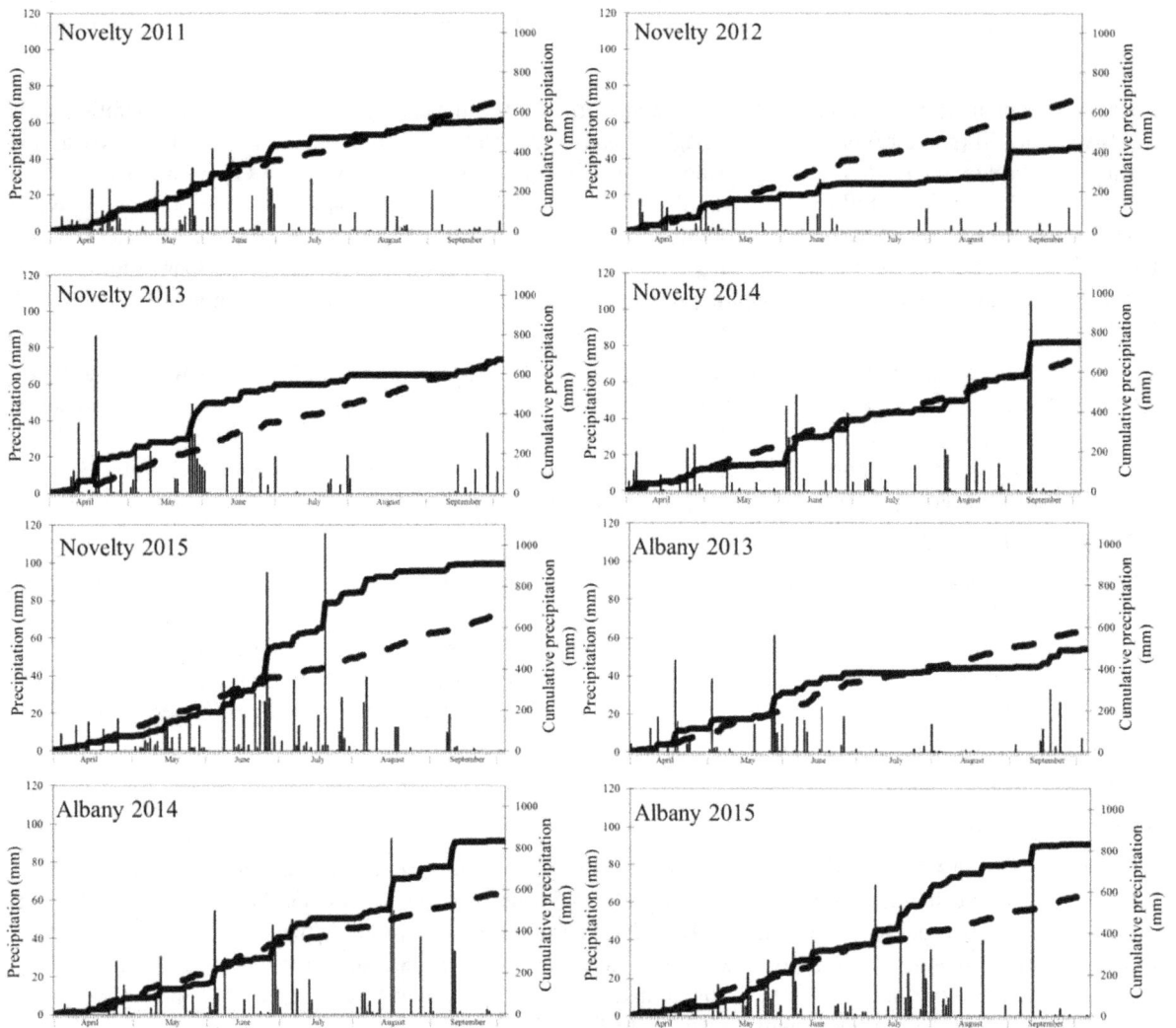

Figure 1. Precipitation over the six site-years for corn (2011-2014) and the following soybean (2012-2015) crop. Bars represent individual precipitation events (left vertical axis). The solid line represents cumulative precipitation throughout the season, and the dashed line represents 10-year average precipitation (right vertical axis)

Table 2. Field and management information for corn sites at Novelty (2011-2014) and Albany (2013-2014)

Management information	2011 Novelty	2012 Novelty	2013		2014	
			Novelty	Albany	Novelty	Albany
Plot size (m)	3 by 12	3 by 12	3 by 15	3 by 11	3 by 15	3 by 11
Hybrid or cultivar	DKC 63-84	DKC 63-84	DKC 63-25 VT3	DKC 64-69	DKC 63-25	DK 64-69
Planting date	12 Apr.	2 Apr.	15 May	14 May	18 Apr.	5 May
Seeding rate (seeds ha^{-1})	76,600	79,100	81,1500	71,700	81,500	74,100
Harvest date	22 Sep.	28 Aug.	7 Oct.	10 Oct.	10 Oct.	16 Oct.
Maintenance fertilizer	31 Mar. 2011	18 Nov. 2011			11 Nov. 2013	NA
Nitrogen	200 kg N ha^{-1} (AA)[†]	213 kg N ha^{-1} (AA) + nitrapyrin at 2.34 L ha^{-1}	200 kg N ha^{-1} (AA)	200 kg N ha^{-1} (AN)	245 kg N ha^{-1}	200 kg N ha^{-1} (AN)
P-S-Zn application	6 May	28 Nov. 2011	29 Apr.	7 May	25 Mar.	5 May
Tillage	No-till	No-till	No-till	Minimum	No-till	Minimum
Weed management						
Burndown/ Pre-emergence[‡]	5 Apr., glyphosate 1.2 kg ae ha^{-1} + saflufenacil 0.03 kg ai ha^{-1} + dimethenameid-P 0.2 kg ai. ha^{-1} + AMS 18 g L^{-1}	19 Mar., saflufenacil 0.03 kg ai. ha^{-1} + dimethenameid-P 0.2 kg ai ha^{-1} + glyphosate 1.2 kg ae ha^{-1} + AMS 18 g L^{-1}	17 May, atrazine 1.5 kg ai ha^{-1} + S-metolachlor 1.5 kg ai ha^{-1} + mesotrione 0.2 kg ai ha^{-1} + MSO 1% v/v + UAN 2.34 L ha^{-1} + Glyphosate 1.2 kg a.e. ha^{-1}	14 May, atrazine 1.5 kg ai ha^{-1} + S-metolachlor 1.5 kg ai ha^{-1} + mesotrione 0.2 kg ai ha^{-1}	13 Nov. 2013, simazine 1.1 kg ai ha^{-1} + glyphosate 0.6 kg a.e. ha^{-1} + 2, 4-D 2.7 kg ae ha^{-1} + COC 2.34 L ha^{-1}	5 May, atrazine 1.5 kg ai ha^{-1}+ s-metolachlor 1.5 kg ai ha^{-1}+ mesotrione 0.2 kg ai ha^{-1} + glyphosate 1.2 kg a.e. ha^{-1}
Postemergence	17 May, acetochlor 3.2 kg ai ha^{-1}	10 May, atrazine 1.5 kg ai ha^{-1} + S-metolachlor 1.5 kg ai ha^{-1} + glyphosate 1.2 kg a.e. ha^{-1} + 0.25% v/v NIS		11 June, glyphosate 1.2 kg a.e. ha^{-1}	24 May, atrazine 1.5 kg ai ha^{-1} + S-metolachlor 1.5 kg ai ha^{-1} + glyphoshate 1.2 kg a.e. ha^{-1} + 0.25% v/v NIS	
Insect management	17 May, Lambda-cyhalothrin 0.02 kg ai ha^{-1}	10 May, Lambda-cyhalothrin 0.02 kg ha^{-1}	NA	NA	NA	NA
Disease management	NA	NA	NA	NA	10 July, azoxystrobin 0.08 kg ai ha^{-1}	NA

Note. [†] Abbreviations: AA, anhydrous ammonia; ae, acid equivalent; ai, active ingredient; AN, ammonium nitrate; COC, crop oil concentrate; MSO, Methylated seed oil; NA, none applied; and UAN, urea ammonium nitrate.

[‡] Chemical Names: acetochlor, 2-chloro-N-(ethoxymethyl)-N-(2-ethyl-6-methylphenyl)acetamide; atrazine, 1-Chloro-3-ethylamino-5-isopropylamino-2,4,6-triazine; azoxystrobin, Methyl (2E)-2-(2-{[6-(2-cyanophenoxy) pyrimidin-4-yl]oxy}phenyl)-3-methoxyacrylate; dimethenameid-P, 2-chloro-N-(2,4-dimethylthiophen-3-yl)-N-[(2S)-1-methoxypropan-2-yl]acetamide; glyphosate, N-(phosphonomethyl)glycine; lambda-cyhalothrin, 3-(2-chloro-3,3,3-trifluoro-1-propenyl)-2,2-dimethyl-cyano(3-phenoxyphenyl)methyl cyclopropanecarboxylate; S-metolachlor, 2-chloro-N-(2-ethyl-6-methylphenyl)-N-[(2S)-1-methoxypropan-2-yl]acetamide; nitrapyrin, 2-chloro-6-(trichloromethyl)pyridine; simazine, 6-chloro-2-N,4-N-diethyl-1,3,5-triazine-2,4-diamine; and sulflufenacil, 2-chloro-4-fluoro-5-[3-methyl-2,6-dioxo-4-(trifluoromethyl)pyrimidin-1-yl]-N-[methyl(propan-2-yl)sulfamoyl]benzamide.

Table 3. Field management information for soybean sites following corn fertilizer treatments at Novelty (2012-2015) and Albany (2014-2015)

Management information[†]	2012 Novelty	2013 Novelty	2014		2015	
			Novelty	Albany	Novelty	Albany
Plot size (cm)	3 by 12	3 by 12	3 by 15	3 by 10	3 by 10	3 by 15
Hybrid or cultivar	Ag3730	Morsoy LL 3759N	Stine 38LE02	AG 3731	Stine 38LEO2	Asgrow 3934
Planting date	25 Apr.	17 May	8 May	15 May	2 Jul.	2 Jul.
Row spacing (cm)	38	19	20	76	19	76
Seeding rate (seeds ha^{-1})	444,800	395,400	444,800	385,500	469,000	370,000
Harvest date	9 Oct.	10 Oct.	18 Oct.	27 Oct.	20 Oct.	12 Nov.
Tillage	No-till	No-till	No-till	Minimum	No-till	No-till
Weed management						
Burndown/PRE[‡]	25 Apr., saflufenacil 0.035 kg ai ha^{-1} + 0.25% v/v NIS + UAN 2.5 L ha^{-1} + glyphosate 1.2 kg ae ha^{-1}	17 May, saflufenacil 0.035 kg ai ha^{-1} + glyphosate 1.2 kg ae ha^{-1} + UAN 2.5 L ha^{-1} + MSO 1% v/v	23 May glufosinate 0.8 kg ai ha^{-1} + AMS 18 g L^{-1}	15 May, S-metolachlor 2.2 kg ai ha^{-1} + metribuzin O.53 kg ai ha^{-1}	3 Apr., duflufenacil 0.3 kg ai ha^{-1} + dimethenamid-P 0.17 kg ai ha^{-1} + glyphosate 0.87 kg ae ha^{-1} + 0.25% v/v NIS+ AMS 18 g L^{-1}	S-metolachor 1.5 kg ai ha^{-1} + metribuzin 0.4 kg ai ha^{-1}
Postemergence	24 May, fomesafen 0.35 kg ha^{-1} + glyphosate 1.2 kg ae ha^{-1} + UAN 2.3 L ha^{-1} + 0.25% v/v NIS	4 June, glufosinate 0.8 kg ai ha^{-1} + AMS 18 g L^{-1} 1 July, glufosinate 0.8 kg ai ha^{-1} + S-metolachor 1.36 kg ai ha^{-1} + fomesafen 0.3 kg ai ha^{-1} + AMS 18 g L^{-1} + 0.25% v/v NIS	25 May S-metolachor 1.36 kg ai ha^{-1} + fomesafen 0.3 kg ai ha^{-1} + AMS 18 g L^{-1} + 0.25% v/v NIS	3 June, glyphosate 1.2 kg ae ha^{-1}	6 Jul., glyphosate 1.58 kg ae ha^{-1} + fomesafen 0.3 kg ai ha^{-1} + S-metolachor 1.2 kg ai ha^{-1}	Glyphosate 1.26 kg ae ha^{-1}
	22 June, glyphosate 1.2 kg ae ha^{-1} + AMS 18 g L^{-1} + 0.25% v/v NIS		9 July, glufosinate 0.8 kg ha^{-1} + flumiclorac-pentyl 0.48 kg ai ha^{-1} + AMS 18 g L^{-1} + 0.25% v/v NIS			
Insect management	NA	NA	NA	NA	NA	NA
Disease management	NA	NA	10 July, azoxystrobin 0.08 kg ai ha^{-1}	NA	NA	NA

Note. [†] Abbreviations: ae, acid equivalent; a.i., active ingredient; COC, crop oil concentrate; MSO, Methylated seed oil; NIS, non-ionic surfactant; NA, none applied; and UAN, urea ammonium nitrate.

[‡] Chemical name: azoxystrobin, Methyl (2E)-2-(2-{[6-(2-cyanophenoxy)pyrimidin-4-yl]oxy}phenyl)-3-methoxyacrylate; flumiclorac-pentyl, pentyl 2-[2-chloro-5-(1,3-dioxo-4,5,6,7-tetrahydroisoindol-2-yl)-4-fluorophenoxy]acetate; dimethenameid-P, 2-chloro-N-(2,4-dimethylthiophen-3-yl)-N-[(2S)-1-methoxypropan-2-yl]acetamide; fomesafen, 5-[2-chloro-4-(trifluoromethyl)phenoxy]-N-methylsulfonyl-2-nitrobenzamide; glufosinate, 2-amino-4-[hydroxy(methyl)phosphoryl]butanoic acid; glyphosate, N-(phosphonomethyl)glycine; metrabuzin, 4-amino-6-tert-butyl-3-methylsulfanyl-1,2,4-triazin-5-one; S-metolachlor, 2-chloro-N-(2-ethyl-6-methylphenyl)-N-[(2S)-1-methoxypropan-2-yl]acetamide; and sulflufenacil, 2-chloro-4-fluoro-5-[3-methyl-2,6-dioxo-4-(trifluoromethyl)pyrimidin-1-yl]-N-[methyl(propan-2-yl)sulfamoyl]benzamide.

In 2014, upstate Missouri experienced record yields due not only to cool summer temperatures (data not presented), but also to uniform distribution of precipitation, with Novelty receiving 771 mm and Albany receiving 751 mm. In 2015, precipitation at Albany and Novelty was above average, which delayed soybean planting.

Initial soil samples were taken from each site (Table 1). Soil test P levels were very low (Novelty in 2011, 2012), low (Novelty in 2013, 2014, and Albany in 2013), and medium (Albany in 2014) according to Buchholz et al. (2004). Soil test Zn levels were low (Novelty in 2011, 2012, and 2013) or medium (Novelty in 2014, Albany in 2013 and 2014) (Buchholz, Brown, Garret, Hanson, & Wheaton, 2004). Finally, soil test S was medium across all site-years (Buchholz et al., 2004).

3.2 Corn Response

Corn plant population was similar at all site-years (Novelty in 2011 and 2012, $P = 0.86$, and Novelty and Albany in 2013 and 2014, $P = 0.92$). Ear leaf tissue concentration at VT for P, S, and Zn were combined over the two years (Novelty in 2013 and 2014) where the measurements were collected due to an absence of a significant interaction between years. These sites had low soil test P and low-to-medium soil test Zn. Ear leaf P concentrations ranged from 2.38 to 3.13 g kg^{-1} (Table 4). All treatments, except the non-treated control, were similar and were within the sufficiency range for ear leaf P concentrations (2.5 to 5.0 g kg^{-1}) (Jones et al., 1967; Bryson et al., 2014). The non-treated control had 2.4 g kg^{-1} P, which was below the plant sufficiency range (Jones et al., 1967; Bryson et al., 2014). Concentrations of ear leaf S and ear leaf P were similar. All treatments were similar, and they were up to 0.2 g kg^{-1} greater than the non-treated no N or no P controls. All treatments including P and N were within the ear leaf sufficiency range (1.5 to 4.0 g kg^{-1}) for Zn (Jones, 1967; Bryson et al., 2014). Ear leaf S concentrations ranged from 1.6 to 2.1 g kg^{-1} and were greater than the no N and no P controls. Similar results for ear leaf P and S concentrations were reported in Iowa (Sawyer & Barker, 2010). There was no clear effect of Zn on P concentration, similar to other research evaluating P-Zn interactions (Keefer et al., 1972; Rehm et al., 1981).

Table 4. Corn ear SO$_4$-S concentration for Zn treatments at Novelty from 2013-2014. Data were combined over years. All P amounts were 80 kg P$_2$O$_5$ ha^{-1}

Zn treatments[†]	Zn amount	P	Zn	SO$_4$-S
	---- kg ha^{-1} ----	---- g kg^{-1} ----	---- mg kg^{-1} ----	---- g kg^{-1} ----
Non-treated control	0	2.4	20.9	1.7
N only control	0	2.9	30.2	2.1
DAP	0	2.8	28.6	2.0
DAP + ZnSO$_4$	2.2	3.0	29.4	2.1
DAP + ZnSO$_4$	5.6	3.0	30.4	2.1
MAP	0	3.0	28.5	2.1
MAP + ZnSO$_4$	2.2	2.9	27.5	2.0
MAP + ZnSO$_4$	5.6	2.8	27.7	1.9
MES10	0	2.8	26.1	1.9
MESZ	2.2	2.9	27.1	2.0
MAP + AS	0	3.1	27.7	2.1
MAP + SuperZn	2.2	3.1	27.7	2.0
MAP + SuperZn	5.6	3.0	29.1	2.0
DAP + AS	0	3.0	27.2	2.1
DAP + SuperZn	2.2	3.0	26.5	2.0
DAP + SuperZn	5.6	3.1	29.2	2.0
LSD ($P = 0.05$)		0.3	2.9	0.2

Note. [†] Abbreviations: AMS, ammonium sulfate; DAP, diammonium phosphate; MAP, monoammonium phosphate.

Ear leaf Zn concentrations were interesting because the N-only control treatment had the second highest Zn concentration, which indicated sufficient plant available Zn in the soil. An application of DAP+ZnSO$_4$ at 5.6 kg Zn ha^{-1} and N-only treatments had 4.13 to 9.43 mg kg^{-1} greater ear leaf Zn concentration than MES10 and the non-treated control. The non-treated, no N control was 5.15 to 9.43 mg kg^{-1} less than all other treatments for ear leaf Zn concentration, but this was probably due to N affecting Zn uptake. These data indicate no clear impact of the Zn treatments on ear leaf Zn concentration over the two years ear leaves were collected since the N-only control had ear leaf Zn concentrations similar to the Zn treatments. A corn fertilizer study in West Virginia with three amounts of Zn (0, 3.36, and 6.72 kg Zn ha^{-1}) also observed no-yield response to Zn treatment (Stout & Bennett, 1983). An efficacy study of ZnO concluded that it and ZnSO$_4$ had similar plant nutrient recovery when incorporated in the soil; however, ZnSO$_4$ had a greater plant recovery than ZnO when the fertilizer was band- or surface-applied (McBeath & McLaughlin, 2014). In Michigan, a three-year study with 220 kg of fertilizer ha^{-1} comparing MAP and DAP observed no change in tissue P or Zn concentration and detected no yield difference between MAP and DAP (Yerokun & Christenson, 1990). In addition, P fertility did not appear to decrease Zn uptake, similar to Halim et al. (1968).

A significant two-way interaction occurred between treatments and site-years for corn grain yield; therefore, data were analyzed by individual site-years and reported separately (Table 5). In a planned comparison, contrasts comparing Zn amounts (2.2 vs 5.6 kg Zn ha^{-1}), Zn sources (SuperZn vs. ZnSO$_4$), and P sources (MAP vs. DAP) showed no significant differences in yield for the four site-years (Novelty and Albany in 2014 and 2015) that were evaluated (data not presented). At Novelty in 2011, all treatments yielded 6,670 to 7,860 kg ha^{-1} greater than the non-treated, no-N control with yields from 9,090 to 10,280 kg ha^{-1} (Table 5). With the extreme drought in 2012, yields were low (1,120 to 1,770 kg ha^{-1}). The non-treated control, N-only, and MESZ yielded 531 to 643 kg ha^{-1} greater than MAP, MAP+ZnSO$_4$ at 2.2 kg Zn ha^{-1}, MAP+ZnSO$_4$ at 5.6 kg Zn ha^{-1}, and MAP+AMS. At Novelty in 2013, MAP+SuperZn at 5.6 kg Zn ha^{-1} and DAP+AMS increased yields 560 to 2,850 over the non-treated control, N-only control, DAP+ZnSO$_4$ at 5.6 kg Zn ha^{-1}, and DAP+ZnSO$_4$ at 2.2 kg Zn ha^{-1}.

Table 5. Corn grain yield response to Zn treatments at Novelty (2011-2014) and Albany (2013-2014). Phosphorus was applied at 90 kg P$_2$O$_5$ ha^{-1}

Zn treatments[†]	Zn amount	Novelty				Albany	
		2011	2012	2013	2014	2013	2014
		---------- kg ha^{-1} ----------					
Non-treated control	0.0	2,420	1,770	6,800	12,700	6,520	7,240
N only control	0.0	9,090	1,710	9,080	16,180	6,990	8,970
DAP	0.0	9,420	1,380	9,420	15,950	6,930	9,200
DAP + ZnSO$_4$	2.2	9,230	1,600	9,040	16,800	6,520	10,420
DAP + ZnSO$_4$	5.6	9,480	1,620	9,010	15,820	6,520	8,760
MAP	0.0	9,740	1,180	9,200	16,480	6,590	10,540
MAP + ZnSO$_4$	2.2	10,270	1,120	9,620	16,340	6,590	9,870
MAP + ZnSO$_4$	5.6	9,660	1,130	9,470	16,800	6,660	10,420
MES10	0.0	9,890	1,420	9,500	16,710	6,590	8,810
MESZ	2.2	10,280	1,750	9,610	16,130	6,790	7,210
MAP + AS	0.0	9,800	1,120	9,350	16,140	6,660	8,710
MAP + SuperZn	2.2	-[‡]	-	9,500	16,280	7,260	10,200
MAP + SuperZn	5.6	-	-	9,650	16,730	6,660	10,230
DAP + AS	0.0	-	-	9,640	16,490	6,660	9,470
DAP + SuperZn	2.2	-	-	9,510	14,790	6,660	10,220
DAP + SuperZn	5.6	-	-	9,400	16,580	6,930	8,560
LSD ($P = 0.05$)		1,260	450	540	806	NS	1,580

Note. [†] Abbreviations: AMS, ammonium sulfate; DAP, diammonium phosphate; MAP, monoammonium phosphate.

[‡] Treatments were not applied at these two locations.

All other treatments had similar grain yields. Grain yields at Albany in 2013 were similar among treatments ($P = 0.57$). In 2014, both Novelty and Albany experienced exceptionally high yields (12,700 to 16,800 kg ha^{-1} and 7,210 to 10,540 kg ha^{-1}, respectively), due to good overall precipitation (Figure 1) and low temperatures during pollination and grain fill (data not presented). At Novelty, DAP+ZnSO$_4$ at 2.2 kg Zn ha^{-1}, MAP+ZnSO$_4$ at 5.6 kg Zn ha^{-1}, MES10, and MAP+SuperZn at 5.6 kg Zn ha^{-1} had 763 to 4098 kg ha^{-1} greater yields than the non-treated control, DAP, DAP+ZnSO$_4$ at 5.6 kg Zn ha^{-1}, or DAP+SuperZn at 2.2 kg Zn ha^{-1}. At Albany in 2014, MAP alone yielded 1,570 to 3,300 kg ha^{-1} greater than the non-treated control, N-only, MES10, MESZ, DAP+SuperZn at 5.6 kg Zn ha^{-1}, DAP+ZnSO$_4$ at 5.6 kg Zn ha^{-1}, and MAP+AMS. Albany had medium soil test P, S, and Zn, which likely resulted in limited yield differences among treatments (Table 1). An increase in corn grain yield has been related to P uptake which has been greater than Zn fertility (Rehm et al., 1983), and there was no apparent Zn-P interaction which was similar to other research (Rehm et al., 1981).

Table 6. Corn grain protein concentrations at Novelty (2011-2014) and Albany (2013-2014). Data combined over years were denoted. Phosphorus was applied at 90 kg P_2O_5 ha^{-1}

Zn treatments[†]	Zn amount	Novelty 2011	Novelty 2013 and 2014[‡]	Albany 2013 and 2014[‡]
	---- kg ha^{-1} ----	\<--- g kg^{-1} ---\>		
Non-treated control	0	70	69	73
N only control	0	94	84	75
DAP	0	96	85	79
DAP + $ZnSO_4$	5.6	96	85	80
DAP + $ZnSO_4$	2.2	95	85	80
MAP	0	89	84	77
MAP + $ZnSO_4$	5.6	94	84	83
MAP + $ZnSO_4$	2.2	96	85	75
MES10	0	95	84	77
MESZ	2.2	94	83	81
MAP + AS	0	95	85	82
MAP + SuperZn	2.2	-	84	82
MAP + SuperZn	5.6	-	83	82
DAP + AS	0	-	83	83
DAP + SuperZn	2.2	-	83	80
DAP + SuperZn	5.6	-	83	83
LSD (P = 0.05)		7	2	5

Note. [†] Abbreviations: AMS, ammonium sulfate; DAP, diammonium phosphate; MAP, monoammonium phosphate.

[‡] Data were combined over years.

[§] Treatments were not applied during these two years.

Grain oil, protein, and starch concentrations were observed at Novelty (2011, 2013, and 2014) and Albany (2013 and 2014), but none of the treatments affected grain oil concentration at any site-year [(Novelty in 2011 and 2012, $P = 0.41$) (Novelty and Albany in 2013 and 2014, $P = 0.16$)] (data not presented). Grain protein at Novelty in 2013 and 2014 as well as Albany in 2013 and 2014 had no treatment-by-year interaction, so data were combined over years (Table 6). At Novelty in 2011, protein concentration ranged from 70 to 96 g kg^{-1}. All fertilizer treatments were similar, but they were 19 to 26 g kg^{-1} greater than the no N, non-treated control. Novelty in 2012 showed no significant difference ($P = 0.11$) among treatments, which likely was due to extreme drought (data not presented). At Novelty and Albany in 2013 and 2014, protein concentration with MAP+AMS was significantly greater than MAP+SuperZn at 5.6 kg Zn ha^{-1} and the non-treated control by 3 and 15 g kg^{-1}, respectively. All other fertilized treatments had similar protein concentrations that ranged from 83 to 85 mg kg^{-1}. At Albany in 2013 and 2014, protein concentrations ranged from 73 to 83 g kg^{-1}. DAP+AMS, MAP+$ZnSO_4$ at 5.6 kg Zn ha^{-1}, and DAP+SuperZn at 5.6 kg Zn ha^{-1} had 6 to 11 g kg^{-1} greater protein concentration than the non-treated control, N-only, MAP, MES10, and MAP+$ZnSO_4$ at 2.2 kg Zn ha^{-1}. At Novelty in 2011, the non-treated control had 64 to 170 g kg^{-1} higher starch concentration than all other treatments (Table 7). MAP had 68 to 124 g kg^{-1} greater starch concentration than all other fertilizer treatments. Novelty in 2012 showed no differences among treatments ($P = 0.87$). At Novelty in 2013 and 2014, the non-treated control was 52 to 96 g kg^{-1} greater than all treatments except DAP+AMS. At Albany in 2013 and 2014, DAP, DAP+$ZnSO_4$ at 5.6 kg Zn ha^{-1}, MAP+$ZnSO_4$ at 2.2 kg Zn ha^{-1}, and DAP+SuperZn at 2.2 kg Zn ha^{-1} had 59 to 100 g kg^{-1} greater grain starch concentration than MAP+SuperZn at 2.2 kg Zn ha^{-1}, DAP+SuperZn at 5.6 kg Zn ha^{-1}, DAP+SuperZn at 2.2 kg Zn ha^{-1}. At Albany in 2013 and 2014, all treatments had similar starch concentrations ($P = 0.17$). Although significance occurred at five site-years for protein and three site-years for starch concentrations, differences were inconsistent. The relationship between nitrogen and corn protein concentration is commonly observed (Uribelarrea, Below, & Moose, 2004); however, Kaiser and Lamb (2008) showed protein contents were lower when P was applied, compared to no P application, especially at 120 kg N ha^{-1} and greater N applications. However, we did not observe this in our research.

3.3 Soil Test P, S, and Zn Following Corn

Treatments with applied P fertilizer had 16 to 37 kg ha^{-1} greater soil test P than the non-treated control and the N-only treatment (Table 8). The largest soil test P was DAP+SuperZn at 5.6 kg Zn ha^{-1}, which had 16 to 33 kg ha^{-1} greater soil test P than the non-treated control, N-only, MAP, DAP, DAP+SuperZn at 2.2 kg Zn ha^{-1}, and MAP+SuperZn at 2.2 kg Zn ha^{-1}. However, the DAP treatment had 14.8 kg ha^{-1} less soil test P than MAP+SuperZn at 5.6 kg Zn ha^{-1}, while all other treatments with P had similar soil test P.

Table 7. Grain starch concentrations at Novelty (2011, 2013-2014) and Albany (2013-2014). Data were combined over years were denoted. Phosphorus was applied at 90 kg P$_2$O$_5$ ha^{-1} for all treatments including P

Zn treatments[†]	Zn amount	Novelty 2011	Novelty 2013-2014[‡]
	kg ha^{-1}	---------------------------- g kg^{-1} ----------------------------	
Non-treated	0	731	739
N-Only	0	716	732
DAP	0	712	730
DAP + ZnSO$_4$	2.2	716	732
DAP + ZnSO$_4$	5.6	716	730
MAP	0	724	732
MAP + ZnSO$_4$	2.2	715	730
MAP + ZnSO$_4$	5.6	714	732
MES10	0	718	732
MESZ	2.2	718	732
MAP + AS	0	714	730
MAP + SuperZn	2.2	-[§]	732
MAP + SuperZn	5.6	-	734
DAP + AS	0	-	734
DAP + SuperZn	2.2	-	730
DAP + SuperZn	5.6	-	730
LSD (*P* = 0.05)		61	49

Note. [†] Abbreviations: AMS, ammonium sulfate; DAP, diammonium phosphate; MAP, monoammonium phosphate.

[‡] Data were combined over years. No differences among treatments were observed in 2012.

[§]Treatments were not applied during these two years.

Table 8. Soil test P, Zn, and SO$_4$-S after corn harvest at Novelty and Albany (2013-2014). Phosphorus was applied at 90 kg P$_2$O$_5$ ha^{-1} for all treatments including P

| Zn treatments[†] | Zn amount | P[‡] | Zn | | SO$_4$-S | |
			Novelty[§] 2013 and 2014	Albany[§] 2013 and 2014	Novelty[§] 2013 and 2014	Albany[§] 2013 and 2014
	--- kg ha^{-1} ---	kg ha^{-1}	------ mg ha^{-1} -----	------ mg ha^{-1} ------	----- mg ha^{-1} -----	------ mg ha^{-1} ------
Non-treated	0	33.8	0.58	0.80	4.48	5.40
N-Only	0	31.6	0.56	0.70	4.08	4.19
DAP	0	48.3	0.61	1.20	4.26	4.61
DAP + ZnSO$_4$	2.2	50.8	0.84	2.20	5.40	4.64
DAP + ZnSO$_4$	5.6	55.1	1.42	1.00	5.02	5.23
MAP	0	49.4	0.66	2.00	4.57	5.88
MAP + ZnSO$_4$	2.2	58.7	1.08	1.30	4.86	4.69
MAP + ZnSO$_4$	5.6	53.3	2.32	2.00	5.34	5.48
MES10	0	55.4	0.62	1.40	6.64	5.24
MESZ	2.2	56.1	1.08	1.10	5.55	4.59
MAP + AS	0	55.5	0.67	0.90	5.40	6.20
MAP + SuperZn	2.2	50.2	0.84	1.10	4.89	5.33
MAP + SuperZn	5.6	61.5	1.37	1.56	4.79	5.93
DAP + AS	0	53.9	0.67	0.90	4.76	5.45
DAP + SuperZn	2.2	48.7	1.10	1.10	5.41	5.50
DAP + SuperZn	5.6	64.8	2.44	1.50	4.80	5.13
LSD		12.9	0.58	NS	0.96	0.75

Note. [†] Abbreviations: AMS, ammonium sulfate; DAP, diammonium phosphate; MAP, monoammonium phosphate.

[‡] Data were combined over sites and years.

[§] Data were combined over years.

Soil test S and Zn had a significant site-year-by-treatment interaction, but due to similar soil series, data were combined over years for individual sites and analyzed. Soil test Zn ranged from 0.56 to 2.44 mg ha^{-1} and 0.71 to 1.56 mg ha^{-1} at Novelty and Albany, respectively. All treatments had at least medium soil test Zn at the end of the growing season (Buchholz et al., 2004). When Zn was applied at Novelty in 2013 and 2014, soil test Zn generally increased compared to fertilizer treatments that had no Zn. DAP+SuperZn at 5.6 kg Zn ha^{-1} and MAP+ZnSO$_4$ at 5.6 kg Zn ha^{-1} had significantly greater soil test Zn than all other treatments, while all other treatments with P, S, or Zn, were similar. At Albany, DAP+ZnSO$_4$ at 2.2 kg Zn ha^{-1} and MAP+ZnSO$_4$ at 5.6 kg Zn ha^{-1}, and MAP had 1.1 to 1.5 mg kg^{-1} greater soil test Zn than DAP+AMS, MAP+AMS, N-only, and the non-treated control, but all other treatments were similar. This indicates that even though there was no crop yield response to Zn, soil buildup occurred.

Table 9. Soybean plant population, grain yield, oil, and protein concentration response to fertilizer treatments applied the previous year at Novelty (2012-2015) and Albany (2014-2015)

Zn treatments[†]	Zn amount	Population		Yield		Oil		Protein	
		2012-13	2014-15	2012-13	2014-15	2012-13	2014-15	2012-13	2014-15
	-- kg ha^{-1} --	------- plants ha^{-1} -------		-------- kg ha^{-1} -------		---------------------- g kg^{-1} ----------------------			
Non-treated	0	307,600	321,700	2,640	2,900	197	189	352	349
N-only	0	314,300	317,000	2,670	3,120	193	189	357	348
DAP	0	310,100	292,300	2,570	3,170	192	188	360	349
DAP + ZnSO4	2.2	331,600	312,100	2,620	3,080	193	188	358	347
DAP + ZnSO4	5.6	122,800	336,800	2,620	3,130	192	189	358	347
MAP	0	340,300	321,000	2,550	3,120	191	188	360	348
MAP + ZnSO4	2.2	316,500	333,600	2,590	3,140	192	188	359	349
MAP + ZnSO4	5.6	280,000	309,400	2,600	3,120	193	188	357	347
MESZ	2.2	333,600	333,100	2,623	3,140	191	189	360	349
MES10	0	340,300	311,600	2,650	3,130	193	188	357	346
MAP + AMS	0	323,000	302,500	2,660	3,130	192	188	358	348
MAP + SuperZn	2.2	-[‡]	317,500	-	3,210	-	189	-	347
MAP + SuperZn	5.6	-	299,700	-	3,190	-	189	-	344
DAP + AMS	0	-	299,700	-	3,230	-	189	-	346
DAP + SuperZn	2.2	-	295,000	-	3,060	-	190	-	347
DAP + SuperZn	5.6	-	333,100	-	3,130	-	189	-	348
P-value		0.5	0.15	0.85	0.16	0.42	0.37	0.33	0.46

Note. [†] Population and yield data were determined for all site-years. Grain oil and protein data were determined at all site-years except Albany in 2014. Data were combined over site-years unless denoted otherwise.

Abbreviations: AMS, ammonium sulfate; DAP, diammonium phosphate; MAP, monoammonium phosphate.

[‡] Treatments were not applied during these two years.

Soil test S ranged from 4.1 to 6.6 mg kg^{-1} and 4.2 to 6.2 mg kg^{-1} at Novelty and Albany, respectively. All treatments had medium soil test S. At Novelty, all treatments with S or Zn had similar soil test S levels (Buchholz et al., 2004). MESZ and MES10 had 0.98 to 2.56 higher soil test S amounts than the non-treated control, N-only, DAP, and MAP. At Albany, MAP+AMS had a 1.07 to 2.01 mg kg^{-1} greater soil test S concentration than DAP+SuperZn at 5.6 kg Zn ha^{-1}, MAP+ZnSO$_4$ at 2.2 kg Zn ha^{-1}, DAP+ZnSO$_4$ at 2.2 kg Zn ha^{-1}, DAP, and N-only. Treatments that included AMS, MES10, and MESZ had similar soil test S concentrations. Post-harvest soil samples showed no significant differences among treatments for potassium ($P = 0.7$), magnesium ($P = 0.75$), calcium ($P = 0.54$), or organic matter ($P = 0.2$) levels (data not presented).

3.4 Soybean Response Following Corn

Soybean plant population ranged from 303,000 to 340,200 plants ha^{-1}, but was non-significant ($P = 0.52$) between treatments (Table 9). At Novelty and Albany in 2014 and 2015, whole-plant samples taken before physiological maturity at R6 (Fehr & Caviness, 1971) were analyzed for P, S, and Zn concentration. At Novelty in 2014 and 2015, no differences were seen in plant P ($P = 0.69$) or S ($P = 0.26$) concentration (Table 10). At Albany in 2015, no significance differences appeared between plant P ($P = 0.64$), S ($P = 0.66$), or Zn ($P = 0.98$) concentration; however, at Albany in 2014, P, S, and Zn concentrations were significantly different. At Albany in 2014, plant P concentration ranged from 2.48 to 3.81 g kg^{-1}. Plant P concentration in the DAP+AMS treatment was 0.57 to 1.33 g kg^{-1} greater than all treatments. The N-only control had a plant Zn concentration similar to 7 of the 16 treatments, indicating there may be no strong effect regarding application of P for soybeans between treatments applied for corn and plant P concentrations observed in soybean the following year. A 14-site study in Iowa showed increased P uptake at eight sites (Borges & Mallarino, 2003).

At Albany in 2014, whole-plant S concentration ranged from 1.95 to 2.99 g kg^{-1} (Table 10). Similar to plant P concentration, DAP+AMS had the greatest plant S concentration, which was 0.33 to 1 g kg^{-1} greater than all treatments except MAP+SuperZn at 5.6 kg Zn ha^{-1} and DAP+SuperZn at 5.6 kg Zn ha^{-1}. Interestingly, treatments including SuperZn generally had a greater plant S concentration than treatments including ZnSO$_4$. Again, this could point to the lack of a good relationship between fertilizer treatments and plant S concentrations. Similarly, when S fertilizer was applied to wheat in a wheat-soybean rotation, no significant differences were reported in soybean S concentrations (Singh et al., 2014).

Plant Zn concentration was significantly affected at both locations in 2014 and Novelty in 2015 (Table 10). At Albany in 2014, plant Zn concentrations ranged from 24.2 to 35.6 mg kg^{-1}. Similar to both plant P and S concentration data, DAP+AMS had the highest plant Zn concentration at 35.6 mg kg^{-1}, which was 5.0 to 11.4 mg kg^{-1} greater than all treatments except DAP+SuperZn at 5.6 kg Zn ha^{-1} and MAP+SuperZn at 2.2 or 5.6 kg Zn ha^{-1}. In a planned contrast comparing Zn rate (2.2 vs. 5.6 kg Zn ha^{-1}) and Zn source (SuperZn vs. ZnSO$_4$), SuperZn had significantly higher plant Zn concentration than ZnSO$_4$ ($P = 0.0005$), while Zn rate showed no significant difference ($P = 0.12$). At Novelty in 2014 and 2015, plant Zn concentration ranged from 26.1 to 34.7 mg kg^{-1}. MAP+SuperZn at 5.6 kg Zn ha^{-1} increased plant Zn concentration by 4.9 to 10.3 mg kg^{-1} over all treatments except DAP+ZnSO$_4$ at 5.6 kg Zn ha^{-1}. In a planned contrast comparing Zn rate (2.2 vs 5.6 kg Zn ha^{-1}) and Zn source (SuperZn vs ZnSO$_4$), Zn at 5.6 kg Zn ha^{-1} increased plant Zn concentration compared to 2.2 kg Zn ha^{-1} ($P = 0.0008$), while Zn source was not significant ($P = 0.72$). These results showed completely different responses at Novelty compared to Albany.

Table 10. Soybean whole plant P, S, and Zn concentrations as affected by Zn treatments at Albany and Novelty in 2014-2015

Zn treatments[†]	Zn amount	Leaf P			Leaf S			Leaf Zn		
		Novelty[‡]	Albany		Novelty[‡]	Albany		Novelty[‡]	Albany	
			2014	2015		2014	2015		2014	2015
	kg ha^{-1}	----------- g kg^{-1} -----------			----------- g kg^{-1} -----------			----------- mg kg^{-1} -----------		
Non-treated control	0	3.1	2.58	3.8	2.2	2.10	2.5	27.3	29.5	24.9
N-only	0	3.1	2.83	3.6	2.2	2.14	2.5	28.6	28.8	24.7
DAP	0	3.1	2.55	3.7	2.1	2.18	2.5	26.1	26.0	25.4
DAP + ZnSO$_4$	2.2	3.3	2.98	3.6	2.4	2.32	2.5	34.7	28.3	24.8
DAP + ZnSO$_4$	5.6	3.1	2.65	3.8	2.3	2.28	2.5	29.6	27.7	25.4
MAP	0	3.0	2.48	3.7	2.1	1.94	2.4	29.1	24.2	24.9
MAP + ZnSO$_4$	2.2	3.2	2.60	3.4	2.3	2.14	2.4	31.5	25.7	26.1
MAP + ZnSO$_4$	5.6	3.2	2.48	3.6	2.2	2.28	2.5	28.9	27.0	25.0
MESZ	2.2	3.2	2.73	3.7	2.3	2.08	2.5	28.2	25.0	26.1
MES10	0	3.2	3.14	3.7	2.2	2.35	2.5	28.5	29.0	25.0
MAP + AMS	0	3.2	3.24	3.6	2.3	2.63	2.5	27.5	30.6	26.1
MAP + SuperZn	2.2	3.0	3.0	3.7	2.2	2.40	2.5	26.3	31.3	25.4
MAP + SuperZn	5.6	3.3	3.59	3.6	2.3	2.80	2.5	36.4	32.0	24.4
DAP + AMS	0	3.3	3.81	3.9	2.3	2.99	2.5	26.5	35.6	25.0
DAP + SuperZn	2.2	3.2	2.86	3.6	2.2	2.67	2.5	30.5	29.0	25.4
DAP + SuperZn	5.6	3.2	3.03	3.5	2.2	2.67	2.4	30.7	34.5	24.6
LSD ($P = 0.05$)		NS	0.04	NS	NS	0.03	NS	3.4	4.7	NS

Note. [†] Abbreviations: AMS; Ammonium sulfate; DAP, diammonium phosphate; MAP, monoammonium phosphate.

[‡] Data were combined over years.

Nutrient concentration and total plant tissue weights were used to calculate total plant uptake of P, S, and Zn (Table 11). Although significant differences were observed in plant nutrient concentrations, all treatments were similar when total plant uptake was calculated [P ($P = 0.52$), S ($P = 0.49$), and Zn ($P = 0.60$)]. In 2012 and 2013, grain yields ranged from 2,380 to 2,490 kg ha^{-1}, and all treatments yielded similarly ($P = 0.85$) (Table 9). Seed oil and protein content were measured to determine any impact of treatments on seed quality. Seed oil ranged from 191 to 197 g kg^{-1}, with no difference between treatments ($P = 0.42$). Protein concentration ranged from 352 to 360 g kg^{-1}, but also showed no effects of the treatments ($P = 0.46$). In 2014 and 2015, soybean population, yield, oil, and protein combined over years and locations (Novelty and Albany). Soybean plant population ranged from 293,300 to 336,800 plants ha^{-1} and showed no difference among treatments ($P = 0.15$). Similarly, yields (2,900 to 3,200 kg ha^{-1}) showed no differences among treatments ($P = 0.16$) (Table 9). Grain quality (oil, protein) was also similar among treatments ($P = 0.37$, $P = 0.46$, respectively). Between treatments, seed oil concentration had a small range, 180 to 190 g kg^{-1}, and protein concentration had a narrower range, 346 to 349 g kg^{-1}.

Fertilizer applications to corn in a corn-soybean rotation had inconsistent effects on soybean yield and seed quality. Anthony et al. (2012) and Buah, Polito, and Killorn (2000) reported no difference in yield between 0 and 56 kg P ha^{-1}. However, others have shown increased yields, but at low soil test P levels (Borges & Mallarino, 2000, 2003). An Iowa study at over 112 locations showed positive yield effects from P fertilizer at only 20 sites (Haq & Mallarino, 2005). Applications of S fertilizer have also had mixed results. Divito, Echeverria, Andrade, and Sadras (2015) reported increased yields with S application, while Singh et al. (2014) reported no response to S fertilizer.

Table 11. Total soybean plant P, S, and Zn uptake at Novelty and Albany in 2014-2015. Data were combined over all site-years

Zn treatments[†]	Zn amount	P uptake	S uptake	Zn uptake
	kg ha^{-1}	---------------------------------- kg ha^{-1} ----------------------------------		
Non-treated control	0	17.5	12.7	1.2
N-only	0	18.1	13.4	1.3
DAP	0	21.8	15.4	1.5
DAP + ZnSO$_4$	2.2	19.0	13.9	1.5
DAP + ZnSO$_4$	5.6	19.8	14.5	1.4
MAP	0	18.0	12.9	1.4
MAP + ZnSO$_4$	2.2	18.7	13.8	1.3
MAP + ZnSO$_4$	5.6	21.1	15.0	1.5
MESZ	2.2	19.8	14.8	1.4
MES10	0	20.5	14.8	1.4
MAP + AMS	0	20.0	14.3	1.4
MAP + SuperZn	2.2	20.8	15.3	1.4
MAP + SuperZn	5.6	19.2	14.2	1.6
DAP + AMS	0	19.0	13.8	1.2
DAP + SuperZn	2.2	21.8	15.8	1.6
DAP + SuperZn	5.6	21.7	15.5	1.6
P-value		0.52	0.49	0.60

Note. [†] Abbreviations: AMS; Ammonium sulfate; DAP, diammonium phosphate; MAP, monoammonium phosphate.

4. Conclusion

Variation in precipitation over site-years strongly affected corn grain yields. Ear leaf P tissue concentration at VT showed fertilized treatments within the sufficiency range, with no significant difference among treatments. Ear leaf S tissue concentration at VT showed no significant differences, except for the non-treated control. All fertilized treatments were within the ear leaf S sufficiency range. All treatments were similar for ear leaf Zn concentration at VT, except for the non-treated control, which was significantly lower than all other treatments and for MES10, which was significantly lower than N-only and DAP+ZnSO$_4$ at 5.6 kg Zn ha^{-1}. Yields showed no significant differences at Novelty in 2013, and all treatments were similar at Novelty in 2011. In 2012, because of a severe drought, yields in the non-treated control were the greatest. At Novelty in 2014 and Albany in 2013 and 2014, adding S and/or Zn had no effect on yield. The rate of Zn fertilizer (2.2 vs 5.6 kg Zn ha^{-1}) also showed no significant effect on yield. When P was determined post-harvest, soil samples reflected the application of P fertilizer. Generally, when Zn was applied, soil test Zn increased. At Novelty, MES10 and MESZ had the greatest increase in soil test S, while at Albany MAP+AMS had the greatest amount of soil test S. Carry-over fertilizer from corn experiments showed differences in plant nutrient concentrations, but this had no effect on total plant nutrient (P, S, or Zn) uptake, grain yield or quality. However, at Novelty plant Zn concentration was increased with 5.6 kg Zn ha^{-1} compared to 2.2 kg Zn ha^{-1}, while Albany showed an increase in soybean Zn concentration with SuperZn when compared to $ZnSO_4$. This indicates that micro-nutrient uptake was affected by Zn rate and source, depending on the soil type.

References

Amijee, F., Stribley, D. P., & Tinker, P. B. (1990). Soluble carbohydrates in roots of leek (*Allium porrum*) plants in relation to phosphorus supply and VA mycorrhizas. *Plant Nutrition-Physiology and Applications*. New York, NY: Springer.

Anthony, P., Maizer, G., Sparrow, S., & Zhang, M. (2012). Soybean yield and quality in relation to soil properties. *Agronomy Journal, 104*, 1443-1458. http://dx.doi.org/10.1007/978-94-009-0585-6_27

Borges, R., & Mallarino, A. P. (2000). Grain yield, early growth, and nutrient uptake of no-till soybean as affected by phosphorus and potassium placement. *Agronomy Journal, 92*, 380-388. http://dx.doi.org/10.2134/agronj2000.922380x

Borges, R., & Mallarino, A. P. (2003). Broadcast and deep-band placement of phosphorus and potassium for soybean managed with ridge tillage. *Soil Science Society America Journal, 67*, 1920-1927. http://dx.doi.org/10.2136/sssaj2003.1920

Briskin, D. P., Bloom, A., Taiz, L., & Zeiger, E. (2010). Mineral Nutrition. *Plant Physiology* (5th ed.). Sunderland, MA: Sinauer Associates Inc.

Broadley, M. R., White, P. J., Hammond, J. P., Zelko, I., & Lux, A. (2007). Zinc in plants. *New Phytologist, 173*, 677-702. http://dx.doi.org/10.1111/j.1469-8137.2007.01996.x

Bryson, G. M., Mills, H. A., Sasseville, D. N., Jones, J. B., & Barker, A. V. (2014). *Plant Analysis Handbook III: A Guide to Sampling, Preparation, Analysis, Interpretation and Use of Results of Agronomic and Horticultural Crop Plant Tissue*. Athens, GA: Micro-Macro Publishing, Inc.

Buah, S. S., Polito, T. A., & Killorn, R. (2000). No-tillage soybean response to banded and broadcast and direct and residual fertilizer phosphorus and potassium applications. *Agronomy Journal, 92*, 657-662. http://dx.doi.org/10.2134/agronj2000.924657x

Buchholz, D. D., Brown, J. R., Garret, J., Hanson, R., & Wheaton, H. (2004). *Soil Test Interpretations and Recommendations Handbook*. Columbia, MO: University of Missouri-College of Agriculture, Division of Plant Sciences.

Camberato, J., & Casteel, S. (2010). Keep an eye open of sulfur deficiency in wheat. *Soil Fertility Update*. Purdue University Extension. Retrieved from https://www.agry.purdue.edu/ext/soybeanArrivals/04-13-10_ JC_SC_Sulfur_deficiency.pdf

Camberato, J., Maloney S., Casteel, S., & Johnson, K. (2012). *Soil Fertility Update*. Purdue University Extension. Retrieved from https://www.agry.purdue.edu/ext/soilfertility/05-03-12Sulfur_deficiency_alfa lfa.pdf

Divito, G. A., Echeverría, H. E., Andrade, F. A., & Sadras, V. O. (2015). Diagnosis of S deficiency in soybean crops: Performance of S and N: S determinations in leaf, shoot and seed. *Field Crops Research, 180*, 167-175. http://dx.doi.org/10.1016/j.fcr.2015.06.006

Fageria, N. K. (2004). Dry matter yield and nutrient uptake by lowland rice at different growth stages. *Journal of Plant Nutrition, 27*, 947-958. http://dx.doi.org/10.1081/PLN-120037529

Fehr, W. R., & Caviness, C. E. (1971). Stages of soybean development. *Crop Sci., 11*, 929-930. http://dx.doi.org/10.2135/cropsci1971.0011183X001100060051x

Freeman, K., Ruffo, M., & Mann, K. (2014). Improving crop yield and nutrient uptake efficiency with premium sulfur enhanced phosphate fertilizer. *American Society of Agronomy*. Abstract, Long Beach, CA.

Graham, R. D., Ascher, J. S., & Hynes, S. C. (1992). Selecting zinc-efficient cereal genotypes for soils of low zinc status. *Plant and Soil, 146*, 241-250. http://dx.doi.org/10.1007/BF00012018

Halim, A. H., Wassom, C. E., & Ellis Jr., R. (1968). Zinc deficiency symptoms and zinc and phosphorous interactions in several strains of corn (*Zea mays* L.). *Agronomy Journal, 60*, 267-271. http://dx.doi.org/ 10.2134/agronj1968.00021962006000030007x

Haq, M. U., & Mallarino, A. P. (2005). Response of soybean grain oil and protein concentrations to foliar and soil fertilization. *Agronomy Journal, 97*, 910-918. http://dx.doi.org/10.2134/agronj2004.0215

Hergert, G. W. (2000). Fertility principals: Sulfur. *Nutrient Management of Agronomic Crops in Nebraska.* University of Nebraska Extension. Retrieved from http://extensionpublications.unl.edu/assets/pdf/ec155.pdf

Huang, C., Barker, S. J., Langridge, P., Smith, F. W., & Graham, R. D. (2000). Zinc deficiency up-regulates expression of high-affinity phosphate transporter genes in both phosphate-sufficient and-deficient barley roots. *Plant Physiology, 124*, 415-422. http://dx.doi.org/10.1104/pp.124.1.415

Jones Jr., J. B. (1967). Interpretation of plant analysis for several agronomic crops. *Soil Testing and Plant Analysis.* Part II. SSSA Special Publ. Series No. 2. Madison, WI: Soil Sci. Soc., Amer.

Kaiser, D. S., Strock, J., & Lamb, J. (2008). *Impact of phosphorus fertilization strategies on efficiency of nitrogen use by corn rotated with soybeans.* Minnesota Department of Agriculture. Retrieved from http://www.mda.state.mn.us/chemicals/fertilizers/afrec/researchprojects/~/media/Files/chemicals/afrec/reports/phosfertstratimpact.ashx

Keefer, R. F., Singh, R. N., Horvath, D. J., & Henderlong, P. R. (1972). Response of corn to time and rate of phosphorous and zinc application. *Soil Science Society of America Journal, 36*, 628-632. http://dx.doi.org/10.2136/sssaj1972.03615995003600040036x

Mahler, R. L. (2004). *Nutrients plants require for growth.* University of Idaho Extension. Retrieved from http://www.cals.uidaho.edu/edComm/pdf/CIS/CIS1124.pdf

McBeath, T. M., & McLaughlin, M. J. (2014). Efficacy of zinc oxides as fertilizers. *Plant and Soil, 374*, 843-855. http://dx.doi.org/10.1007/s11104-013-1919-2

Mosaic. (2007). *A New Vision of Phosphate from Mosaic.* Plymouth, MN: The Mosaic Company.

Nathan, M. V., Stecker, J. A., & Sun, Y. (2012). *Soil Testing in Missouri.* Univ. Extension, Division of Plant Sciences, College of Agriculture Food and Natural Resources, University of Missouri, Columbia, MO.

Olsen, S. (1982). Micronutrient interaction. In J. M. Mortved, & W. L. Lindsay (Eds.), *Micronutrients in agriculture* (pp. 243-264). Soil Science Society America, Madison, WI.

Pierzynski, G. M., McDowell, R. W., Sims, J. T., & Sharpley, A. N. (2005). Chemistry, cycling, and potential movement of inorganic phosphorus in soils. *Phosphorus: Agriculture and the Environment* (pp. 53-86). ASA, Madison, WI.

Rehm, G. W., Sorensen, R. C., & Wiese, R. A. (1981). Application of phosphorous, potassium and zinc to corn grown for grain or silage: early growth and yield. *Soil Sci. Soc. Am. J., 45*, 523-528. http://dx.doi.org/10.2136/sssaj1981.03615995004500030017x

Rehm, G. W., Sorensen, R. C., & Wiese, R. A. (1983). Application of phosphorous, potassium and zinc to corn grown for grain or silage: Nutrient concentration and uptake. *Soil Sci. Soc. Am. J., 47*, 697-700. http://dx.doi.org/10.2136/sssaj1983.03615995004700040019x

Robson, A. D., & Pitman, M. G. (1983). Interactions between nutrients in higher plants. *Inorganic plant nutrition* (pp. 147-180). Springer. http://dx.doi.org/10.1007/978-3-642-68885-0_6

Sawyer, J., & Barker, D. (2009). *Evaluation of Mosaic MicroEssentials sulfur fertilizer products for corn production* (p. 5). Iowa State University Department of Agronomy, Ames, IA.

Sawyer, J., & Barker, D. (2010). *Evaluation of combination phosphorus-sulfur fertilizer products for corn production* (p. 4). Iowa State University Department of Agronomy, Ames, IA.

Sawyer, J., & Barker, D. (2012). Sulfur fertilization response in Iowa corn and soybean production. *Proceedings of the 2012 Wisconsin Crop Management Confer*ence (pp. 39-48). Madison, WI: University of Wisconsin-Madison.

Schachtman, D. P., Reid, R. J., & Ayling, S. M. (1998). Phosphorus uptake by plants: From soil to cell. *Plant Physiology, 116*, 447-453. http://dx.doi.org/10.1104/pp.116.2.447

Schnappinger, M. G., Martens, D. C., & Hawkins, G. W. (1969). Response of corn to Zn-EDTA and $ZnSO_4$ in field investigations. *Agron. J., 61*, 834-836. http://dx.doi.org/10.2134/agronj1969.00021962006100060002x

Schulte, E. E. (2004). *Soil and applied zinc* (p. 2). University of Wisconsin Extension.

Schulte, E. E., & Kelling, K. A. (1992). *Understanding plant nutrients: Soil and applied phosphorus* (No. A2520, p. 32). Univ. Wis. Extn. Pub.

Singh, J. P., Karamanos, R. E., & Stewart, J. W. B. (1986). Phosphorus-induced zinc deficiency in wheat on residual phosphorus plots. *Agronomy Journal, 78*, 668-675. http://dx.doi.org/10.2134/agronj1986.0002196 2007800040023x

Singh, J. P., Karamanos, R. E., & Stewart, J. W. B. (1988). The mechanism of phosphorus-induced zinc deficiency in bean (*Phaseolus vulgaris* L.). *Canadian Journal of Soil Science, 68*, 345-358. http://dx.doi.org/10.4141/cjss88-032

Singh, S. P., Singh, R., Singh, M. P., & Singh, V. P. (2014). Impact of sulfur fertilizer on different forms and balance of soil sulfur and the nutrition of wheat in wheat-soybean cropping sequence in Tarai soil. *Journal of Plant Nutrition, 37*, 618-632. http://dx.doi.org/10.1080/01904167.2013.867987

Slaton, N. A., Norman, R. J., Roberts, T. L., DeLong, R. E., Massey, C., Clark, S., & Branson, J. (2010). Evaluation of new fertilizers and different methods of application for rice production. *B.R. Wells Rice Research Studies* (pp. 266-277).

Stout, W. L., & Bennett, O. L. (1983). Effect of Mg and Zn fertilization on soil test levels, ear leaf composition, and yields of corn in northern West Virginia. *Communications in Soil Science & Plant Analysis, 14*, 601-613. http://dx.doi.org/10.1080/00103628309367392

Tabatabai, M. A. (1984). Importance of Sulphur in Crop Production. *Biogeochemistry, 1*, 45-62. http://dx.doi.org/10.1007/BF02181120

Uribelarrea, M., Below, F. E., & Moose, S. P. (2004). Grain composition and productivity of maize hybrids derived from the Illinois protein strains in response to variable nitrogen supply. *Crop Sci., 44*, 1593-1600. http://dx.doi.org/10.2135/cropsci2004.1593

USDM. (2015). *United States Drought Monitor*. Retrieved from http://doughtmonitor.unl.edu

Vance, C. P., Uhde-Stone, C., & Allan, D. L. (2003). Phosphorus acquisition and use: critical adaptations by plants for securing a nonrenewable resource. *New Phytologist, 157*, 423-447. http://dx.doi.org/10.1046/j.1469-8137.2003.00695.x

Woolfork, C., Olson, R., Mann, K., & Perez, O. (2014). *Canola (Brassica napus) fertilizer seed safety*. ASA, CSSA, SSSA. Abstract. Long Beach, CA.

Yerokun, O. A., & Christenson, D. R. (1990). Relating High Soil Test Phosphorus Concentrations to Plant Phosphorus Uptake. *SSSAJ, 54*, 796-799. http://dx.doi.org/10.2136/sssaj1990.03615995005400030029x

4

Socioeconomic Analysis of Rural Credit and Technical Assistance for Family Farmers in the Transamazonian Territory, in the Brazilian Amazon

Galdino Xavier de Paula Filho[1], Miquéias Freitas Calvi[2] & Roberta Rowsy Amorim de Castro[3]

[1] Education Faculty, Federal University of Amapá, Brazil

[2] Forestry Faculty, Federal University of Pará, Brazil

[3] Exact Sciences and Technology Faculty, Federal University of Pará, Brazil

Correspondence: Galdino Xavier de Paula Filho, Education Faculty, Federal University of Amapá, Mazagão, State of Amapá, Brazil. E-mail: galdinoxpf@gmail.com

Abstract

In Brazil, Rural Credit and Technical Assistance policies for family farming were formulated with the goal of promoting rural development in a sustainable and integrated manner. This study is the result of the *Monitoring and assessment of public policies for territory management in the Pará Amazon* project, undertaken by the Federal University of Pará (UFPA), aimed to evaluate the main socioeconomic impacts and limitations for the execution of these policies in the Transamazonian Territory. It is characterized as qualitative and exploratory, developed from bibliographic research and field research, based on data obtained through interviews conducted with 22 families of farmers who are beneficiaries of Rural Credit, the B modality of the National Programme for Strengthening Family Agriculture (PRONAF) and of the Technical Assistance Policy, whose sample corresponds to 10% of total contracts made effective within that Territory, between the years of 2013 and 2014. In addition to these farmers, for the analysis of the Technical Assistance service, interviews were conducted with extension workers from eight organizations, one of which is a state public company and seven of which are outsourced companies hired by the Federal Government to provide this service. The descriptive analysis shows that PRONAF B focuses on areas that produce short cycle food crops and on fishing activities. The technical assistance service provided by the public company is carried out in all the cities within the Territory, but only meets 10% of the demand; the service provided by the outsourced companies also occurs in all cities and its greatest setback is the delay in the release of funds by the Federal Government, which generates delays in the agricultural calendar and discontinuity in the productive activities, due to the end of the term of the companies' contracts.

Keywords: Brazilian Amazon, family farming, rural credit, technical assistance

1. Introduction

Although the importance of family farming in food production is unquestionable, in many countries this category is still developing mainly with its own resources, with a low level of access to rural credit and technical assistance policies to foment these production systems (Reardon et al., 2009; Poulton et al., 2010). Worldwide, there are approximately 2.5 to 3 billion people living in rural areas that are directly and indirectly involved with food productions, for subsistence and/or for supplying urban centers (Borras & Franco, 2012). One may observe that regions with developed agriculture contribute towards local development and the development of society as a whole. For this reason, this theme is a recurring topic in several countries. In the Brazilian Amazon, family farming exhibits reduced productivity due to low soil fertility in most of the territory, lack of technical orientation and precarious infrastructure in the rural zone (Bicalho & Hoefle, 2010). Many establishments reach productive exhaustion due to the practice of shifting cultivation, which degrades the soil and the forest, making the continuity of many production systems impossible (Prates-Clark et al., 2009).

Among the rural development policies, one may observe that in the last years, rural credit and technical assistance for family farming have been contemplated in many governmental actions. However, these services are still insufficient to meet the actual demand of the farmers, especially in developing countries, as the state or private institutions consider credits for farmers to be a risky, difficult and expensive business, and therefore offer different interest rates with respect to large-scale agriculture (Papias & Ganesan, 2009). Technical assistance contributes towards a greater efficiency of production systems and towards the integration of several agricultural functions in different levels of development, resulting in the accumulation and validation of technical knowledge among farmers (Labarthe, 2009). All of these services are essential to potentiate economic growth and to improve the quality of life within the rural zone (Rahman et al., 2011).

Kumar et al. (2010) point out that credit is one of the essential inputs for agricultural development, because it capitalizes and guides farmers towards proceeding with new investments and/or adopting new technologies. In relation to technical assistance, one notes that, due to environmental imperatives faced by many countries, this service has been guided, in the last years, by agricultural production programs and actions with environmental conservation, such as the *Conservation Technical Assistance Program* in the United States (Brinson & Eckles, 2011) and the National Technical Assistance and Rural Extension Policy (PNATER) in Brazil.

However, the conclusion that emerges is that there is a positive correlation between rural credit and technical assistance (Taveira & Oliveira, 2008; Uaiene et al., 2009). In a study on technical efficiency and potential for productivity of cocoa culture (*Theobroma cacao* L.) among farmers in countries in Central Africa, Binam et al. (2010) observed that access to credit had a significant impact on the technical efficiency of producers, especially in Cameroon, Nigeria, Ivory Coast and Ghana. These authors reveal that the role of credit, accompanied by technical assistance, cannot be underestimated, as when farmers have access to these services, they display greater efficiency from a productive point of view. Access to rural credit and technical assistance can enhance the ability of poor households to acquire agricultural raw materials, and as such, the development of rural credit institutions is a necessary condition for increasing work and soil productivity, and is a crucial factor for encouraging the development of low income farmers (Binam et al., 2010).

In Brazil, especially in the Brazilian Amazon, rural credit and technical assistance policies are widely discussed in a way that is integrated into strategies to improve the productivity of family farmers, as restrictions in the use of new areas for agriculture became a great concern in the last decade due to governmental strategies aimed at the reduction of deforestation (Balbino et al., 2011; Martins & Pereira, 2012). These policies, along with the diffusion of technology, use of agricultural inputs and fertilizers, mechanization, supply of information to farmers and social infrastructure investments in the rural areas are extremely important for the development of the agriculture (Browder et al., 2008). There are cases in which, even though application of these policies takes place, there are no results which involve structural changes in the production system of the farmers, so as to indicate that many of them need improvement (Gomes et al., 2009). According to the last Agricultural Census carried out in Brazil (IBGE, 2006), in the Brazilian states that correspond to the Amazon Region, there are 413,101 family agricultural establishments (9.4% of family agricultural establishments in Brazil), in an area of 16,647,328 hectares (20.7% of the area of family agricultural establishments in Brazil). However, there is a vast literature regarding the absence or inadequacy of public policies for this category, especially those related to rural credit and technical assistance (Tourneay & Bursztyn, 2010; Marchan, 2012; Martins & Pereira, 2012; Silveira & Wiggers, 2013).

The PRONAF B benefits farmers linked to social organizations (cooperative organizations, associations, fisherman colonies etc.) with annual income until R$ 20,000.00 and possess the PRONAF Declaration of Aptitude. The farmers benefited by the outsourced technical assistance are exclusively residents of the areas of the new modalities of land reform, such as the Settlement Project (PA); the State Extractivist Settlement Project (PEAEX); the Sustainable Development Project (PDS); and the Extractivist Reserve (RESEX); it is also necessary for them to be in the list of beneficiaries of the land reform carried out by the National Institute of Colonization and Land Reform (INCRA) and that they are considered family farmers, according to Law n° 11326/2006 (DOU, 2006).

Considering this situation, this study has objective to evaluate the rural credit policies, the B modality of PRONAF and the public technical assistance executed in the Transamazonian Territory, in the Southwest Region of the State of Pará, within the Brazilian Amazon. The main impacts of these policies on socioeconomic and productive aspects were analyzed, as well as in the local and regional dimensions, in addition to the main limitations for the full effectuation and execution these policies.

One must note that outsourced technical assistance, although a service provided by private companies, is a public service in nature in this case, as the outsourced companies are hired by the Federal Government. However, for the purpose of this study, this modality of service will be named outsourced technical assistance. Additionally, the technical assistance studied in this work relates to the *Technical Assistance and Rural Extension*, which results from PNATER, directed towards the family farming segment in Brazil.

The conception of this work is contemplated in the actions of the *Monitoring and Assessment of Public Policies for Territory Management in the Pará Amazon Project*, executed by way of a partnership between the Ministry of Agrarian Development (MDA) and the National Council of Scientific and Technological Development (CNPq) with the Federal University of Pará (UFPA).

2. Methodology

2.1 Characteristics of the Study Area

The integration of the Territory is the Transamazonian Highway (BR 230) and the Xingu River. It is formed by 10 cities: Altamira, Anapu, Brasil Novo, Medicilândia, Pacajá, Placas, Porto de Moz, Senador José Porfírio, Uruará and Vitória do Xingu (Figure 1). The Territory possesses a total area of 250,973 km² and an approximate population of 331,770 inhabitants, with a population density of 2.62 inhabitants/km² (IBGE, 2014).

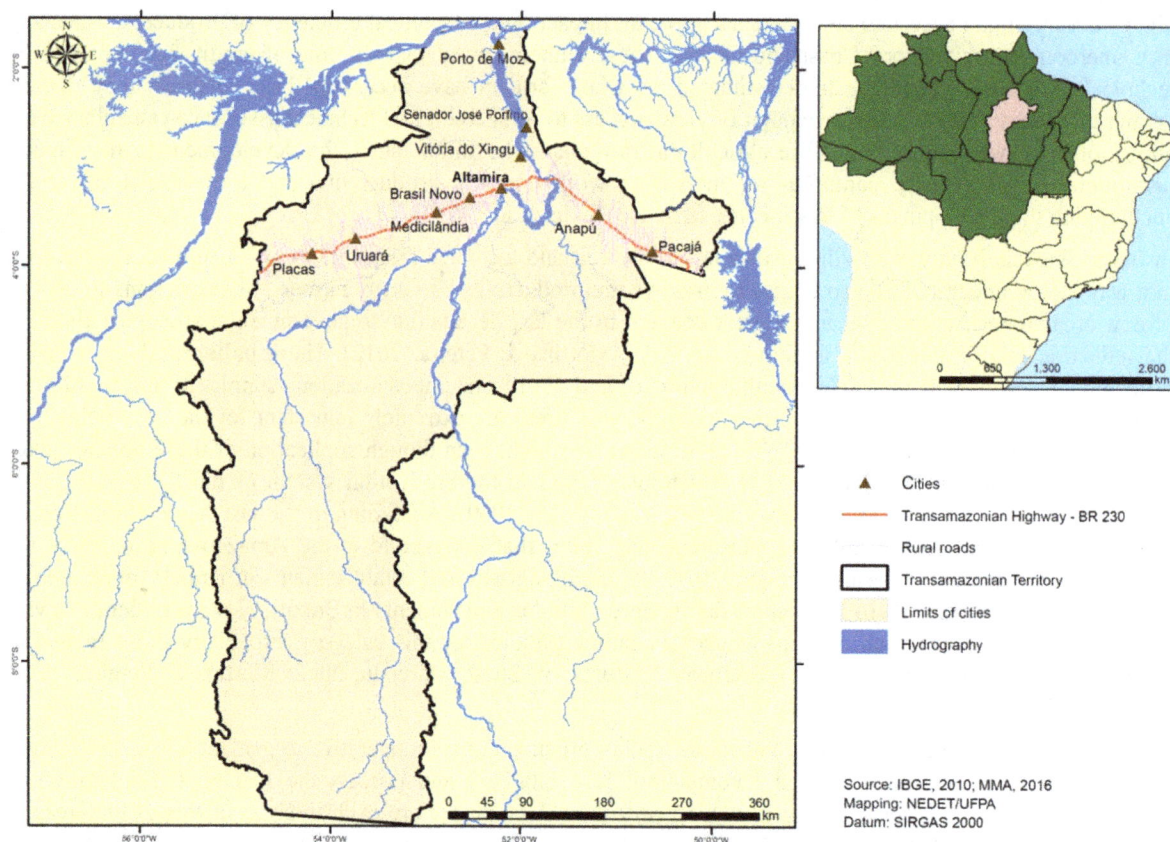

Figure 1. Transamazonian Territory, state of Pará, Brazil

In this region reside family farmers, settlers from land reform programs, fishermen, riverside communities, indigenous peoples and *quilombolas*. The most notable rural activities within the Territory are cattle rising, practiced extensively, cocoa production and the cultivation of food crops which are consumed by these family groups, who then commercialize the surplus (Santos et al., 2014; Paula Filho et al., 2016).

The Transamazonian Territory is located in the Central Region of the Brazilian Amazon; its economic is characterized by vegetable extractives, and which have been transformed into agricultural frontiers for the production and expansion of the *commodity* market, a process which intensified by way of public funding towards the implantation of agropecuary projects, increasing the levels of deforestation, as stated in a vast literature available on this theme (Escada et al., 2005; Soares-Filho et al., 2005; Scouvart et al., 2008; Rodrigues et al., 2009; Pacheco, 2012; Souza et al., 2013).

Currently, the Territory is entering a new cycle of natural resource exploitation, aimed at making use of the hydroelectric power of the Xingu River, in the Altamira Region, where recently the Belo Monte Hydroelectric Dam, the third largest in the planet, was constructed. This enterprise is one of the main works of the Growth Acceleration Program carried out by the Federal Government, which justifies it as being strategic for the resumption of economic growth in the country (Chernela, 2010), based on the national security discourse. This fact has significantly altered the agricultural and agrarian conformation and reality within the Territory, culminating in new challenges and opportunities in sight for family farms, which demand new studies, searches for market strategies and supply of information to farmers and technicians inserted in this space (Randell, 2015).

2.2 Data Collection

This study was conducted in 2013 and 2014. The target demographic was composed of family farmers who lived in the Transamazonian Territory and were beneficiaries of rural credit in the PRONAF B modality, as well as of the public technical assistance actions. The study utilized primary data collected from the farmers by way of structured questionnaires, and secondary data from academic literature and official documents.

The sampling technique was utilized and interviews were conducted with 22 farmers who are beneficiaries of rural credit (PRONAF B) and of technical assistance in seven cities of the Territory (Altamira, Brasil Novo, Medicilândia, Placas, Senador José Porfírio, Vitória do Xingu and Porto de Moz), which is equivalent to 10% of total contracts existent during that period (Table 1). Interviews were also conducted with employees of the Pará State Technical Assistance Company (EMATER), which acts in all the cities within the Territory, as well as with employees or directors of seven outsourced companies hired by the Brazilian government to provide services in three cities (Anapu, Pacajá and Porto de Moz).

2.3 Data Analysis

One encourages an analysis of the Transamazonian Territory that particularizes the cities, especially regarding their physical and productive characteristics, as the cities that are located along the Xingu River (Porto de Moz, Senador José Porfírio, Vitória do Xingu and part of Altamira), have an economy based on artisanal fishing, chestnut collection and annual crop cultivation. The cities located along the Transamazonian Highway (Placas, Uruará, Medicilândia, Brasil Novo, Anapu, Pacajá and the other part of Altamira) the system of production is based on cocoa production and extensive cattle raising, as analyzed by Bratman (2011), Godar et al. (2012) and Hetrick et al. (2013).

2.4 Analytical Procedures

The analysis of performance to the public policies for family farming was carried out in a particularized and individual form between the farmers of the cities that compose the Territory, verifying the performance level of a city in relation to another by way of descriptive and inferential statistics. In this sense, a hypothesis was elaborated in which cities with a higher degree of access to information, local development or schooling of its farmers may be more susceptible to the execution of these policies. At the same time, the social environment and the socioeconomic dynamic associated with the production possibilities of the cities are factors that may contribute towards greater access to the policies.

3. Results and Discussions

3.1 Characteristics and Performance of PRONAF B among Beneficiary Farmers

There are only seven cities (out of ten) within the Transamazonian Territory in which there are farmers who are benefited by PRONAF B. By analyzing Table 1, one may observe that this credit modality was applied predominantly in fishing activities (58%), food culture production (37%) and small animal raising (5%). In other words, PRONAF B was applied mainly in fishing activities of residents of the Xingu River margins and in small subsistence food culture systems (horticulture, tubers and grains), located in cities along the Transamazonian Highway. It is worth noting that the city of Altamira presents both characteristics (Table 1).

Table 1. Activities funded by rural credit (PRONAF B) in the Transamazonian Territory

Municipality	Interviewed farmers	Funded activity	Percentage of the sample	Percentage by funded activity	Region of the beneficiaries
Medicilândia	01	Small animals	5%	Small animals: 5%	
Brasil Novo	03	Food crops	16%		Highway: 42% of the funded activity
Placas	03	Food crops	16%	Food crops: 37%	
Altamira *(highway)*	01	Food crops	5%		
Altamira *(river)*	02	Fishing	11%		
Vitória do Xingu	03	Fishing	11%	Fishing: 58%	River: 58% of the funded activity
Senador José Porfírio	03	Fishing	5%		
Porto de Moz	06	Fishing	31%		
Transamazonian Territory	22	-	100%	100%	100%

The study revealed that PRONAF B faces several hindrances to its execution in the Transamazonian Territory. Although fishing is one of the most funded activities, EMATER, the public institution that guides the beneficiaries, has no professionals in this area to carry out the technical assistance on this raising system. Another difficulty is the absence of banking agencies in some cities, which forces the beneficiaries to dislocate themselves to other cities in order to receive the financial resources of their funding. This dislocation elevates costs and diminishes the buying power of the received credit.

Literature shows that rural credit for small farmers has been essential in structuring their production systems, as observed by Reardon et al. (2009) in a revision work regarding this policy in countries of Latin America; and in a work conducted with asymmetric information developed by Boucher et al. (2008) in Peru, Nicaragua and Honduras, whose results are similar to those found in this study, in which it was found that the main benefit and positive aspect of access to credit was the organization of the families' productive systems. In this study, the received credit was invested in the acquisition of equipment and construction of service structures, which contributed directly to the productive sovereignty of the families and to the food security strategies, seeing as the activities performed by these farmers, in addition to being aimed at commercialization, also had food self-sufficiency as a goal.

3.2 Technical Assistance in the Vision of the Farmers and the Policy's Executors

The technical assistance executed in the Transamazonian Territory is stratified in two forms:

i) Technical assistance executed by a public agency (EMATER). Although the national directive recommended by PNATER is to universalize this service for farmers, the company prioritizes service mainly to the beneficiaries of PRONAF. EMATER acts in all the cities within the Territory, but admits that the service provided falls short of the demand, serving only 10% of the farmers who need these services. For years, the company has been demanding from the State of Pará Government the hiring of more professionals in order to meet the demand for this service.

ii) Technical assistance executed by outsourced companies hired by INCRA. This action is articulated within the scope of the *Plano Brasil Sem Miséria* (Brazil Without Misery Plan), a Federal Government Program, by way of public tenders for specific areas that are the new land reform modalities, such as in the example of settlement projects in the cities of Anapu and Pacajá, and in the *Verde Para Sempre* (Forever Green) RESEX, in the city of Porto de Moz.

The beneficiaries of the Brazil Without Misery Plan are families who find themselves in a low income situation. The selection is carried out by the outsourced company's technicians, together with the Social Assistance Municipal Offices. All cities within the Territory were contemplated by this policy, but it was mostly applied in the cities with a smaller Human Development Index (HDI), as there is a strong correlation between the socioeconomic situation of the families and the level of development of the cities where they reside (Table 2).

This study identified 1,300 families being benefited by the technical assistance provided by the outsourced companies in the Anapu and Pacajá settlement projects and 1,849 families in the *Verde Para Sempre* RESEX, in the city of Porto de Moz (Table 2). One may note that the fact that the city of Pacajá holds the largest number of families benefited by the technical assistance policies may be justified by it being the city with the highest number of settlement projects and of families settled in land reform areas among the cities within the Territory.

Table 2. Data from the technical assistance provided to the farming families in the Territory (2014)

Municipality	N° of families served by PBSM[1]	N° of families served in the land reform areas[2]	N° of families served by EMATER[3]	HDI of the cities[4]
Altamira	225	0	230	0.665 (medium)
Anapu	325	1,979	172	0.548 (low)
Brasil Novo	75	0	204	0.613 (medium)
Medicilândia	150	0	313	0.582 (low)
Pacajá	375	1,300	382	0.515 (low)
Placas	75	0	135	0.552 (low)
Porto de Moz	75	1,849	201	0.503 (low)
Senador José Porfírio	150	0	91	0.514 (low)
Uruará	225	0	269	0.589 (low)
Vitória do Xingu	150	0	87	0.597 (low)
Transamazonian Territory	**1,825**	**5,128**	**2,084**	**0.567 (low)**

Note. [1] *Plano Brasil Sem Miséria*; [2] According with the Agricultural Census (IBGE, 2006); [3] Estimated value based on EMATER's 10% service capacity. The demand takes into consideration the current number of agropecuary establishments in the cities, based on the Brazilian Institute of Geography and Statistics' Agricultural Census (IBGE, 2006); [4] Based on data from the United Nations Development Program (PNUD, 2010).

Among the main difficulties encountered for the consolidation of the technical assistance policy, one may highlight those related to the logistics of access to the farmers' properties, especially those located in the settlement projects and in the RESEX, which require hours of travel to reach, as observed previously in works conducted by Ludewigs et al. (2009), and Pereira and Sauer (2011). However, the greatest difficulty, which persists over the years, has been the lack of effectuation of the public policies and actions by INCRA toward the consolidation of the land reform areas, especially the settlement projects, such as the actions of road construction and maintenance of roads throughout the year (especially in the rainy period); the issuance of property titles of rural lots; and, especially, the release of funds for technical assistance within an adequate timeframe. The delay in the release of financial resources by INCRA for the providers of these services is a recurring problem that affects the agricultural calendar, generating a series of problems that become irreversible. This situation, regrettably, takes place throughout the entire Amazon, as may be observed in works carried out by Pacheco (2009), Tourneau and Bursztyn (2010), and Silveira and Wiggers (2013).

The present study found that there is a deficit in the provision of technical assistance services, both by the public company and by the outsourced companies. In addition to this deficit, in the settlement projects there is the further problem of discontinuity of the actions, especially due to the fact that the outsourced companies operate by public calls, with a specified deadline for the execution of the activities, and after the end of the contract, they leave the settlement, leaving the farmers to wait for the next contract, which is not always executed by the same company or initiated in sequence. This public policy model has generated interruption, discontinuity and repetition of stages that have already been carried out due to lack of communication, increasing public spending and producing few effective results.

The privatization of technical assistance has not been positively evaluated by national and international experts. Diesel et al. (2008) argue that the initial justification for the privatization of services was that agriculture had achieved a satisfactory level of production, requiring advanced technology. However, this is not the case of the region being studied, where the production systems of family farmers still have a number of limitations, especially those related to the absence of the many structural public policies, such as lack of electrification in a large part of rural homes, lack of recovery of local roads, among others, as observed by Hostiou et al. (2006) and Brondizio and Moran (2008) in works conducted with family farmers in the Transamazonian Territory. It is possible that this scenario may change over the next few years, as there are other land reform areas in these cities that may be contemplated by the public calls for technical assistance.

3.3 Performance of the Cities upon Accessing the Rural Credit and Technical Assistance Policies

Considering the public policies are aimed at family farming, it is assumed that the cities with the best performance in access and operationalization of the policies are the ones with the highest number of agricultural establishments. However, the higher number of family farmers is not enough to ensure the execution of these policies, as it is necessary that the family farmers are duly apt to access them. Relating the information of Tables 1 and 3, one finds that the cities with the largest amount of agropecuary establishments were not necessarily those that most accessed the credit and technical assistance policies conceded to family farming in the Transamazonian Territory.

Pacheco (2009), Leite et al. (2011), and Castro and Singer (2012) consider the condition that the farmers find themselves in to be fundamental in determining access to these policies; they analyze that, although, in the last years, many settlements for family farming have been created in the Brazilian Amazon, many of these are not duly consolidated with structural policies, such as roads and rural electrification. Ludewigs et al. (2009), Pers et al. (2011), and Bratman (2011) point out that this situation hinders the execution of the other public policies for settlements, which may lead to the abandonment of these areas by the families who were settled there.

Table 3. Distribution of agropecuary establishments in the Transamazonian Territory (2014)

Municipality	N° of Agropecuary Establishments	% of Agropecuary Establishments
Altamira	2,305	11
Anapu	1,729	8
Brasil Novo	2,044	10
Medicilândia	3,139	15
Pacajá	3,825	18
Placas	1,351	7
Porto de Moz	2,012	10
Senador José Porfírio	911	5
Uruará	2,693	12
Vitória do Xingu	874	4
Transamazonian Territory	**20,883**	**9.3***
Brazil	**5,219,504**	

Note. * Percentage in relation to the number of establishments in the State of Pará. Source: Adapted of IBGE (2006).

Regarding PRONAF B, one notes a predominance of small farmers and fishermen in Porto de Moz, Altamira and Brasil Novo, the cities that have the highest number of agropecuary establishments in the entire Transamazonian Territory (Table 3). This analysis refers only to rural credit in the B modality of PRONAF, although there are at least ten other credit modalities. However, due to the heterogeneity exhibited by these segments, as well as the local opportunities that present themselves (public policies, market, productive structure), which in turn are related to physical and socioeconomic characteristics of the cities, it is possible that in analyses of other modalities, access is concentrated in other groups of cities.

The public technical assistance, conducted by EMATER, which follows (or should follow) PNATER's directives, is lacking, as the company cannot meet its demand, due to a reduced technical staff. The last hiring of extension workers took place in the year of 2006. This deficit, regrettably, is a reality across Brazil, as previously observed in works conducted by Diesel et al. (2008), and Taveira and Oliveira (2008). In a study conducted on the performance of family farming in Brazil between the years of 1996 and 2006, Guanziroli et al. (2012) highlighted that this segment is diversified and requires specific policies, and that the absence or inadequacy of policies for family farming contributes towards the existence of negative indicators and little increase in agricultural production.

As for the technical assistance executed by the outsourced companies, directed towards the new modalities of land reform, the cities with the highest number of families settled in these areas are likely to have a higher number of beneficiaries, a fact which may justify the city of Pacajá being the one with the highest number of beneficiaries (Table 2), since it is the city with the highest number of settlement projects and settled families, as observed in Table 4.

Table 4. Modalities of land reform and number of families settled in the Transamazonian Territory (2014)

Municipality	Modalities of land reform	Nº of settled families	Families settled by city
Altamira	05 PA	1,718	4,367
	05 PDS	2,516	
	03 RESEX	133	
Anapu	01 PA	1,658	2,107
	03 PDS	449	
Brasil Novo	04 PA	773	773
Medicilândia	02 PA	1,705	2,876
	01 PDS	1,171	
Pacajá	24 PA	6,591	7,519
	01 PDS	928	
Placas	06 PA	1,381	2,002
	04 PDS	621	
Porto de Moz	01 PA	90	2,202
	01 PDS	46	
	01 RESEX	2,002	
	01 PEAEX	64	
Senador José Porfírio	06 PA	2,083	2,149
	01 PDS	66	
Uruará	09 PA	2,146	2,819
	03 PDS	673	
Transamazonian Territory		**26,814**	
Brazil		**969,296**	

Note. PA: Settlement Project; PDS: Sustainable Development Project; PEAEX: State Agroextractive Settlement Project; RESEX: Extractive Reserve.

Source: INCRA (2015).

Outsourced technical assistance services in Extractive Reserves in the Amazon are recent, and scientific literature does not yet contain works with results that show the efficiency of these services, which makes a deeper discussion regarding this theme and the establishment of comparative parameters for the results impossible. The experience encountered in the city of Porto de Moz is one of the pioneers within the Amazon. This study revealed that the main difficulties encountered by the technical assistance services in this conservation unit are the delay in the transfer of funds for the companies that provide the services (which was also observed in settlement projects), and the logistics surrounding the provision of services to the families, due to the geographical dimensions and the difficulty of access to the various communities of the RESEX. Although there are 2002 families in this unit, the company's hiring tender foresaw service to 1849 families, or 92% of the demand. According to the company that provided this service in the RESEX, the financial and technical limitations made service to all families impossible.

4. Conclusions

The PRONAF B rural credit was concentrated in areas whose production is based on short cycle food cultures and on fishing activities, playing an important contribution to the sovereignty and food security of the benefited families, as they invest the funds in the acquisition and/or improvement of the productive infrastructure. However, the absence of banking agents in some cities and of skilled professionals to carry out the technical orientation on the funded activities hinder the effectiveness of this policy.

Public technical assistance is carried out by EMATER in all the cities, but this institution can only meet 10% of the existing demand. The main limitation encountered is the reduced number of technicians, as professionals have not been hired for at least ten years.

Outsourced technical assistance takes place in all cities by way of actions articulated within the scope of the Federal Government's Brazil Without Misery Plan, aimed at farmers who find themselves in a low income situation; at family farmers and extractives situated in a RESEX; and two settlement projects. The greatest difficulty for the implementation of these actions is the delay in the release of funds to the providers of services by INCRA, resulting in the delay in technical assistance activities related to the agricultural calendar, as well as in the interruption of the activities, due to the end of the contracts with the companies that operate this policy.

In a general way, it is observed that these public policies for family farming are poorly accessed among farmers in the study region, the Transamazonian Territory, however, this work does not exhaust the debate regarding the assessment of productive inclusion programs for family farming. It may contribute in the evaluation of access to these programs, from the specificities of each city, considering physical and productive s factors and, mainly, the socioeconomic and political dynamic of these cities.

References

Balbino, L. C., Cordeiro, L. A. M., Porfírio-da-Silva, V., Moraes, A., Martínez, G. B., Alvarenga, R. C., … Galerani, P. R. (2011). Evolução tecnológica e arranjos produtivos de sistemas de integração lavoura – pecuária – floresta no Brasil. *Pesquisa Agropecuária Brasileira, 46*(10), 1-12. http://dx.doi.org/10.1590/S0100-204X2011001000001

Bicalho, A. M. S. M., & Hoefle, S. W. (2010). Economic Development, Social Identity and Community Empowerment in the Central and Western Amazon. *Geographical Research, 48*(3), 281-296. http://dx.doi.org/10.1111/j.1745-5871.2009.00626.x

Binam, J. N., Gockowski, J., & Nkamleu, G. B. (2008). Technical efficiency and productivity potential of cocoa farmers in West African countries. *The Developing Economies, 46*(3), 242-263. http://dx.doi.org/10.1111/j.1746-1049.2008.00065.x

Borras, S. M., & Franco, J. C. (2012). Global Land Grabbing and Trajectories of Agrarian Change: A Preliminary Analysis. *Journal of Agrarian Change, 12*(1), 34-59. http://dx.doi.org/10.1111/j.1471-0366.2011.00339.x

Boucher, S. R., Carter, M. R., & Guirkinger, C. (2008). Risk Rationing and Wealth Effects in Credit Markets: Theory and Implications for Agricultural Development. *American Journal of Agricultural Economics, 90*(2), 409-423. http://dx.doi.org/10.1111/j.1467-8276.2007.01116.x

Bratman, E. Z. (2011). Villains, Victims, and Conservationists? Representational Frameworks and Sustainable Development on the Transamazon Highway. *Human Ecology, 39*(4), 441-453. http://dx.doi.org/10.1007/s10745-011-9407-x

Brinson, M. M., & Eckles, S. D. (2011). U.S. Department of Agriculture conservation program and practice effects on wetland ecosystem services: A synthesis. *Ecological Applications, 21*(3), 116-127. http://dx.doi.org/10.1890/09-0627.1

Brondizio, E. S., & Moran, E. F. (2008). Human dimensions of climate change: the vulnerability of small farmers in the Amazon. *Philosophical Transactions of the Royal Society, 363*(1), 1803-1809. http://dx.doi.org/10.1098/rstb.2007.0025

Browder, J. O., Pedlowski, M. A., & Walker, R. (2008). Revisiting Theories of Frontier Expansion in the Brazilian Amazon: A Survey of the Colonist Farming Population in Rondônia's Post-frontier, 1992-2002. *World Development, 36*(8), 1469-1492. http://dx.doi.org/10.1016/j.worlddev.2007.08.008

Castro, M. C., & Singer, B. H. (2012). Agricultural settlement and soil quality in the Brazilian Amazon. *Population and Enviroment, 34*(1), 22-43. http://dx.doi.org/10.1007/s11111-011-0162-0

Chernela, J. (2010). Opposition in the Time of Avatar: Belo Monte Dam in the Brazilian Amazon. *Anthropology News, 51*(8), 26-27. http://dx.doi.org/10.1111/j.1556-3502.2010.51826_2.x

Diesel, V., Froehlich, J. M., Neumann, P. S., & Silveira, P. R. C. (2008). Privatização dos serviços de extensão rural: uma discussão (des) necessária. *Revista de Economia e Sociologia Rural, 46*(4), 1155-1188. http://dx.doi.org/10.1590/S0103-20032008000400010

DOU (Diário Oficial da União). (2006). *Lei 11.326/2006* (Edition of 27/07/2016, Section 01, p. 1).

Escada, M. I. S., Vieira, I. C. G., Kampel, S. A., Araújo, R., Veiga, J. B., Aguiar, A. P. D., ... Câmara, G. (2005). Processos de ocupação nas novas fronteiras da Amazônia: o interflúvio do Xingu/Iriri. *Estudos Avançados, 19*(54), 9-23. http://dx.doi.org/10.1590/S0103-40142005000200002

Godar, J., Tizardo, E. J., Pokorny, B., & Johnson, J. (2012). Typology and Characterization of Amazon Colonists: A Case Study Along the Transamazon Highway. *Ecology, 40*(2), 251-267. http://dx.doi.org/10.1007/s10745-012-9457-8

Gomes, E. G., Mello, J. C. C. B. S., Souza, G. S., Mezza, L. A., & Mangabeira, J. A. C. (2009). Efficiency and sustainability assessment for a group of farmers in the Brazilian Amazon. *Annals of Operations Research, 169*(1), 167-181. http://dx.doi.org/10.1007/s10479-008-0390-6

Guanziroli, C. E., Buainain, A. M., & Di Sabbato, A. (2012). Dez anos de evolução da agricultura familiar no Brasil: (1996 e 2006). *Revista de Economia e Sociologia Rural, 50*(2), 351-370. http://dx.doi.org/10.1590/S0103-20032012000200009

Hetrick, S., Chowdhury, R. R., Brondizio, E., & Moran, E. (2013). Spatiotemporal Patterns and Socioeconomic Contexts of Vegetative Cover in Altamira City, Brazil. *Land, 2*(4), 774-796. http://dx.doi.org/10.3390/land2040774

Hostiou, N., Veiga, J. B., & Tourrand, J. F. (2006). Dinâmica e evolução de sistemas familiares de produção leiteira em Uruará, frente de colonização da Amazônia brasileira. *Revista de Economia e Sociologia Rural, 44*(2), 295-311. http://dx.doi.org/10.1590/S0103-20032006000200007

IBGE (Instituto Brasileiro de Geografia e Estatística). (2006). *Censo Agropecuário 2006*. Rio de Janeiro. Available at: http://www.cidades.ibge.gov.br/xtras/temas.php?lang=&codmun=150060&idtema=3&search=para|censo-agropecuario-2006 Accessed in: 16/07/2015.

IBGE (Instituto Brasileiro de Geografia e Estatística). (2014). *Cidades 2014*. Rio de Janeiro. Retrieved July 22, 2015, from http://www.cidades.ibge.gov.br/xtras/uf.php?lang=&coduf=15&search=para

INCRA (Instituto Nacional de Colonização e Reforma Agrária). (2015). *Projetos de Reforma Agrária Conforme Fases de Implementação*. Brasília. Retrieved September 11, 2015, from http://www.incra.gov.br/sites/default/files/uploads/reforma-agraria/questao-agraria/reformaagraria/projetos_criados-geral.pdf

Kumar, A., Singh, K. M., & Sinha, S. (2010). Institutional Credit to Agriculture Sector in India: Status, Performance and Determinants. *Agricultural Economics Research Review, 23*(2), 253-264. http://dx.doi.org/10.5958/0974-0279.2010.00001.9

Labarthe, P. (2009). Extension services and multifunctional agriculture. Lessons learnt from the French and Dutch contexts and approaches. *Journal of Environmental Management, 90*(2), 193-202. http://dx.doi.org/10.1016/j.jenvman.2008.11.021

Leite, F. L., Caldas, M., Simmons, C., Pers, S. G., Aldrich, S., & Walker, R. T. (2011). The social viability and environmental sustainability of direct action land reform settlements in the Amazon. *Environment, Development and Sustainability, 13*(4), 773-788. http://dx.doi.org/10.1007/s10668-011-9289-5

Ludewigs, T., D'Antona, A. O., Brondízio, E. S., & Hetrick, S. (2009). Agrarian Structure and Land-cover Change Along the Lifespan of Three Colonization Areas in the Brazilian Amazon. *Development, 37*(8), 1348-1359. http://dx.doi.org/10.1016/j.worlddev.2008.08.018

Marchand, S. (2012). The relationship between technical efficiency in agriculture and deforestation in the Brazilian Amazon. *Ecological Economics, 77*(1), 166-175. http://dx.doi.org/10.1016/j.ecolecon.2012.02.025

Martins, P. F. S., & Pereira, T. Z. S. (2012). Cattle-raising and public credit in rural settlements in Eastern Amazon. *Ecological Indicators, 20*(1), 316-323. http://dx.doi.org/10.1016/j.ecolind.2012.02.031

Pacheco, P. (2009). Agrarian Reform in the Brazilian Amazon: Its Implications for Land Distribution and Deforestation. *Word Development, 37*(8), 1337-1347. http://dx.doi.org/10.1016/j.worlddev.2008.08.019

Pacheco, P. (2012). Actor and frontier types in the Brazilian Amazon: Assessing interactions and outcomes associated with frontier expansion. *Geoforum, 43*(1), 864-874. http://dx.doi.org/10.1016/j.geoforum.2012.02.003

Papias, M. M., & Ganesan, P. (2009). Repayment behaviour in credit and savings cooperative societies. *International Journal of Social Economics, 36*(5), 608-625. http://dx.doi.org/10.1108/03068290910954059

Paula Filho, G. X., Calvi, M. F., & Castro, R. R. A. (2016). Institutional Markets for Family Agriculture: Analysis of the Food Acquisition Program (PAA) and the National School Feeding Program (PNAE) within a Territory in the Brazilian Amazon. *International Journal of Research Studies in Agricultural Sciences, 2*(4), 12-23. http://dx.doi.org/10.20431/2454-6224.0204002

Pereira, J. M. M., & Sauer, S. (2011). A "reforma agrária assistida pelo mercado" do Banco Mundial no Brasil: dimensões políticas, implantação e resultados. *Revista Sociedade e Estado, 26*(3), 587-612. http://dx.doi.org/10.1590/S0102-69922011000300009

Pers, S. G., Leite, F., Simmons, C., Walker, R., Aldrich, S., & Caldas, M. (2010). Intraregional Migration, Direct Action Land Reform, and New Land Settlements in the Brazilian Amazon. *Bulletin of Latin American Research, 29*(4), 459-476. http://dx.doi.org/10.1111/j.1470-9856.2010.00384.x

PNUD (Programa das Nações Unidas para o Desenvolvimento). (2010). *Atlas do Desenvolvimento Humano: Ranking decrescente do IDH-M dos municípios do Brasil.* Brasília: PNUD. Retrieved September 14, 2015, from http://www.pnud.org.br/arquivos/ranking-idhm-2010.pdf

Poulton, C., Dorward, A., & Kydd, J. (2010). The Future of Small Farms: New Directions for Services, Institutions, and Intermediation. *World Development, 38*(10), 1413-1428. http://dx.doi.org/10.1016/j.worlddev.2009.06.009

Prates-Clark, C. C., Lucas, R. M., & Santos, J. R. (2009). Implications of land-use history for forest regeneration in the Brazilian Amazon. *Canadian Journal of Remote Sensing: Journal Canadien de Télédétection, 35*(6), 534-553. http://dx.doi.org/10.5589/m10-004

Rahman, M. W., Luo, J., & Cheng, E. (2011). Policies and performances of agricultural/rural credit in Bangladesh: What is the influence on agricultural production? *African Journal of Agricultural Research, 6*(31), 6440-6452. http://dx.doi.org/10.5897/AJAR11.1575

Randell, H. (2015). Structure and agency in development-induced forced migration: The case of Brazil's Belo Monte. *Population and Enviroment, 37*(168), 1573-1596. http://dx.doi.org/10.1007/s11111-015-0245-4

Reardon, T., Barret, C. B., Berdegué, J. A., & Swinnen, J. F. M. (2009). Agrifood Industry Transformation and Small Farmers in Developing Countries. *World Development, 37*(11), 1717-1727. http://dx.doi.org/10.1016/j.worlddev.2008.08.023

Rodrigues, A. S. L., Ewers, R. M., Parry, L., Souza Jr., C., Veríssimo, A., & Balmford, A. (2009). Boom-and-Bust Development Patterns Across the Amazon Deforestation Frontier. *Science, 324*(5933), 1435-1437. http://dx.doi.org/10.1126/science.1174002

Santos, I. V., Porro, N. M., & Porro, R. (2014). Mobilidade de camponeses entre assentamentos de reforma agrária: territorialidades em cheque no desenvolvimento local da Transamazônica, Pará, Brasil. *Interações, 16*(1), 43-53. http://dx.doi.org/10.1590/151870122015103

Scouvart, M., Adams, R. T., Caldas, M., Dale, V., Mertens, B., Nedelec, V., ... Lambin, E. F. (2008). Causes of deforestation in the Brazilian Amazon: a qualitative comparative analysis. *Journal of Land Use Science, 2*(4), 257-282. http://dx.doi.org/10.1080/17474230701785929

Silveira, L. B., & Wiggers, R. (2013). Protegendo a floresta, reconfigurando espaços na Amazônia: O caso do Projeto de Assentamento Extrativista Santa Maria Auxiliadora, Humaitá (AM). *Revista de Administração Pública, 47*(3), 671-693. http://dx.doi.org/10.1590/S0034-76122013000300007

Soares-Filho, B. S., Nepstad, B. C., Curran, L., Cerqueira, G. C., Garcia, R. A., Ramos, C. A., ... McGrath, D. (2005). Cenários de desmatamento para a Amazônia. *Estudos Avançados, 19*(54), 137-152. http://dx.doi.org/10.1590/S0103-40142005000200008

Souza, R. A., Miziara, F., & De Marco Junior, P. (2013). Spatial variation of deforestation rates in the Brazilian Amazon: A complex the assistência técnica e extensão rural for agrarian technology, agrarian structure and governance by surveillance. *Land Use Policy, 30*(1), 915-924. http://dx.doi.org/10.1016/j.landusepol.2012.07.001

Taveira, L. R. S., & Oliveira, J. T. A. (2008). A extensão rural na perspectiva de agricultores assentados do Pontal do Paranapanema. *Revista de Economia e Sociologia Rural, 46*(1), 9-30. http://dx.doi.org/10.1590/S0103-20032008000100001

Tourneay, F. M., & Bursztyn, M. (2010). Assentamentos rurais na Amazônia: Contradições entre a política agrícola e a política ambiental. *Ambiente e Sociedade, 13*(1), 111-130. http://dx.doi.org/10.1590/S1414-753X2010000100008

Uaiene, R. N., Arndt, C., & Masters, W. A. (2009). *Determinants of Agricultural Technology adoption in Mozambique*. Discussion Papers N°. 67E, Maputo, National Directorate of Studies and Policy Analysis.

5

Confirmation of Glyphosate- and Acetolactate Synthase (ALS)-Inhibitor-Resistant Kochia (*Kochia scoparia*) in Nebraska

Neha Rana[1] & Amit J. Jhala[1]

[1] Department of Agronomy and Horticulture, University of Nebraska-Lincoln, Lincoln, NE, USA

Correspondence: Amit J. Jhala, Department of Agronomy and Horticulture, University of Nebraska-Lincoln, Lincoln, NE 68583, USA. E-mail: amit.jhala@unl.edu

Abstract

Kochia is an early emerging weed of increasing concern across the Great Plains region of the United States due to the evolution of resistance to herbicides. Greenhouse studies were conducted to confirm and characterize the level of glyphosate and acetolactate synthase (ALS)-inhibiting herbicide resistance in kochia biotype collected from a field in Sheridan County in Nebraska. The response of kochia biotype to 9 rates (0 to 16×) of tribenuron and glyphosate was evaluated in a whole plant dose-response bioassay. On the basis of the values at the 90% effective dose (ED_{90}), the putative-resistant kochia biotype had a 6- and 15-fold level of resistance to glyphosate and tribenuron, respectively. Future research will evaluate strategies for the management of glyphosate- and ALS-resistant kochia under field conditions.

Keywords: glyphosate, acetolactate synthase (ALS)-inhibiting herbicides, resistance management

1. Introduction

Kochia [*Kochia scoparia* (L.) schrad], an annual broadleaf weed species from the Chenopodiaceae family, is native to Eurasia and was introduced to the western hemisphere in the early 1900s as an ornamental plant (Dodd & Moore, 1993; Friesen et al., 2009). Commonly referred to as Fireweed (Casey, 2009), it is an economically important weed in cropland and non-crop areas including pastures, roadsides, wasteland, and ditchbanks (Forcella, 1985; Beckie et al., 2013). Kochia is a C4 species characterized by early season emergence at low soil temperatures, rapid growth, and high tolerance to heat, drought, and salinity. It is well adapted to the semiarid and arid regions of the Canadian Prairies and the Great Plains of North America (Nussbaum et al., 1985; Schwinghamer & Van Acker, 2008; Friesen et al., 2009).

In the Midwestern and western United States, kochia has become an increasingly problematic weed with detrimental effects on yields in annual crops, including wheat (*Triticum aestivum* L.), sorghum [Sorghum bicolor (L.)], sunflower (*Helianthus annuus* L.), sugar beet (*Beta vulgaris* L.), corn (*Zea mays* L.), and soybean [*Glycine max* (L.) Merr.]. It is estimated that the interference and competition from kochia reduced crop yields by 30% to as much as 95% depending on the density of the kochia and the crop being investigated (Durgan et al., 1990; Forcella, 1985; Weatherspoon & Schweizer, 1969; Wicks et al., 1993, 1994). Furthermore, kochia is a weed of major concern because it has evolved resistance to number of herbicide chemistries (Friesen et al., 2009). Kochia seed viability is short-lived in the soil, with low seed dormancy impacting the turnover time for favorable fitness traits to become predominant in the population (Anderson & Nielsen, 1996; Schwinghamer & Van Acker, 2008). In addition, kochia is wind-pollinated and seed dispersal from resistant plants tumble across fields, allowing for the rapid spread of resistance. Thus, herbicide-resistant kochia has become prevalent in areas where herbicide(s) with the same mode of action have been used repeatedly (Crespo et al., 2014).

Kochia is a weed species that has evolved resistance as early as 1970s; the first case of triazine-resistant kochia was reported in 1976 in a Kansas corn field (Heap, 2016). A decade later, 5 years after commercialization of the first sulfonylurea herbicide, a chlorsulfuron-resistant biotype of kochia was documented in wheat (Primiani et al., 1990). Since then, there has been a widespread occurrence of kochia biotypes resistant to triazines, sulfonylureas (SU), and imidazolinone (IMI) herbicides (Beckie et al., 2013; Heap, 2016; Kumar et al., 2014). In 1994, dicamba (synthetic auxin)-resistant kochia was first documented in Montana, but it has since been confirmed in North Dakota, Idaho, Nebraska, and Colorado (Cranston et al., 2001; Crespo et al., 2014; Heap, 2016; Preston et al., 2009). Furthermore, glyphosate-resistant (GR) kochia was first reported in Kansas in 2007, then in South

Dakota in 2009, followed by Nebraska and Colorado in 2011, and Montana and North Dakota in 2012 (Heap, 2016; Kumar et al., 2014; Waite et al., 2013). Additionally, multiple herbicide-resistant kochia biotypes have been reported with resistance to acetolactate synthase (ALS)-and photosystem (PS) II-inhibitors in Illinois and Indiana, and resistance to ALS- and 5-enolpyruvylshikimate-3-phosphate (EPSP) synthase-inhibitors in Montana and western Canada in 2013 (Beckie et al., 2013; Heap, 2016). In this respect, Kochia is similar to Palmer amaranth (*Amaranthus palmeri* S. Wats) and common waterhemp (*Amaranthus rudis* Sauer) biotypes that have evolved resistance to atrazine and HPPD-inhibiting herbicides in Nebraska (Jhala et al., 2014; Rana et al., 2013).

Kochia is not a new weed species in Nebraska. It is usually found in the western and southwestern part of the state, but over the past several years it has been reported as problematic further east. Kochia populations resistant to PS II- (triazine) and ALS-inhibiting herbicides (sulfonylureas and imidazolinones) occur in many corn fields in Nebraska (Heap, 2016), and recently, glyphosate- and dicamba-resistant kochia were confirmed in western Nebraska in 2010 and 2011, respectively (Heap, 2016; Crespo et al., 2014). Thus, kochia has evolved resistance to herbicide active ingredients belonging to four different sites of action in Nebraska.

Nebraska ranks 4th in the United States in terms of market value of agricultural products sold, with livestock, corn, and soybean being the top three agricultural products in terms of revenue generated (USDA-NASS, 2012). Corn and soybean cropping systems have a huge impact on the state's economy, and weed control is an important component of the crop production system. In the late 1990s, due to the widespread and repeated use of ALS-inhibiting herbicides, several weed species—including both grasses and broadleaf weeds—evolved resistance to ALS-inhibiting herbicides in several Midwestern states, including Nebraska (Patzoldt et al., 2002; Tranel & Wright, 2002). With the commercialization of GR crops in 1996, selection pressure from ALS herbicides lessened as glyphosate provided new and effective management of resistant weed species (Shaner, 2014). However, with the increased use of glyphosate, fields in Iowa, Illinois, and Missouri began reporting GR weed species (Culpepper, 2006; Owen, 2002; Rosenbaum & Bradley, 2013). Additionally, glyphosate and ALS-inhibiting herbicides are predominant chemistries used to control kochia in chemical-fallow, pre-plant burndown, and post-harvest in wheat (Donald & Prato, 1991; Lloyd et al., 2011; Mickelson et al., 2004; Kumar et al., 2014; Kumar & Jha, 2015a, 2015b). In a recent research Kumar et al. (2015) reported the underlying mechanisms of glyphosate and ALS resistance in kochia accessions from Montana.

Growers from western Nebraska reported kochia control failure following repeated applications of glyphosate and ALS-inhibiting herbicides, justifying the need to confirm the existence of glyphosate and ALS inhibitor-resistant kochia in Nebraska. The objective of this study was to confirm a kochia biotype resistant to glyphosate and ALS-inhibiting herbicide (tribenuron) by quantifying the level of resistance in a whole plant dose response study. We hypothesized that the putative kochia biotype would be resistant to both glyphosate- and ALS-inhibiting herbicides because of their control failure under field conditions.

2. Method

2.1 Plant Materials

In 2012, a grower from Sheridan County in western Nebraska reported failure to control kochia following repeated applications of glyphosate and tribenuron methyl. The field in question had been under glyphosate-tolerant corn, conventional spring wheat, and grass seed production for at least eight years, mostly relying on glyphosate and tribenuron for weed control. During fall of 2012, seeds of surviving kochia plants were collected from the same field and considered as a putative-glyphosate and ALS-inhibitors-resistant kochia biotype. Kochia seeds collected in 2010 from a field near Grant, Nebraska with a known history of effective control with the recommended rate of glyphosate and tribenuron were considered as a susceptible biotype and used for comparison in this study.

Seeds were cleaned thoroughly and stored separately in airtight polythene bags at 4 °C until their use in this study. The seeds were planted in germination trays containing potting mix (Berger BM1 All-Purpose Mix, Berger Peat Moss Ltd., Saint-Modeste, Quebec, Canada). Seedlings were transplanted at the first true leaf stage to square plastic pots (10 × 10 × 12-cm) containing a 3:1 mixture of potting mix to soil. Plants were supplied with adequate water and nutrients and kept in a greenhouse maintained at 28 °C and 24 °C day and night temperatures, respectively. Artificial lighting was provided using metal halide lamps with 600 μmol photon m^{-2} s^{-1} light intensity to ensure a 16 h photoperiod.

2.2 Whole Plant Dose-Response Study

Greenhouse whole plant dose-response bioassays were conducted in 2013 at the University of Nebraska-Lincoln to determine the level of resistance in the putative glyphosate-and ALS-inhibitor-resistant kochia biotype. The experiments were arranged in a randomized complete block design with four replications. Separate experiments were conducted for glyphosate and ALS-inhibiting herbicides (tribenuron). Single kochia plant per pot was considered as an experimental unit. Glyphosate (Touchdown HiTech®, Syngenta Crop Protection, LLC, P.O. Box 18300, Greensboro, NC 27419-8300) treatments included 9 rates (0, 0.125×, 0.25×, 0.5×, 1×, 2×, 4×, 8×, and 16×), where 1× = the recommended field rate of glyphosate (870 g ae ha^{-1}). Tribenuron-methyl (Express herbicide, DuPont, 1007 Market Street, Wilmington, Delaware 19898) treatments included 9 rates (0, 0.125, 0.25×, 0.5×, 1×, 2×, 4×, 8×, and 16×), where 1× = the recommended field rate of tribenuron methyl (17.5 g ai ha^{-1}). The 8- to 10-cm tall kochia seedlings were treated with glyphosate or tribenuron treatments in a single-tip chamber sprayer (DeVries Manufacturing Corp, Hollandale, MN 56045) fitted with an 8001 E nozzle (TeeJet, Spraying Systems Co., Wheaton, IL 60187) calibrated to deliver 140 L ha^{-1} spray volume at 207 kPa at a speed of 4 kmph. Each herbicide treatment was prepared in distilled water and mixed with ammonium sulfate (AMS, DSM Chemicals North America Inc., Augusta, GA) at 2.5% wt/v and nonionic surfactant (NIS, Induce, Helena Chemical Co., Collierville, TN) at 0.25% v/v.

Visual control estimates were recorded at 7, 14, and 21 d after treatment (DAT) using a scale ranging from 0 to 100%, with 0% meaning no control and 100% meaning complete death or control of kochia. Percent control was assessed on the basis of chlorosis, necrosis, and stunting in plant height compared with nontreated control plants. Aboveground biomass of each kochia plant was harvested at 21 DAT and oven-dried at 65 °C until it reached a constant weight. The biomass data were converted into percent biomass reduction as compared to the nontreated control (Wortman, 2014).

$$Percent\ biomass\ reduction = [(C-B)/C] \times 100 \qquad (1)$$

Where, C is the mean biomass of the four nontreated control replicates, and B is the biomass of an individual treated experimental unit.

2.3 Statistical Analysis

Data were subjected to ANOVA using the PROC MIXED procedure in SAS to test the significance of experimental run, treatment, replication, and treatment-by-experiment interaction. Data from both experiments were pooled on the basis of non-significant experiment run and treatment by experiment interaction; therefore, combined data are presented. For the dose response studies, visual injury estimates and biomass reduction data at 21 DAT were analyzed using a nonlinear regression model with *drc* 1.2 version package in R 2.3.0 (R statistical software, R Foundation for Statistical Computing, Vienna, Austria; http://www.R-project.org) (Knezevic et al., 2007). Data were subjected to dose response models using a four-parameter log-logistic equation (Seefeldt et al., 1995; Streibig et al., 1993).

$$Y = C + \{D - C/1 + \exp[B\ (\log X - \log E)]\} \qquad (2)$$

Where, Y is the response variable based on the visual injury estimate or biomass reduction, C is the lower limit, D is the upper limit, X is the glyphosate or tribenuron dose, E is the glyphosate or tribenuron dose required for 50% response (e.g., an effective dose, ED_{50}, required to produce 50% control or biomass reduction), and B is the slope of each curve. Finally, the effective doses of glyphosate or tribenusron required to cause 50 and 90% injury and biomass reduction at 21 DAT were calculated.

3. Results and Discussion

The labeled rate of glyphosate (870 g ha^{-1}) completely controlled the susceptible kochia biotype, with 670 g ha^{-1} of glyphosate required for 90% control (Table 1). The labeled rate of glyphosate made little to no injury on the putative-resistant kochia biotype (Figure 1), which required 2,516 and 3,781 g ha^{-1} of glyphosate to achieve 50 and 90% control, respectively. The putative-resistant biotype thus exhibited a 6-fold level of resistance compared to the glyphosate-susceptible biotype (Table 1). Similarly, Beckie et al. (2013) reported the first occurrence of three GR kochia biotypes in western Canada exhibiting resistance levels ranging from 4- to 6-fold. A similar study in Montana on four GR kochia accessions required glyphosate doses of 2350 to 3640 g ha^{-1} for 50% response (Kumar et al., 2014).

Table 1. Values of ED_{50} and ED_{90} for control of kochia biotypes in a dose response to glyphosate and tribenuron at 21 d after treatment in a greenhouse study conducted at the University of Nebraska-Lincoln

Herbicide	Kochia biotypes	ED_{50} (±SE) (g ae ha^{-1})	ED_{90} (±SE) (g ae ha^{-1})	Resistance Level[a]	Regression Parameters[b]	
					B	D
Glyphosate						
	Susceptible	126 (5)	670 (53)	-	-1 (0.1)	102 (1)
	Resistant	2516 (91)	3781 (190)	6×	-5 (0.5)	99 (2)
Tribenuron						
	Susceptible	16 (0.1)	30 (0.2)	-	-4 (0.05)	100 (0.2)
	Resistant[c]	150 (5)	443 (34)	15×	-2 (0.1)	98 (5)

Note. ED_{50} = effective dose required to control 50% population, ED_{90} = effective dose required to control 90% population, SE = standard error. The value in parenthesis is standard error.

[a] Regression parameters B and D for 3-parameter log-logistic model obtained using the nonlinear least-squares function of the statistical software R.

[b] Resistance level was calculated by dividing ED_{90} value of resistant biotype by that of the susceptible biotype.

[c] These values have limited biological meaning because 90% control was not achieved even with the highest rate of glyphosate and tribenuron used in this study.

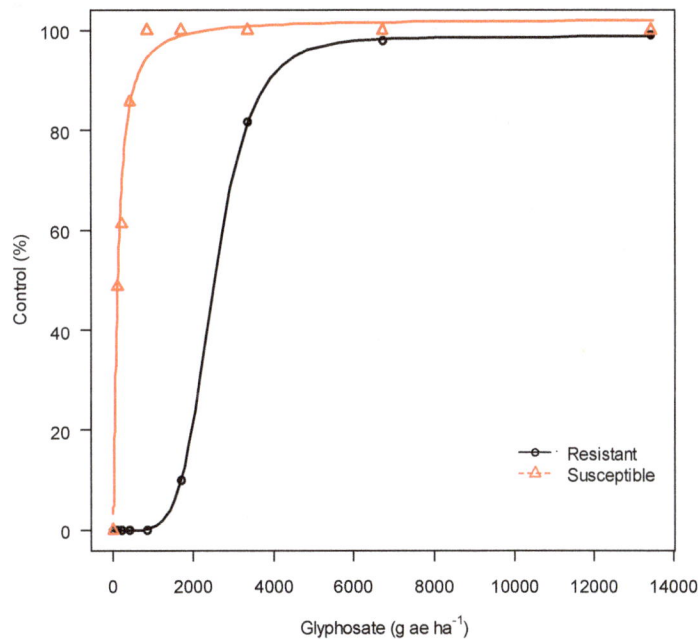

Figure 1. Control of susceptible and resistant kochia biotypes in a dose response to glyphosate at 21 d after treatment in a greenhouse study conducted at the University of Nebraska-Lincoln

The higher level of resistance to the ALS-inhibiting herbicide (tribenuron) was also observed in the putative-resistant kochia biotype. For example, the ED_{90} value of the putative-resistant kochia biotype was achieved at 443 g ha^{-1} compared with 30 g ha^{-1} for the susceptible biotype, thus achieving a 15-fold level of resistance (Figure 2; Table 1). Beckie et al. (2013) also reported a high frequency of ALS-inhibitor herbicide resistance in all confirmed GR kochia populations collected from Alberta, Canada. On the basis of the dose-response curve of shoot biomass reduction, ED_{50} values for the glyphosate and tribenuron-resistant kochia biotype were 2,329 and 117 g ha^{-1} (Figure 2; Table 2). Furthermore, ED_{90} values were 6- and 15-fold greater than the susceptible biotype and 7 and 10 times the labeled rate of glyphosate and tribenuron, respectively. Tribenuron was never able to provide > 65% control of the resistant kochia biotype, even with the highest rate (280 g ha^{-1}) tested (Figure 2; Table 2).

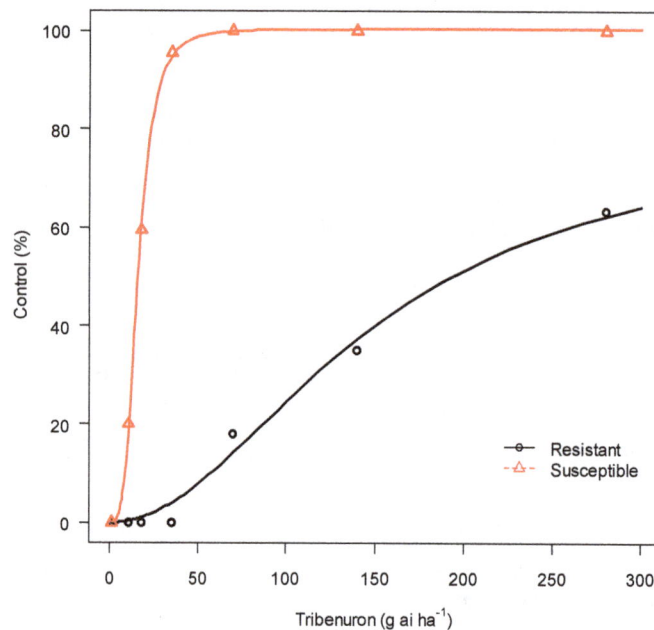

Figure 2. Control of susceptible and resistant kochia biotypes in a dose response to tribenuron at 21 d after treatment in a greenhouse study conducted at the University of Nebraska-Lincoln

Table 2. Values of ED_{50} and ED_{90} for percent biomass reduction of kochia biotypes in a dose response to glyphosate and tribenuron at 21 d after treatment in a greenhouse study at the University of Nebraska-Lincoln

Herbicide	Kochia biotypes	ED_{50} (±SE) (g ae ha^{-1})	ED_{90} (±SE) (g ae ha^{-1})	Regression Parameters[a]	
				B	D
Glyphosate					
	Susceptible	639 (157)	719 (192)	-2.1 (0.5)	101 (2)
	Resistant	2329 (353)	4845 (734)	-1.5 (0.1)	74 (5)
Tribenuron					
	Susceptible	20 (2)	47 (8)	-2.5 (0.5)	102 (3)
	Resistant	117 (17)	463 (68)	-1.1 (0.1)	95 (2)

Note. ED_{50} = effective dose required for 50% biomass reduction, ED_{90} = effective dose required for 90% biomass reduction, SE = standard error. The values in parenthesis are standard errors.

[a] Regression parameters B and D for 3-parameter log-logistic model obtained using the nonlinear least-squares function of the statistical software R.

Overall results of the dose response curves suggested that the susceptible kochia biotype was effectively controlled with the labeled rates of glyphosate and tribenuron. The percent visual control dose response analysis showed the GR kochia biotype with a 6-and 15-fold resistance to glyphosate and tribenuron, respectively, relative to the susceptible biotype (Table 1). Biomass reduction data for glyphosate and tribenuron supported the percent visual injury estimates at 21 DAT and suggested a similar level of resistance at the ED_{50} and ED_{90} levels (Figure 3; Figure 4).

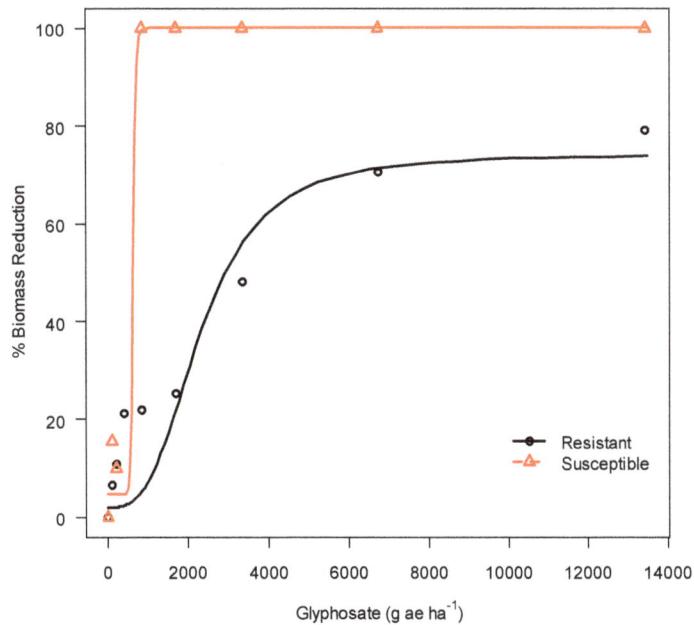

Figure 3. Biomass reduction of susceptible and resistant kochia biotypes in a dose response to glyphosate at 21 d after treatment in a greenhouse study conducted at the University of Nebraska-Lincoln

The resistant kochia biotype had a 15-fold level of resistance to POST-applied tribenuron. The level of ALS-resistance is likely much higher, since the highest dose only provided 63% control (Figure 2) and 46% biomass reduction (Figure 4).

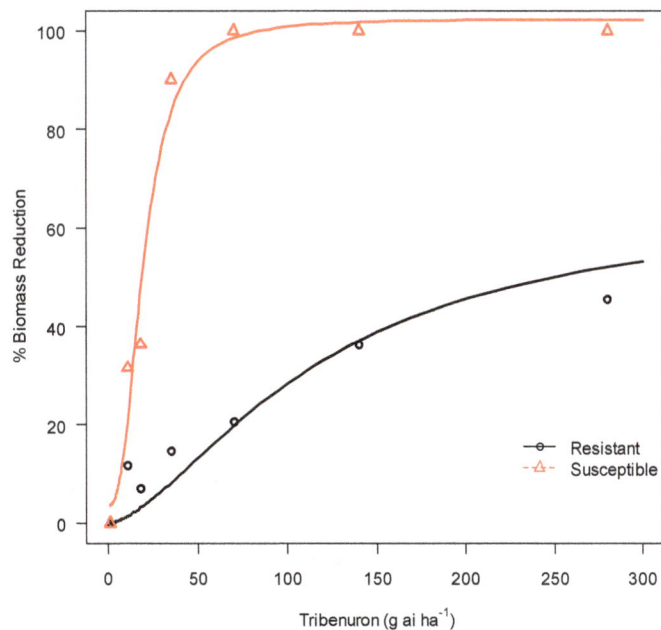

Figure 4. Biomass reduction of susceptible and resistant kochia biotypes in a dose response to tribenuron at 21 d after treatment in a greenhouse study conducted at the University of Nebraska-Lincoln

Weed management and containment efforts to control herbicide-resistant weeds are highly influenced by the adoption of best management practices by growers (Norsworthy et al., 2012). The putative-resistant kochia biotype from the production field in Nebraska is resistant to both glyphosate and ALS-inhibiting herbicides. In the Midwestern United States, kochia was one of the first weed species to evolve resistance to several herbicide chemistries, including synthetic auxins and EPSP synthase inhibitors (Primiani et al., 1990; Beckie et al., 2013). Glyphosate and tribenuron are important weed control options that can be applied POST or pre-seeding (burndown), in-crop, or in chemical-fallow situations in glyphosate-resistant crops and most cereal crops. With the evolution of multiple-herbicide resistance there would be no registered in-crop herbicides to control the multiple-herbicide resistant biotype in sunflower (*Helianthus annuus* L.), lentil (*Lens culinaris* Medik), chickpea (*Cicer arietinum* L.), dry bean (*Phaseolus vulgaris* L.), soybean, or potato (*Solanum tuberosum* L.). Glyphosate and tribenuron could still be used to control other troublesome broadleaf and grass weed species, but the growers would have to modify their practices to control kochia using other cultural, physical or chemical practices. Due to the evolution of EPSP synthase and ALS-inhibitor-resistant kochia biotype, weed management in these crop production systems will be challenging. Therefore, integrated weed management strategies including crop rotation, tillage, and a PRE followed by POST herbicide program with effective sites of action will be required for the control of EPSP synthase and ALS-inhibiting herbicide-resistant kochia biotype.

References

Anderson, R. L., & Nielsen, D. C. (1996). Emergence pattern of five weeds in the central Great Plains. *Weed Technology, 10*, 744-749.

Beckie, H. J., Blackshaw, R. E., Low, R., Hall, L. M., Sauder, C. A., Martin, S., ... Shirriff, S. W. (2013). Glyphosate- and acetolactate synthase inhibitor–resistant kochia (*Kochia scoparia*) in western Canada. *Weed Science, 61*, 310-318. http://dx.doi.org/10.1614/WS-D-12-00140.1

Casey, P. A. (2009). *Plant Guide for Kochia (Kochia scoparia)*. USDA-Natural Resources Conservation Service, Kansas Plant Materials Center, Manhattan, KS.

Cranston, H. J., Kern, A. J., Hackett, J. L., Miller, E. K., Maxwell, B. D., & Dyer, W. E. (2001). Dicamba resistance in Kochia. *Weed Science, 49*, 164-170. http://dx.doi.org/10.1614/0043-1745(2001)049[0164:DRIK]2.0.CO;2

Crespo, R. J., Bernards, M. L., Sbatella, G. M., Kruger, G. R., Lee, D. J., & Wilson, R. G. (2014). Response of Nebraska Kochia (*Kochia scoparia*) Accessions to Dicamba. *Weed Technology, 28*, 151-162. http://dx.doi.org/10.1614/WT-D-13-00109.1

Culpepper, A. S. (2006). Glyphosate-induced weed shifts. *Weed Technology, 20*, 277-281. http://dx.doi.org/10.1614/WT-04-155R.1

Dodd, J., & Moore, J. H. (1993). Introduction and status of *Kochia scoparia* in western Australia. *Proceedings of the 10th Australian and 14th Asian-Pacific Weeds Conference, 1*, 496-500.

Donald, W. W., & Prato, T. (1991). Profitable, effective herbicides for planting-time weed control in no-till spring wheat (*Triticum aestivum*). *Weed Science, 39*, 83-90.

Durgan, B. R., Dexter, A. G., & Miller, S. D. (1990). Kochia (*Kochia scoparia*) interference in sunflower (*Helianthus annuus*). *Weed Technology, 4*, 52-56.

Forcella, F. (1985). Spread of kochia in the northwestern United States. *Weeds Today, 16*, 4-6.

Friesen, L. F., Beckie, H. J., Warwick, S. I., & Van Acker, R. C. (2009). The biology of Canadian weeds. 138. *Kochia scoparia* (L.) Schrad. *Canadian Journal of Plant Science, 89*, 141-167. http://dx.doi.org/10.4141/CJPS08057

Heap, I. (2016). *International Survey of Herbicide Resistant Weeds*. Retrieved April, 2016, from http://www.weedscience.org

Jhala, A. J., Sandell, L. D., Rana, N., Kruger, G. R., & Knezevic, S. Z. (2014). Confirmation and control of triazine and 4-Hydroxyphenylpyruvate dioxygenase-inhibiting herbicide-resistant Palmer amaranth (*Amaranthus palmeri*) in Nebraska. *Weed Technology, 28*, 28-38. http://dx.doi.org/10.1614/WT-D-13-00090.1

Knezevic, S. Z., Streibig, J. C., & Ritz, C. (2007). Utilizing R software package for dose-response studies: the concept and data analysis. *Weed Technology, 21*, 840-848. http://dx.doi.org/10.1614/WT-06-161.1

Kumar, V., & Jha, P. (2015a). Effective preemergence and postemergence herbicide programs for kochia control. *Weed Technology, 29*, 24-34. http://dx.doi.org/10.1614/WT-D-14-00026.1

Kumar, V., & Jha, P. (2015b). Influence of herbicides applied postharvest in wheat stubble on control, fecundity, and progeny fitness of *Kochia scoparia* in the US Great Plains. *Crop Protection, 71*, 144-149. http://dx.doi.org/10.1016/j.cropro.2015.02.016

Kumar, V., Jha, P., & Reichard, N. (2014). Occurrence and characterization of kochia (*Kochia scoparia*) accessions with resistance to glyphosate in Montana. *Weed Technology, 28*, 122-130. http://dx.doi.org/10.1614/WT-D-13-00115.1

Kumar, V., Jha, P., Giacomini, D., Westra, E. P., & Westra, P. (2015). Molecular basis of evolved resistance to glyphosate and Acetolactate Synthase-Inhibitor herbicides in kochia (*Kochia scoparia*) accessions from Montana. *Weed Science, 63*, 758-769. http://dx.doi.org/10.1614/WS-D-15-00021.1

Lloyd, K. L., Johnson, J. M., Gover, A. E., & Sellmer, J. C. (2011). Preemergence and postemergence suppression of kochia on rights-of-way. *Weed Technology, 25*, 292-297. http://dx.doi.org/10.1614/WT-D-10-00117.1

Mickelson, J. A., Bussan, A. J., Davis, E. S., Hulting, A. G., & Dyer, W. E. (2004). Postharvest kochia (*Kochia scoparia*) management with herbicides in small grains. *Weed Technology, 18*, 426-431. http://dx.doi.org/10.1614/WT-03-164R1

Norsworthy, J. K., Ward, S. M., Shaw, D. R., Llewellyn, R. S., Nichols, R. L., Webster, T. M., ... Barrett, M. (2012). Reducing the risks of herbicide resistance: best management practices and recommendations. *Weed Science, 60*, 31-62. http://dx.doi.org/10.1614/WS-D-11-00155.1

Nussbaum, E. S., Wiese, A. F., Crutchfield, D. E., Chenault, E. W., & Lavake, D. (1985). The effect of temperature and rainfall on emergence and growth of eight weeds. *Weed Science, 33*, 165-170.

Owen, M. D. K. (2002). Glyphosate resistant waterhemp in Iowa. *Proceedings of North Central Weed Science Society, 57*, 210.

Primiani, M. M., Cotterman, J. C., & Saari, L. L. (1990). Resistance of *Kochia scoparia* to sulfonylurea and imidazolinone herbicides. *Weed Technology, 4*, 169-172.

Rana, N., Knezevic, S. Z., & Scott, J. (2013). HPPD-resistant waterhemp in Nebraska. *Proceedings of 2013 Crop Production Clinics* (pp. 165-171). Lincoln, NE: University of Nebraska-Lincoln Extension.

Rosenbaum, K. K., & Bradley, K. W. (2013). A Survey of Glyphosate-Resistant Waterhemp (*Amaranthus rudis*) in Missouri Soybean Fields and Prediction of Glyphosate Resistance in Future Waterhemp Populations Based on In-Field Observations and Management Practices. *Weed Technology, 27*, 656-663. http://dx.doi.org/10.1614/WT-D-13-00042.1

Schwinghamer, T. D., & Van Acker, R. C. (2008). Emergence timing and persistence of kochia (*Kochia scoparia*). *Weed Science, 56*, 37-41. http://dx.doi.org/10.1614/WS-07-098.1

Seefeldt, S. S., Jensen, J. E., & Fuerst, E. P. (1995). Log-logistic analysis of herbicide dose-response relationships. *Weed Technology, 9*, 218-227.

Shaner, D. L. (2014). Lessons learned from the history of herbicide resistance. *Weed Science, 62*, 427-431. http://dx.doi.org/10.1614/WS-D-13-00109.1

Streibig, J. C., Rudemo, M., & Jensen, J. E. (1993). Dose-response curves and statistical models. In J. C. Streibig, & P. Kudsk (Eds.), *Herbicide Bioassays* (pp. 29-55). Boca Raton, FL: CRC.

Waite, J., Thompson, C. R., Peterson, D. E., Currie, R. S., Olson, B. L. S., Stahlman, P. W., & Khatib, K. A. (2013). Differential Kochia (*Kochia scoparia*) Populations Response to Glyphosate. *Weed Science, 61*, 193-200. http://dx.doi.org/10.1614/WS-D-12-00101.1

Weatherspoon, D. M., & Schweizer, E. E. (1969). Competition between kochia and sugarbeets. *Weed Science, 17*, 464-467.

Wicks, G. A., Martin, A. R., & Mahnken, G. W. (1993). Control of triazine-resistant kochia (*Kochia scoparia*) in conservation tillage corn (*Zea mays*). *Weed Science, 41*, 225-231.

Wicks, G. A., Martin, A. R., Haack, A. E., & Mahnken, G. W. (1994). Control of triazine-resistant kochia (*Kochia scoparia*) in sorghum (*Sorghum bicolor*). *Weed Technology, 8*, 748-753.

Wortman, S. E. (2014). Integrating weed and vegetable crop management with multifunctional air propelled abrasive grits. *Weed Technology, 28*, 243-252. http://dx.doi.org/10.1614/WT-D-13-00105.1

Evaluation of Serenade Max to Control Fruit Rot of Grapes

T. Thomidis[1], S. Pantazis[2] & K. Konstantinoudis[3]

[1] Alexander Technological Educational Institute of Thessaloniki, Sindos, Macedonia, Greece

[2] ANADIAG Hellas, Veria, Macedonia, Greece

[3] Technological Educational Institute of Central Macedonia, Serres, Greece

Correspondence: T. Thomidis, Alexander Technological Educational Institute of Thessaloniki, Sindos, 57400, Macedonia, Greece. E-mail: thomi-1@otenet.gr

Abstract

The effectiveness of the biopesticide Serenade Max to control fruit rot in grape was investigated. The experiments were conducted in commercial vineyard (cv Xinomavro Naousa). The results showed that this product was relatively effective to control fruit rots on grapes. No significant difference was observed between Serenade Max and Switch25/37.5 WG in both years of experiment. It was found that the fungus *Botrytis cinerea* commonly caused fruit rots (at a percentage of 96%) on the control grape fruits.

Generally, Serenade Max could be an alternative method against fruit rots of grape in biological fruit production system.

Keywords: *Bacillus subtilis*, biological control, *Botrytis cinerea*, fruit rot, vineyard

1. Introduction

Bunch rot of grapes is a very important disease occurring in vineyards worldwide. *Botrytis* has been found as the main pathogen causing fruit rot on grapes (Thomidis et al., 2015). Several other genera and species such as *Colletotrichum*, *Rhizopus*, *Aspergillus* and *Alternaria* have been reported to cause fruit rots on grape fruits, but in low percentage (Latinovic et al., 2012). The disease is most severe during prolonged rainy and cloudy periods just before or during harvest. Use of effective fungicides at appropriate times during the growing season can provide significant control. The use of biological products reduces potential risk to farm workers and the environment and to promote public confidence in food safety. Previous works showed the antagonistic bacterial strains Bacillus C6, Brevibacterium MFD-47, Enterobacter MFD-81 and Pantoea MFD-232 were effective against *B. cinerea* (Donmez et al., 2011). Serenade Max is a bio-fungicide/bactericide (contains a unique strain of *Bacillus subtilis* (QST 713 strain)) to aid in the control and suppression of many plant pathogens. This product is suitable for use in organic production systems and has BioGro certification (Anonymous, 2011). The main mode of action of Serenade is based on the production of lipopeptides disrupt pathogen cell membranes. These modes of action are different from other fungicides and therefore represent an opportunity for resistance management and synergy with single-site fungicides (Sereno et al., 2013). In addition, lipopeptides induce systemic acquired resistance (SAR) and other defence responses against a number of pathogens in host plants (Niu et al., 2011; Ongena & Jacques, 2008).

The main objective of this study was the evaluation of the biopesticide Serenade Max (BAYER Hellas, Sorou 18-22, 15125, Marousi Athens) to control fruit rots on grape.

2. Method

Field experiments were conducted in a commercial vineyard (cv. Xinimavro Naoussas, 8-yr-old) located in Strantza Naousas, the Prefecture of Imathia, Greece, for two consecutive years (2014, 2015). Plants were grown under commercial cultural practices and no fungicide was applied against fruit rots. A back pack sprayer (Boom ID-GR T19cr) was used with a spray volume 1000 l/ha. The date of applications was at 29th August and 15th September 2014 and again 2015. Necrotic tissues from the margin of the diseased portion of the unsprayed fruit were aseptically removed and placed on potato dextrose agar plates. In 2014, one concentration (2.5 mg Serenade Max/ml) was evaluated, while, in 2015, two concentrations were evaluated (2.5 mg Serenade Max/ml and 4 mg Serenade Max/ml). The experiment was laid out in randomized complete block design. There were 5 replicates, each with three plants. Plants sprayed with the fungicide fludioxonil/cyprodinil (1 g/l) recommended by producer (Switch 25/37.5 WG-Syngenta Hellas) and untreated plants were used as control. Thirty fruits/bunches were randomly collected from each treatment at harvesting period and the percentage of rotted fruits was recorded by using a rate: 0 = healthy, 100 = totally rotted.

Data for fruit rots expressed in percentages were analysed after angular transformation to active normality (data presented in the tables are back-transformed). For testing of differences, the Duncan's Multiple Range Test was applied at the significant level a = 0.05.

3. Results and Discussion

The results showed that this product was relatively effective to control fruit rots on grapes (Table 1). No significant difference was observed between Serenade Max and Switch 25/37.5 WG in both years of experiment. Also, there is no significant difference between the two concentrations of Serenade Max used. Previous works showed that the *B. subtilis* strains B-3 and B-16 were effective against *B. cinerea* in apple (Peighami-Ashnaei et al., 2009). In other study, the *B. subtilis* strain B-28 inhibited *B. cinerea* of tomato at a rate of 71.1% *in vitro* and 52.4% *in vivo* (Wang et al., 2009). However, Anonymous (2011) reported that the symptoms of Botrytis bunch rot were evident in both treatments of Serenade Max (incidence 21.5% and 24.5% respectively) under high disease pressure.

Isolation of pathogens from the control grape fruits showed that the fungus *B. cinerea* was mainly responsible at a percentage of 96%. Fungi of genus *Colletotrichum*, *Alternaria*, *Aspergillus* were also isolated from rotted fruits. Similarly, Latinovic et al. (2012) isolated the fungi *Colletotrichum*, *Rhizopus*, *Aspergillus* and *Alternaria* from grape fruit with symptoms of rot.

Generally, Serenade Max could be an alternative method against fruit rots of grape in biological fruit production system. In addition, a spray programme including Serenade Max and fungicide application is recommended in integrated fruit production system to reduce the percentage of fruit rots.

Table 1. Effectiveness of Serenade Max to control fruit rot of grapes

2014			
Treatment	Rate (mg/ml)	Percentage of Fruit rots	
Serenade Max	2.5	1.40	a[z]
Switch 25/37.5 WG	7.5	0.70	a
Control	-	5.55	b
2015			
Treatments	Rate (mg/ml)	Percentage of Fruit rots	
Serenade Max	2.5	3.71	a
	4	2.75	a
Switch 25/37.5 WG	7.5	1.63	a
Control	-	14.20	b

Note. [z]Values in the same column followed by the same letter are not significantly different according to Duncan's Multuple Range Test (P < 0.05).

References

Anonymous. (2011). *An evaluation of the efficacy of two Bacillus subtilis based products against Botrytis when used in a full season programme during 2010/2011 season.* Retrieved from http://www.grochem.co.nz/Portals%5C537%5CPDF/Clarity_WineGrapeFieldTrial2011.pdf

Donmez, M. F., Esitken, A., Yildiz, H., & Ercisli, S. (2011). Biocontrol of *Botrytis cinerea* on strawberry fruit by plant growth promoting bacteria. *Journal of Animal Plant Science, 21*, 758-763.

Latinovic, J., Latinovic, N., Tiodorovic, J., & Odalovic, A. (2012). First report of Anthracnose fruit rot of strawberry caused by *Colletotrichum acutatum* in Montenegro. *Plant Disease, 96*, 1066. http://dx.doi.org/10.1094/PDIS-02-12-0108-PDN

Niu, D., Liu, H., Jiang, C., Wang, Y., Wang, Q., Jin, H., & Guo, J. (2011). The plant growth-promoting rhizobacterium *Bacillus cereus* AR156 induces systemic resistance in *Arabidopsis thaliana* by simultaneously activating salicylate- and jas-monate/ethylene-dependent signaling pathways. *Molecular Plant-Microbe Interactions, 24*, 533-542. http://dx.doi.org/10.1094/MPMI-09-10-0213

Ongena, M., & Jacques, P. (2008). Bacillus lipopeptides: versatile weapons for plant disease biocontrol. *Trends in Microbiology, 16*,115-125. http://dx.doi.org/10.1016/j.tim.2007.12.009

Peighami-Ashnaei, S., Sharifi-Tehrani, A., Ahmadzadeh, M., & Behboudi, K. (2009). Selection of bacterial antagonists for the biological control of *Botrytis cinerea* in apple (*Malus domestica*) and in comparison with application of thiabendazole. *Communications in Agricultural and Applied Biological Sciences, 74*, 739-743.

Serrano, L., Manker, D., Brandi, F., & Cali, T. (2013). The use of *bacillus subtilis* QST 713 and *bacillus pumilus* qst 2808 as protectant fungicides in conventional application programs for black leaf streak control. *Acta Horticulturae, 986*, 149-156. http://dx.doi.org/10.17660/ActaHortic.2013.986.15

Thomidis, T., Pantazis, S., Navrozidis, E., & Karagiannidis, N. (2015). Biological control of fruit rots on strawberry and grape by BOTRY-Zen. *New Zealand Journal of Crop and Horticultural Science, 43*, 68-72. http://dx.doi.org/10.1080/01140671.2014.958502

Wang, S., Hu, T., Jiao, Y., Wie, J., & Cao, K. (2009). Isolation and characterization of *Bacillus subtilis* EB-28, an endophytic bacterium strain displaying biocontrol activity against *Botrytis cinerea* Pers. *Frontiers Agricultural China, 3*, 247-252. http://dx.doi.org/10.1007/s11703-009-0042-x

7

Nutritional Value of Crisphead 'Iceberg' and Romaine Lettuces (*Lactuca sativa* L.)

Moo Jung Kim[1,2], Youyoun Moon[1], Dean A. Kopsell[3], Suejin Park[1], Janet C. Tou[2] & Nicole L. Waterland[1]

[1] Division of Plant and Soil Sciences, West Virginia University, Morgantown, WV, USA

[2] Division of Animal and Nutritional Sciences, West Virginia University, Morgantown, WV, USA

[3] Department of Plant Sciences, The University of Tennessee, Knoxville, TN, USA

Correspondence: Nicole L. Waterland, Division of Plant and Soil Sciences, West Virginia University, Morgantown, WV 26506, USA. E-mail: nicole.waterland@mail.wvu.edu

The research is financed by the West Virginia University Foundation and West Virginia University Agriculture and Forestry Experimental Station (Hatch Grant WVA00665 and WVA00640). Scientific Article No. 3282 of the West Virginia Agricultural and Forestry Experiment Station, Morgantown.

Abstract

Lettuce (*Lactuca sativa* L.) is one of the most popular vegetables worldwide, but is often viewed as low in nutritional value. However, lettuce contains health-promoting nutrients and biosynthesis of such phytochemicals varies depending on cultivar, leaf color and growing conditions. Studies of such parameters on the nutritional value have not been conclusive because the lettuce samples were collected from heterogeneous growing conditions and/or various developmental stages. In our study nutritional composition was evaluated in the two most popular lettuce types in Western diets, romaine and crisphead 'Iceberg', with red or green leaves grown under uniform cultivating conditions and harvested at the same developmental stage. In the investigated lettuce cultivars, insoluble fiber content was higher ($P \leq 0.05$) in romaine than crisphead lettuces. α-linolenic acid (omega-3 polyunsaturated fatty acid) was the predominant fatty acid and was higher in romaine than crisphead. Iron and bone health-promoting minerals (Ca, Mg and Mn) were significantly higher ($P \leq 0.001$) in romaine. The content of β-carotene and lutein in romaine (668.3 μg g^{-1} dry weight) was ~45% higher than in crisphead (457.3 μg g^{-1}dry weight). For leaf color comparison, red cultivars provided higher amount of minerals (Ca, P, Mn and K), total carotenoids, total anthocyanins and phenolics than green cultivars. Based on our study results, romaine was generally higher in nutrients analyzed, especially red romaine contained significantly higher amount of total carotenoids, total anthocyanins and phenolics. Therefore, romaine type lettuces with red rather than green leaves may offer a better nutritional choice.

Keywords: carotenoid, dietary mineral, fatty acid, insoluble fiber, lettuce, phenolic compound

1. Introduction

Epidemiological studies have reported a correlation between fresh vegetable consumption and reduced risk of chronic diseases (Rodriguez-Casado, 2016). Lettuce (*Lactuca sativa* L.) is one of the most popular vegetables worldwide. In countries with Western diets such as the United States (US), the most serious public health threat is chronic diseases partially due to poor dietary habits (Cordain et al., 2005). Even though salad consumption is increasing, lettuce is regarded as low in nutritional value. However, nutritional value of lettuce greatly varies with lettuce types and depending on lettuce type, and nutrient composition can be equivalent to other "nutritious" vegetables (Kim, Moon, Tou, Mou, & Waterland, 2016).

Lettuce contains several dietary minerals important for human health such as iron (Fe), zinc (Zn), calcium (Ca), phosphorus (P), magnesium (Mg), manganese (Mn), and potassium (K) and other health-promoting bioactive compounds (Kim et al., 2016). Yet, few human clinical studies have investigated lettuce consumption in disease prevention. A case-control study reported an inverse association between colorectal cancer and lettuce consumption (Fernandez, LaVecchia, Davanzo, Negri, & Franceschi, 1997). Investigating potential nutrients in lettuce contributing to reduced colorectal cancer found no relationship to Ca, vitamin E, and folate, while β-carotene showed a significant relationship. Beneficial health properties of lettuce have mainly been attributed to carotenoids and other phytochemicals such as phenolic compounds (López, Javier, Fenoll, Hellin, & Flores, 2014). High quantities of carotenoids (*i.e.* β-carotene and lutein) were reported for several lettuce types including crisphead, butterhead, romaine, green and red leaf lettuces (Mou, 2005; Nicolle et al., 2004). Phenolic acids and anthocyanins were reported in red and green butterhead, crisphead (subtype Batavia), and green and red oak leaf lettuces (García-Macías et al., 2007; Nicolle et al., 2004). In these studies, carotenoids, phenolic acids, and anthocyanin contents varied depending on lettuce types. Although low in fat, α-linolenic acid, an omega-3 polyunsaturated fatty acid, was found to be the major fatty acid in lettuce (Le Guedard, Schraauwers, Larrieu, & Bessoule, 2008; Pereira, Li, & Sinclair, 2001) but to date, fatty acid composition in different lettuce types has not been well investigated.

In the US, crisphead and romaine are the most popular lettuce types, however, their nutritional value has not been extensively evaluated. The United States Department of Agriculture (USDA) nutrient database (2015) includes crisphead and romaine lettuce; however, few bioactive compounds were measured and cultivars with different leaf colors (*i.e.* green and red) were not specified. Nutritional composition is greatly affected by environmental (light, fertilizers, growing conditions) and biological (cultivar, leaf color, developmental stage) factors (Kim et al., 2016). In many studies samples were prepared from lettuces grown under different growing conditions and/or various developmental stages. Thus, the evaluation of the nutritional value in lettuces among different researchers have been often inconsistent. Particularly, the influence of type/cultivar and leaf pigmentation on nutritional value merits further investigation.

Our hypothesis is that genetic difference controlling characteristics of lettuce (type/cultivar) and pigmentation of leaf (red or green) affect phytochemical contents. Although a number of studies reporting nutritional composition of lettuce are available, there are only a few studies comparing both essential and non-essential nutrients of popular lettuce types that are grown under the same cultural conditions. Therefore the objective of this study was to investigate the effect of type/cultivar and leaf color on nutritional value. In our study nutritional composition, including fatty acid, essential dietary minerals, phytonutrients carotenoids and phenolic compounds, and insoluble dietary fiber was evaluated in two most popular lettuce types in the Western diets, crisphead 'Iceberg', and romaine, with red or green leaves grown under uniform conditions.

2. Materials and Methods

2.1 Plant Materials and Growing Conditions

Seeds of green and red cultivars of crisphead (*L. sativa* L. var. *capitata*, 'Ithaca' and 'Red Iceberg') and romaine (*L. sativa* L. var. *longifolia*, 'Coastal Star' and 'Outredgeous') lettuces were sown to plug trays filled with commercial media (Sunshine Mix #1, Sun Gro Horticulture, Agawam, MA), and germinated under a misting system in a greenhouse. Three-week old seedlings were transplanted to 15-cm pots and grown for additional 60 d. A total of four plants were grown for each cultivar with one plant as an individual biological replication. Plants were grown under natural sunlight with supplemental high pressure sodium lighting when the light intensity was below 50 W m^{-2} with 14 h d^{-1} of photoperiod.

Mean greenhouse temperature was 22.6/18.0 ± 1.7/2.5 °C (mean ± SD) day/night, and plants were irrigated as needed with a 20% Hoagland modified nutrient solution #2 (PhytoTechnology Lab., Shawnee Mission, KS). After harvest plants were freeze-dried (VirTis Freezemobile 12SL, SP Scientific, Warminster, PA) and ground to powder, and then stored at -80 °C until analyses.

2.2 Insoluble Fiber Content

Insoluble fiber content was analyzed using neutral detergent method to determine cellulose, hemicellulose, and lignin following the method of Goering and Soest (1970) with modifications. Freeze-dried powdered lettuce (0.5 g) sample were placed into individual fiber filter bag (ANKOM Technology, Macedon, NY) and sealed. Bags were placed in a fiber analyzer (ANKOM^{200} Fiber Analyzers, ANKOM Technology, Macedon, NY) and agitated with 2 L of neutral detergent solution (ANKOM Technology, Macedon, NY) and 4 mL of heat stable α-amylase at 100 °C for 1 h. Neutral detergent solution was drained and samples were washed four times with boiling water (2 L) for 5 min. Bags were rinsed with 100% acetone and then dried in a drying oven at 100 °C overnight. Individual bags were weighed to quantify total insoluble fiber. Each sample was analyzed in duplicate.

2.3 Fatty Acids Analysis

For lipid extraction, a ground sample (50 mg) was mixed with 5 mL of Triz/EDTA buffer and an internal standard (50 μL of nonadecanoic acid) (Nu Chek Prep, Inc., Elysion, MN). Then, samples were mixed with 20 mL chloroform:methanol:acetic acid solution (2:1:0.015, v/v/v) and centrifuged at 900 g_n for 10 min at 10 °C.

The chloroform layer was collected and filtered through a pre-rinsed 1-phase separation filter paper (Whatman 1-PS, GE Healthcare Bio-Sciences, Pittsburgh, PA). Samples were re-extracted with 10 mL of chloroform:methanol solution (4:1, v/v). Extracted lipids were transmethylated into fatty acid methyl esters (FAMEs) according to Fritsche and Johnston (1990). Briefly, samples were dried under a nitrogen gas at 55 °C and extracted fatty acids were methylated by adding 4 mL of 4% H_2SO_4 in anhydrous methanol and incubated at 90 °C for 1 h. After samples were cooled to room temperature, 3 mL of deionized distilled water and 8 mL of chloroform were added. Samples were centrifuged at 900 g_n for 10 min at 10 °C and chloroform layer was filtered through Na_2SO_4. Samples were dried under a nitrogen gas at 55 °C then dissolved in 2 mL of iso-octane. To analyze FAMEs, a gas chromatography (Varian CP-3800 GC system, Agilent Technologies, Inc., Santa Clara, CA) equipped with an autosampler, a flame ionization detector, and a CP-Sil 88 capillary column (Agilent Technologies, Inc., Santa Clara, CA) was used. Initial temperature was 140 °C and held for 5 min and then increased 4 °C per min to the final temperature of 220 °C. Total running time was 85 min per sample. Carrier gas was nitrogen at 0.4 mL min^{-1} of flow rate. Fatty acids were identified by retention time and quantified using a standard curve made with external standards (16A, Nu-Chek Prep, Inc., Elysian, MN).

2.4 Mineral Content

Lettuce powder (0.5 g) was ashed at 550 °C overnight. Ashed samples were dissolved in 2% nitric acid and adjusted to total volume of 20 mL with deionized distilled water. Mineral content of samples was analyzed by inductively coupled plasma mass spectrometry (Optima 2100DV, Perkin Elmer Corp., Waltham, MA), and each sample was analyzed in triplicate.

2.5 Carotenoid Analysis

Carotenoids were analyzed according to Kopsell, Barickman, Sams, and McElroy (2007). Freeze-dried samples (0.1 g) were re-hydrated in 0.8 mL of ultrapure water for 10 min. Ethyl-β-8'-apo-carotenoate (Carotenature, Lupsingen, Switzerland) was added as an internal standard (0.8 mL). Then, tetrahydrofuran (THF) (2.5 mL) stabilized with 2,6-di-*tert*-butyl-4-methoxyphenol (25 mg L^{-1}) was added. Samples were homogenized then centrifuged at 500 g_n for 3 min. The supernatant was collected and samples were re-extracted in 2 mL of THF as described above. The combined supernatants were evaporated and total volume was adjusted to 5 mL with acetone. Final extracts of 2 mL was filtered through a 0.2-μm polytetrafluoroethylene filter (Model Econofilter PTFE 25/20, Agilent Technologies, Wilmington, DE). A high performance liquid chromatography (Agilent 1200 series, Agilent Technologies, Palo Alto, CA) equipped with a photodiode array detector was used for carotenoid analysis. Carotenoids were identified and quantified using an external standard (ChromaDex Inc., Irvine, CA).

2.6 Total Anthocyanin and Phenolic Analysis

Total anthocyanin and phenolic compounds were extracted using the method of Nicolle et al. (2004) and analyzed following the method of Olsen, Aaby, and Borge (2010) with modifications. Absorbance was measured at 510 and 700 nm using a spectrophotometer (Spectronic® Genesys™ 5, Thermo Fisher Scientific Inc., Waltham, MA). The total anthocyanin content was expressed as cyanidin 3-glucoside equivalent.

To determine total phenolic content, 0.3 mL of extract or gallic acid standard was mixed with 1.5 mL of 0.5 N Folin-Ciocalteu reagent (Sigma-Aldrich, St. Louis, MO) and the mixture was incubated for 5 min. Then 1.5 mL of sodium carbonate solution (0.5 M) was added and the mixture was incubated for 1 h at room temperature in the dark before measuring absorbance at 765 nm using a spectrophotometer (Spectronic® Genesys™ 5, Thermo Fisher Scientific Inc., Waltham, MA). The total phenolic content was determined using a gallic acid standard curve.

2.7 Statistical Analysis

The experimental design was a randomized complete block with four replications (n = 4) per cultivar. Each cultivar within a block was randomly placed. All results were expressed as mean ± standard error. Two-way analysis of variance (ANOVA) was conducted using SAS (SAS 9.2, SAS Institute, Inc., Cary, NC, USA) with lettuce type (crisphead vs. romaine) and leaf color (red vs. green) as the main factors. Fisher's least significant difference test (LSD) was performed at the 95% significance level for mean separation.

3. Results and Discussion

3.1 Insoluble Fibers

Fiber consumption is low in the US (Dahl & Stewart, 2015). Fresh lettuce (100 g) can provide up to 10% of the daily recommended intake of fiber for adults (Institute of Medicine, 2002). Eichholz, Förster, Ulrichs, Schreiner, and Huyskens-Keil (2014) reported the dietary fiber in lettuce consisted mainly of insoluble fiber, with 80% as cellulose. In our study, insoluble fiber content depended on lettuce type (Figure 1, Table 1). Insoluble fiber content was higher ($P \leq 0.05$) in the romaine than in the crisphead cultivars. The USDA (2015) reported fiber content of 21 mg g^{-1} FW (420 mg g^{-1} DW) in romaine lettuce and 12 mg g^{-1} FW (240 mg g^{-1} DW) in 'Iceberg' lettuce, but total fiber, rather than insoluble fiber, was measured and leaf color of these cultivars were not mentioned. Eichholz et al. (2014) reported lower fiber content in red butterhead 'Teodore' than in green butterhead 'Wiske', suggesting leaf color can affect fiber content depending on lettuce type. Our study showed no significant difference in insoluble fiber content between green and red pigmented cultivars of romaine and crisphead lettuces. Based on the results, fiber intake in the diet may be increased by choosing romaine lettuce types over others. Among lettuce cultivars examined in our study, insoluble fiber content was higher in romaine than crisphead lettuce types. This is significant since crisphead ('Iceberg') is the most commonly used lettuce in fast food restaurants (Mulabagal et al., 2010).

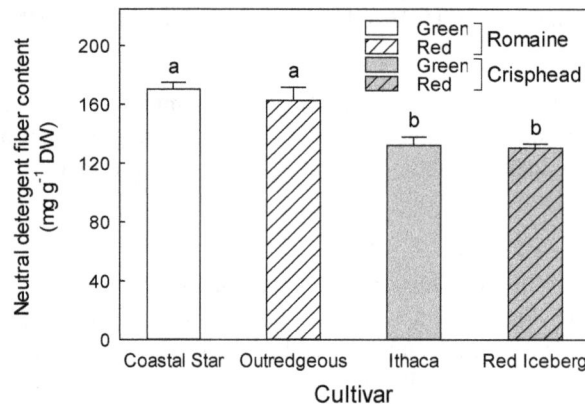

Figure 1. Neutral detergent fiber content in green and red cultivars of romaine (*L. sativa* L. var *longifolia*) and crisphead (*L. sativa* L. var *capitata*) lettuces

Note. Vertical bars are standard errors of the means with four replications (n = 4). Different letters indicate significant difference at $P \leq 0.05$ by Fisher's LSD.

Table 1. ANOVA results for neutral detergent fiber, total anthocyanin and total phenolic contents in green and red cultivars of romaine (*L. sativa* L. var *longifolia*) and crisphead (*L. sativa* L. var *capitata*) lettuces

Source of variation	Insoluble fiber	Total anthocyanin	Total phenolic
Type (T)	*	***	***
Color (C)	NS	***	***
T × C	NS	***	***

Note. [NS], *, *** Nonsignificant or significant at $P \leq 0.05$ or 0.001, respectively.

3.2 Fatty Acid Profile

Major fatty acids present in romaine and crisphead lettuces were the essential fatty acids, α-linolenic acid (ALA) and linoleic acid (LA), which comprise over 75% of the total fatty acids (Table 2). Similarly, the major fatty acids reported for 'Trocadéro' butterhead lettuce (Le Guedard et al., 2008) and Cos romaine lettuce (Pereira, Li, & Sinclair, 2001) were ALA and LA. In our study, lettuce type, but not leaf color influenced total fatty acids content. Green romaine had greater ($P \leq 0.01$) ALA content than crisphead cultivars. The USDA (2015) reported higher ALA for romaine compared to crisphead ('Iceberg'). However, leaf color of these types were not identified. It has been reported that low omega-3 to omega-6 PUFA intake contributes to the increased burden of chronic diseases (Innis, 2014). Modern Western diets typified by the US are low in omega-3 PUFAs. Based on our results, green romaine lettuce provides a higher amount of omega-3 PUFA due to its higher ALA content. Although lettuce provides only a minor contribution to omega-3 PUFA intake in the human diet due to its low fat content, consumption of green romaine would fulfill some of the omega-3 fatty acid requirement.

Table 2. Fatty acid composition of green and red cultivars of romaine (*L. sativa* L. var *longifolia*) and crisphead (*L. sativa* L. var *capitata*) lettuces

Type	Color (Cultivar)	α-Linolenic acid (18:3n-3)	Linoleic acid (18:2n-6)	Palmitic acid (16:0)	Stearic acid (18:0)	Oleic acid (18:1n-9)	Total
		--mg g^{-1} DW--					
Romaine	Green (Coastal Star)	12.9±0.4a[z] (58.0%)[y]	4.5±0.2a (20.3%)	3.7±0.1a (16.5%)	0.8±0.0a (3.5%)	0.5±0.1a (1.7%)	22.2±0.7a
	Red (Outredgeous)	11.3±1.0ab (58.4%)	4.1±0.3a (21.4%)	3.1±0.2a (15.8%)	0.5±0.1a (2.8%)	0.3±0.1a (1.7%)	19.3±1.2a
Crisphead	Green (Ithaca)	7.3±1.1c (47.1%)	4.2±0.3a (28.1%)	3.0±0.4a (19.4%)	0.5±0.2a (3.0%)	0.3±0.1a (2.5%)	15.2±1.8a
	Red (Red Iceberg)	8.7±2.1bc (53.4%)	3.9±0.6a (24.8%)	2.7±0.4a (17.1%)	0.3±0.1a (2.2%)	0.4±0.1a (2.5%)	16.0±3.0a
Significance							
Type (T)		**	NS	NS	NS	NS	*
Color (C)		NS	NS	NS	NS	NS	NS
T × C		NS	NS	NS	NS	NS	NS

Note. [z] Data are presented as mean ± standard error (n = 4). Mean separation within columns by Fisher's LSD at $P \leq 0.05$; [y] Relative ratio of each fatty acid to the total fatty acid content; [NS], [*], [**] Nonsignificant or significant at $P \leq 0.05$ or 0.01, respectively.

3.3 Dietary Minerals

Inadequate intake of Zn and Fe is a concern in vegetarian diets (Hunt, 2003), in part due to a relatively low bioavailability of these minerals in plants compared to meat. Lettuce (100 g fresh weight) provided only 1-4% of the recommended Zn intake of 8-11 mg/day for adults (Kim et al., 2016). In our study, there was no significant difference in Zn content regardless of lettuce type and leaf color (Table 3). Fe content was higher ($P \leq 0.001$) in the romaine than the crisphead cultivars. Similarly, the USDA (2015) reported higher Fe content in romaine (9.7 µg g^{-1} FW) compared to crisphead ('Iceberg') lettuce (4.1 µg g^{-1} FW). Mou and Ryder (2004) also reported higher Fe content in romaine than in crisphead. Fresh lettuce (100 g) depending on lettuce type can provided up to 18% of recommended Fe intake of 8-18 mg/day for adults (Institute of Medicine, 2005; Kim et al., 2016).

Vegetarians may be at greater risk of low bone density and bone fracture due to low Ca in vegetarian diets (Tucker, 2014). The Ca content was higher ($P \leq 0.05$) in romaine than crisphead lettuces (Table 3). There was an interaction of lettuce type and leaf color with higher ($P \leq 0.05$) Ca content in the green than red crisphead cultivars, but not romaine lettuce cultivars. Another important mineral for bone health, P was higher ($P \leq 0.01$) in red than in green romaine and crisphead cultivars. Koudela & Petříková (2008) analyzed five cultivars of leaf lettuces with different leaf colors and reported a significant influence of cultivar on Ca content; however, influence of pigmentation on Ca content was not analyzed. Good-quality vegetarian diets are high in Mg and Mn which are also important to bone health (Tucker, 2014). In our study, romaine had significantly higher Mg content than crisphead lettuces. Red 'Outredgeous' romaine had higher Mn content than crisphead lettuces (Table 3). Based on our results, minerals important for bone health were generally higher in the romaine than the crisphead lettuces studied.

In our study, sodium (Na) content in lettuces ranged from 0.4-0.8 mg g^{-1} DW (20-40 µg g^{-1} FW) which is low based on the daily recommended intake of Na of 1.2-1.5 g/day for adults (Kim et al., 2016). Reducing Na and increasing K intake lowers the risk of hypertension (Adrogué & Madias, 2014). Red 'Outredgeous' romaine had the highest ($P \leq 0.05$) K content among the different lettuce cultivars evaluated. And mineral analysis within the same lettuce types revealed significantly higher amount of K as well as P content in red than green cultivars. While the role of leaf pigmentation on mineral content has not been intensively studied, our result indicated that leaf color might be associated with certain mineral compositions of romaine and crisphead lettuces. The mineral content of our study was generally lower than the result by Baslam, Morale, Garmendia, and Goicoechea (2013) or similar or higher in Zn and Mg content, respectively, reported by Lyons, Goebel, Tikai, Stanley, and Taylor (2015), indicating that mineral content in lettuce can greatly vary depending on lettuce type, cultivar, and cultural conditions.

Table 3. Mineral content in green and red cultivars of romaine (*L. sativa* L. var *longifolia*) and crisphead (*L. sativa* L. var *capitata*) lettuces

Type	Color (Cultivar)	Zn	Fe	Ca	P	Mg	Mn	Na	K
		---------μg g^{-1} DW--------		----------------mg g^{-1} DW--------------			--μg g^{-1} DW--	---------mg g^{-1} DW---------	
Romaine	Green (Coastal Star)	24.6±2.7a^z	38.8±2.1a	9.2±0.3a	2.5±0.1bc	5.5±0.1a	19.0±0.5ab	0.8±0.2a	21.9±0.8b
	Red (Outredgeous)	25.1±0.7a	37.6±4.1a	9.1±0.3a	3.0±0.3a	5.8±0.2a	21.8±1.5a	0.7±0.1a	28.3±0.7a
Crisphead	Green (Ithaca)	22.5±3.0a	26.8±2.2b	8.1±0.4b	2.4±0.2c	5.0±0.2b	14.9±0.6c	0.5±0.1a	18.9±0.2c
	Red (Red Iceberg)	26.1±4.9a	27.9±2.0b	6.5±0.1c	2.9±0.2ab	4.6±0.1b	16.5±0.9bc	0.4±0.0a	21.9±0.6b
Significance									
Type (T)		NS	***	***	NS	***	***	*	***
Color (C)		NS	NS	*	**	NS	*	NS	***
T × C		NS	NS	*	NS	*	NS	NS	*

Note. ^{z} Data are presented as mean ± standard error (n = 4). Mean separation within columns by Fisher's LSD at $P \leq 0.05$; ^{NS}, *, **, *** Nonsignificant or significant at $P \leq 0.05$, 0.01 or 0.001, respectively.

3.4. Carotenoids

Several carotenoids including β-carotene have been studied for their ability to reduce risk of chronic diseases. Oxygenated carotenoids (xanthophylls) have been studied for its protective effects against macular degeneration (Cooper, Eldridge, & Peters, 1999). Some carotenoids such as β-carotene have a provitamin A activity with β-carotene to retinol (animal form of vitamin A) conversion rate ranging 3.6-28.1 by weight in humans (Tang, 2010). In our study β-carotene, lutein, and antheraxanthin were the major carotenoids ranging from 63.2% of total carotenoids in green romaine 'Coastal Star' to 70.6% in red crisphead 'Red Iceberg' (Table 4). The β-carotene and lutein were higher ($P \leq 0.01$) in red 'Outredgeous' romaine compared to the crisphead lettuces. According to the USDA database (2015), romaine contained a higher amount of β-carotene (52.3 μg g^{-1} FW) compared to crisphead ('Iceberg') lettuce (3.0 μg g^{-1} FW). It was difficult to make a direct comparison since the values in USDA database were reported from two studies, and growing conditions, leaf color, and cultivar were not provided.

Biosynthesis of carotenoids is regulated by light. Differences in carotenoid content in lettuce types was suggested to be related to head structure. Crisphead lettuce forms a closed head that obstructs light penetration into the head. In contrast, romaine lettuce has an open head structure which allows more light penetration, resulting in higher amount of carotenoids accumulated (Mou, 2005; Mou & Ryder, 2004). In these studies carotenoid contents were higher in green than red cultivars. In another study, the contents of carotenoids were evaluated in two differently pigmented lettuce cultivars; 'Blonde of Paris Batavia' (green) and 'Oak Leaf' (Pérez-López, Miranda-Apodaca, Muñoz-Rueda, & Mena-Petite, 2015). The carotenoids were higher in red than in green cultivar by 130%. In our study red leaf 'Outredgeous' contained the highest amount of total carotenoids ($P \leq 0.05$), and there was no difference between leaf colors within the same lettuce type. Of the major carotenoids measured in our study, there was an interaction of type and color on antheraxanthin content. Red romaine was higher ($P \leq 0.001$) in antheraxanthin than in green cultivars.

Other carotenoids detected in our lettuce samples were neoxanthin and violaxanthin (~13-18%). There was a significant effect of lettuce type, but not leaf color. Romaine was higher in neoxanthin ($P \leq 0.01$) than in crisphead lettuce. Violaxanthin was higher ($P \leq 0.05$) in red romaine than in the green crisphead, but no difference was found in cultivars differing leaf colors within the same type. Baslam et al. (2013) analyzed one romaine and two crisphead type lettuces and reported significantly higher levels of both neoxanthin and violaxanthin in romaine lettuce. Advantages of higher consumption of neoxanthin and violaxanthin in humans is unknown. We detected only small quantities of zeaxanthin and α-carotene at 1-1.8% of the total carotenoid content. Among the lettuce types evaluated in our study, total carotenoid content was highest in red 'Outredgeous' romaine lettuce.

Table 4. Carotenoid content in green and red cultivars of romaine (*L. sativa* L. var *longifolia*) and crisphead (*L. sativa* L. var *capitata*) lettuces

Type	Color (Cultivar)	α-Carotene	β-Carotene	Lutein	Antheraxanthin	Neoxanthin	Violaxanthin	Zeaxanthin	Total
		$\mu g\ g^{-1}$ DW							
Romaine	Green (Coastal Star)	$16.5\pm2.9a^z$ $(1.4\%)^y$	$280.6\pm7.8ab$ (23.6%)	$338.5\pm11.1ab$ (28.5%)	$132.5\pm5.1c$ (11.1%)	$215.2\pm6.6a$ (18.1%)	$184.2\pm6.5ab$ (15.5%)	$22.0\pm1.1a$ (1.8%)	$1189.4\pm38.0b$
	Red (Outredgeous)	$16.0\pm2.2a$ (1.0%)	$344.2\pm35.1a$ (20.7%)	$373.4\pm34.7a$ (22.5%)	$450.8\pm44.8a$ (27.1%)	$228.5\pm23.8a$ (13.8%)	$226.2\pm27.7a$ (13.6%)	$21.4\pm5.3a$ (1.3%)	$1660.4\pm167.3a$
Crisphead	Green (Ithaca)	$12.7\pm1.3a$ (1.3%)	$193.3\pm9.8c$ (19.1%)	$220.3\pm14.3c$ (21.8%)	$296.7\pm18.9b$ (29.3%)	$135.6\pm8.4b$ (13.4%)	$138.8\pm8.5b$ (13.7%)	$14.4\pm0.9a$ (1.4%)	$1011.7\pm60.6b$
	Red (Red Iceberg)	$12.2\pm0.6a$ (1.0%)	$232.0\pm18.0bc$ (18.7%)	$268.9\pm21.2bc$ (21.7%)	$374.2\pm19.1ab$ (30.2%)	$160.6\pm12.6b$ (13.0%)	$177.7\pm12.9ab$ (14.3%)	$13.5\pm1.6a$ (1.1%)	$1239.1\pm84.4b$
Significance									
Type (T)		*	**	**	NS	**	*	*	*
Color (C)		NS	NS	NS	***	NS	NS	NS	*
T × C		NS	NS	NS	**	NS	NS	NS	NS

Note. [z] Data are presented as mean ± standard error (n = 4). Mean separation within columns by Fisher's LSD at $P \leq 0.05$; [y] Relative ratio of each carotenoid to the total carotenoid content; [NS], [*], [**], [***] Nonsignificant or significant at $P \leq 0.05$, 0.01, or 0.001, respectively.

3.5 Total Phenolic and Anthocyanin

Total phenolic and anthocyanin contents were significantly higher in romaine than in crisphead lettuce in both green and red cultivars (Figure 2, Table 1). Biosynthesis of phenolic compounds might be light dependent as suggested by Tsormpatsidis et al. (2008). As discussed above, romaine lettuces have an open head structure that allows greater light penetration compared to crisphead lettuces. Therefore, romaine lettuce could accumulate higher levels of phenolic compounds including anthocyanins.

Total phenolic content was significantly higher in red compared to green cultivars of romaine and crisphead (Figure 2, Table 4). Llorach, Martínez-Sánchez, Tomás-Barberán, Gil, and Ferreres (2008) reported higher total phenolic content in red oak leaf and lollo rosso than in green crisphead ('Iceberg') and romaine lettuces. Liu et al. (2007) also reported higher total phenolic content in red compared to green cultivars of romaine lettuces as well as leaf type lettuces. We evaluated anthocyanin content since this subgroup of phenolic compound is the main plant pigment responsible for the red color in lettuce (Ferreres, Gil, Castañer, & Tomás-Barberán, 1997). Total anthocyanin content in red 'Outredgeous' romaine and crisphead 'Red Iceberg' was 3418.2 and 442.5 $\mu g\ g^{-1}$ DW, respectively (Figure 2). No anthocyanin was detected in green cultivars of either type lettuce. Similarly, anthocyanin in red oak leaf and lollo rosso lettuce was reported, but anthocyanin was not detected in green cultivars of crisphead ('Iceberg') and romaine (Llorach et al., 2008). Based on the results, red pigmentation was indicative of total anthocyanin and phenolic content. There was also an interaction between leaf color and lettuce type (Table 4). Total phenolic and anthocyanin was significantly higher in red 'Outredgeous' romaine than crisphead 'Red Iceberg' (Figure 2). Red lettuce is increasing in popularity due to public perception that red color is associated with greater health benefits. Among lettuces evaluated in our study, choosing red leaf lettuce such as 'Outredgeous' romaine lettuce may result in greater intake of anthocyanin and phenolic compounds which have been associated with diverse health benefits (Kris-Etherton et al., 2002).

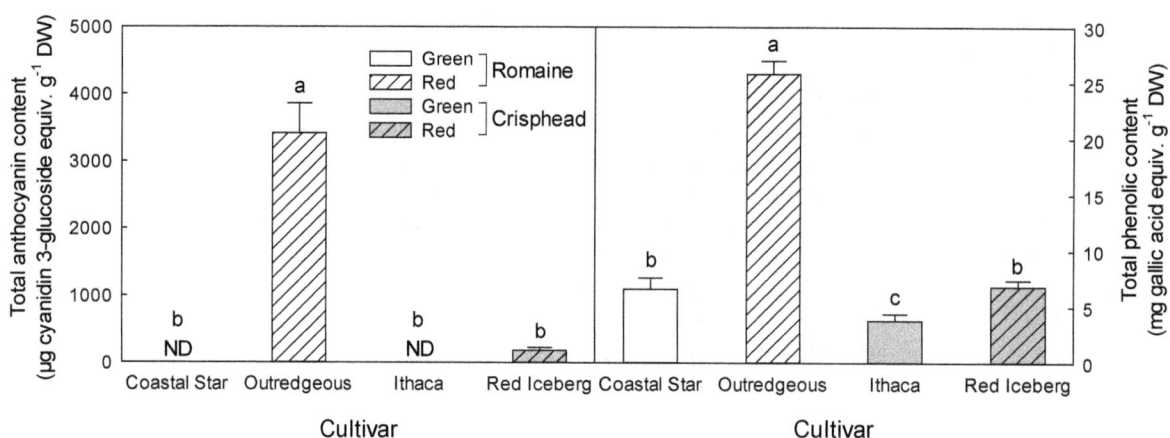

Figure 2. Total anthocyanin and phenolic contents in green and red cultivars of romaine (*L. sativa* L. var *longifolia*) and crisphead (*L. sativa* L. var *capitata*) lettuces

Note. Vertical bars are standard errors of the means with four replications (n = 4). Different letters indicate significant difference at $P \leq 0.05$ by Fisher's LSD.

4. Conclusions

Although lettuce is considered low in nutritional value, the results of the current study showed that nutrient content depended on lettuce type, as well as the morphological factor of leaf color. Generally, the crisphead lettuces were lower in all nutrients analyzed compared to romaine lettuces. Romaine cultivars contained higher amount of insoluble fiber and the essential fatty acids, ALA and LA, which comprise over 75% of the total fatty acids regardless of leaf color. Red cultivars had higher phenolic content than green cultivars with the highest amount in red 'Outredgeous' romaine lettuce. The results of this study provide comparative nutrient data of several popularly consumed lettuce cultivars that were grown under the same growing condition. This information will assist consumers to make food choices that provide higher nutritional value.

References

Adrogué, H. J., & Madias, N. E. (2014). Sodium surfeit and potassium deficit: keys to the pathogenesis of hypertension. *Journal of the American Society of Hypertension, 8*(3), 203-213. http://dx.doi.org/10.1016/ j.jash.2013.09.003

Baslam, M., Morale, F., Garmendia, I., & Goicoechea, N. (2013). Nutritional quality of outer and inner leaves of green and red pigmented lettuces (*Lactuca sativa* L.) consumed as salads. *Scientia Horticulturae, 151*, 103-111. http://dx.doi.org/10.1016/j.scienta.2012.12.023

Cooper, D., Eldridge, A., & Peters, J. (1999). Dietary carotenoids and certain cancers, heart disease, and age-related macular degeneration: a review of recent research. *Nutrition Reviews, 57*(7), 201-214. http://dx.doi.org/10.1111/j.1753-4887.1999.tb06944.x

Cordain, L., Eaton, S. B., Sebastian, A., Mann, N., Lindeberg, S., Watkins, B. A., & Brand-Miller, J. (2005). Origins and evolution of the Western diet: health implications for the 21st century. *The American Journal of Clinical Nutrition, 81*(2), 341-354.

Dahl, W. J., & Stewart, M. L. (2015). Position of the academy of nutrition and dietetics: Health implications of dietary fiber. *J Acad Nutr Diet, 115*, 1861-1870. http://dx.doi.org/10.1016/j.jand.2015.09.003

Eichholz, I., Förster, N., Ulrichs, C., Schreiner, M., & Huyskens-Keil, S. (2014). Survey of bioactive metabolites in selected cultivars and varieties of *Lactuca sativa* L. under water stress. *Journal of Applied Botany and Food Quality, 87*, 265-273. http://dx.doi.org/10.5073/JABFQ.2014.087.037

Fernandez, E., LaVecchia, C., Davanzo, B., Negri, E., & Franceschi, S. (1997). Risk factors for colorectal cancer in subjects with family history of the disease. *British Journal of Cancer, 75*(9), 1381-1384. http://dx.doi.org/10.1038/bjc.1997.234

Ferreres, F., Gil, M. I., Castañer, M., & Tomás-Barberán, F. A. (1997). Phenolic metabolites in red pigmented lettuce (*Lactuca sativa*). Changes with minimal processing and cold storage. *Journal of Agricultural and Food Chemistry, 45*(11), 4249-4254. http://dx.doi.org/10.1021/jf970399j

Fritsche, K. L., & Johnston, P. V. (1990). Effect of dietary α-linolenic acid on growth, metastasis, fatty acid profile and prostaglandin production of two murine mammary adenocarcinomas. *Journal of Nutrition, 120*(12), 1601-1609.

García-Macías, P., Ordidge, M., Vysini, E., Waroonphan, S., Battey, N. H., Gordon, M. H., ... Wagstaffe, A. (2007). Changes in the flavonoid and phenolic acid contents and antioxidant activity of red leaf lettuce (Lollo Rosso) due to cultivation under plastic films varying in ultraviolet transparency. *Journal of Agricultural and Food Chemistry, 55*(25), 10168-10172. http://dx.doi.org/10.1021/jf071570m

Goering, H. K., & Soest, P. J. V. (1970). Forage fiber analysis. *Agriculture Handbook No 379*. USDA, Washington, D.C.

Hunt, J. R. (2003). Bioavailability of iron, zinc, and other trace minerals from vegetarian diets. *The American Journal of Clinical Nutrition, 78*(3), 633S-639S.

Innis, S. (2014). Omega-3 fatty acid biochemistry: Perspectives from human nutrition. *Military Medicine, 179*, 82-87. http://dx.doi.org/10.7205/MILMED-D-14-00147

Institute of Medicine. (2002). *Dietary reference intakes for energy, carbohydrate, fiber, fat, fatty acids, cholesterol, protein, and amino acids*. Washington D.C.: Institute of Medicine. http://dx.doi.org/10.17226/10490

Institute of Medicine. (2005). *Dietary reference intakes for water, potassium, sodium, chloride, and sulfate*. Washington, D.C.: Institute of Medicine. http://dx.doi.org/10.17226/10925

Kim, M. J., Moon, Y, Tou, J. C., Mou, B., & Waterland, N. L. (2016). Nutritional value, bioactive compounds and health benefits of lettuce (*Lactuca sativa* L.). *Journal of Food Composition and Analysis, 49*, 19-34. http://dx.doi.org/10.1016/j.jfca.2016.03.004

Kopsell, D. A., Barickman, T. C., Sams, C. E., & McElroy, J. S. (2007). Influence of nitrogen and sulfur on biomass production and carotenoid and glucosinolate concentrations in watercress (*Nasturtium officinale* R. Br.). *Journal of Agricultural and Food Chemistry, 55*(26), 10628-10634. http://dx.doi.org/10.1021/jf072793f

Koudela, M., & Petříková, K. (2008). Nutrients content and yield in selected cultivars of leaf lettuce (*Lactuca sativa* L. var. *crispa*). *Horticultural Science, 35*(3), 99-106.

Kris-Etherton, P. M., Hecker, K. D., Bonanome, A., Coval, S. M., Binkoski, A. E., Hilpert, K. F., Griel, A. E., & Etherton, T. D. (2002). Bioactive compounds in foods: their role in the prevention of cardiovascular disease and cancer. *American Journal of Medicine, 113*(Suppl. 9B), 71S-88S. http://dx.doi.org/10.1016/S0002-9343(01)00995-0

Le Guedard, M., Schraauwers, B., Larrieu, I., & Bessoule, J.-J. (2008). Development of a biomarker for metal bioavailability: The lettuce fatty acid composition. *Environmental Toxicology and Chemistry, 27*(5), 1147-1151. http://dx.doi.org/10.1897/07-277.1

Liu, X., Ardo, S., Bunning, M., Parry, J., Zhou, K., Stushnoff, C., ... Kendall, P. (2007). Total phenolic content and DPPH˙ radical scavenging activity of lettuce (*Lactuca sativa* L.) grown in Colorado. *LWT - Food Science and Technology, 40*(3), 552-557. http://dx.doi.org/10.1016/j.lwt.2005.09.007

Llorach, R., Martínez-Sánchez, A., Tomás-Barberán, F. A., Gil, M. I., & Ferreres, F. (2008). Characterisation of polyphenols and antioxidant properties of five lettuce varieties and escarole. *Food Chemistry, 108*(3), 1028-1038. http://dx.doi.org/10.1016/j.foodchem.2007.11.032

López, A., Javier, G. A., Fenoll, J., Hellin, P., & Flores, P. (2014). Chemical composition and antioxidant capacity of lettuce: Comparative study of regular-sized (Romaine) and baby-sized (Little Gem and Mini Romaine) types. *Journal of Food Composition and Analysis, 33*(1), 39-48. http://dx.doi.org/10.1016/j.jfca.2013.10.001

Lyons, G., Goebel, R. G., Tikai, P., Stanley, & K.-J., Taylor, M. (2015). Promoting nutritious leafy vegetables in the Pacific and Northern Australia. *Acta Horticulturae, 1102*, 253-260. http://dx.doi.org/10.17660/ActaHortic.2015.1102.31

Mou, B. (2005). Genetic variation of beta-carotene and lutein contents in lettuce. *Journal of the American Society for Horticultural Science, 130*(6), 870-876.

Mou, B., & Ryder, E. J. (2004). Relationship between the nutritional value and the head structure of lettuce. *Acta Horticulturae, 637*, 361-367. http://dx.doi.org/10.17660/ActaHortic.2004.637.45

Mulabagal, V., Ngouajio, M., Nair, A., Zhang, Y., Gottumukkala, A. L., & Nair, M. G. (2010). *In vitro* evaluation of red and green lettuce (*Lactuca sativa*) for functional food properties. *Food Chemistry, 118*, 300-306. http://dx.doi.org/10.1016/j.foodchem.2009.04.119

Nicolle, C., Carnat, A., Fraisse, D., Lamaison, J. L., Rock, E., Michel, H., ... Remesy, C. (2004). Characterisation and variation of antioxidant micronutrients in lettuce (*Lactuca sativa* folium). *Journal of the Science of Food and Agriculture, 84*(15), 2061-2069. http://dx.doi.org/10.1002/jsfa.1916

Olsen, H., Aaby, K., & Borge, G. I. A. (2010). Characterization, quantification, and yearly variation of the naturally occurring polyphenols in a common red variety of curly kale (*Brassica oleracea* L. convar. *acephala* var. *sabellica* cv. 'Redbor'). *Journal of Agricultural and Food Chemistry, 58*(21), 11346-11354. http://dx.doi.org/10.1021/jf102131g

Pereira, C., Li, D., & Sinclair, A. J. (2001). The alpha-linolenic acid content of green vegetables commonly available in Australia. *International Journal for Vitamin and Nutrition Research, 71*(4), 223-228. http://dx.doi.org/10.1024/0300-9831.71.4.223

Pérez-López, U., Miranda-Apodaca, J., Muñoz-Rueda, A., & Mena-Petite, A. (2015). Interacting effects of high light and elevated CO_2 on the nutraceutical quality of two differently pigmented *Lactuca sativa* cultivars (Blonde of Paris Batavia and Oak Leaf). *Scientia Horticulturae, 191*, 38-48. http://dx.doi.org/10.1016/j.scienta.2015.04.030

Rodriguez-Casado, A. (2016). The health potential of fruits and vegetables phytochemicals: Notable examples. *Critical Reviews in Food Science and Nutrition, 56*(7), 1097-1107. http://dx.doi.org/10.1080/10408398.2012.755149

Tang, G. (2010). Bioconversion of dietary provitamin A carotenoids to vitamin A in humans. *American Journal of Clinical Nutrition, 91*(5), 1468/s-1473/s. http://dx.doi.org/10.3945/ajcn.2010.28674G

Tsormpatsidis, E., Henbest, R. G. C., Davis, F. J., Battey, N. H., Hadley, P., & Wagstaffe, A. (2008). UV irradiance as a major influence on growth, development and secondary products of commercial importance in Lollo Rosso lettuce 'Revolution' grown under polyethylene films. *Environmental and Experimental Botany, 63*(1-3), 232-239. http://dx.doi.org/10.1016/j.envexpbot.2007.12.002

Tucker, K. L. (2014). Vegetarian diets and bone status. *The American Journal of Clinical Nutrition, 100*(Suppl. 1), 329S-335S. http://dx.doi.org/10.3945/ajcn.113.071621

USDA. (2015). *National Nutrient Database for Standard Reference Release 28*. Retrieved September 6, 2015, from http://ndb.nal.usda.gov/ndb

8

Few Journal Article Organizational Structure Characteristics Affect Article Citation Rate: A Look at Agricultural Economics Articles Using Regression Analysis

Joe L. Parcell[1], Glynn T. Tonsor[2] & Jason V. Franken[3]

[1] Department of Agricultural and Applied Economics, University of Missouri, Columbia, MO, USA

[2] Department of Agricultural Economics, Kansas State University, Manhattan, KS, USA

[3] School of Agriculture, Western Illinois University, MaComb, IL, USA

Correspondence: Joe L. Parcell, Department of Agricultural and Applied Economics, College of Agriculture, Food and Natural Resources, University of Missouri, 200 Mumford, Columbia, MO 65201, USA. E-mail: parcellj@missouri.edu

Abstract

When reporting research findings, a journal article's organizational structure influences whether others can easily assess the published research's procedures, interpret the results, and synthesize the implications. Organizational structure characteristics include sufficiently explained variables, data format, number of exhibits, and presence of an appendix. This study endeavors to empirically test whether journal article organizational structure influences citation rates. Citations are used for ranking academic fields, evaluating faculty for promotion, and assessing faculty performance for merit-based salary increases. Journal editors desire higher citation rates to enhance journal exposure, and faculty target publishing in journals with higher impact factors, which reflect citation rates. To assess whether journal article organization affects citation rates, this study uses data from a survey of 68 *Journal of Agricultural and Resource Economics* articles published between 1994 and 1998, and it uses citation rates between February 2010 and the publication date as the dependent variable. These articles were selected because they used regression methods and had all information necessary for this analysis. Using Tobit and truncated ordinary least squares regressions, this study evaluated the marginal effects of variables, including organizational structure characteristics, influencing citation rates. The results indicated a lack of statistical significance for most organizational structure variables affecting citation rates. The use of panel data use and presence of an appendix were the two only organizational structure variables that had significant effects on journal article organizational structure. They had respective positive and negative effects. Thus, little evidence supports that a professional impact, measured as citations, will result from at least this particular journal making efforts to improve article format structure. The current study may motivate future research that replicates the methods and examines other journals and article characteristics.

Keywords: article organization structure, citation rate, impact factor, publication frequency

1. Introduction

The frequency at which a particular journal article is cited within peer journal articles is used as a measure of impact for ranking academic and non-academic research fields and determining faculty promotions and merit-based pay raises (Hamermesh, Johnson, & Weisbrod, 1982; C. E. Hilmer & M. J. Hilmer, 2005; Kalaitzidakis, Mamuneas, & Stengos, 2003; Perry, 1994). Some journal articles become seminal research that stands the test of time. Others enter into obscurity soon after publication. Prior research identifies several factors that contribute to citation rates for agricultural economics journals, including author region, author status, article length, and article placement in the journal (e.g., Hilmer & Lusk, 2009; Lusk & Hudson, 2009; Petrolia & Hudson, 2013). One previously unexamined factor that seemingly may influence citation rates is an article's organizational structure. For purposes of this study, organizational structure refers to the components that a researcher chooses to share in a journal article – those include data format, variable explanation, equations, exhibits, appendices, and interpretation – and their presentation and format. It impacts the ease at which other researchers can assess the procedures, interpret the results, and synthesize the implications from published research. This study's objective is to empirically test whether journal article organizational structure affects citation rate.

The reader's ability to assess implications of research results depends on the author's ability to explain procedures, extract generalities, and quantify effects. If the reader cannot clearly follow the author's outline, which is a product of the journal article's organizational structure, then the research's value is reduced greatly because overarching implications will likely be missed and replicating results will undoubtedly be more difficult. Tomek (1993) has suggested that research confirmation and replication involve a quality-quantity trade-off. Ladd (1991, p. 8) noted, "When I was a student, we were taught replication was a necessary process". Surely, making research easier to confirm and replicate improves quality, perhaps without loss of quantity.

We believe several stakeholders stand to benefit if relationships between an article's organizational structure and its citation rate can be identified. Citation rates are components of journal impact factors. Thus, enhancing citation rates would appeal to journal editors who desire to increase a journal's exposure and improve its academic stature. If organizational structure affects citation rates, then editors are likely to accept journal articles that follow structures thought to lead to increased citation rates. Additionally, journal reviewers and editors could provide authors more constructive and consistent feedback about preferred article formats. Similarly, researchers could better control their ability to effectively organize, describe, and synthesize research findings. Understanding if and which organizational factors contribute to higher citation rates is important for researchers as they seek to share their work and have it be referenced by others in the research community.

The present study assesses whether journal article organizational structure affects article citation rates using data from a survey of 68 *Journal of Agricultural and Resource Economics* (*JARE*) articles that Parcell et al. (2000) used to determine best publication practices and consistency in methodologies, reporting practices, and clarity presented in published journal articles. As reported by Parcell et al. (2000), their survey measures of organization style and structure included the type of data analysis, use of simulation, results interpretation, and results presentation. The survey data provide a base for assessing the impact of organizational structural factors on citation rates for these articles between the publication date and through 2010. This paper proceeds with summarizing related literature and describing the data and modeling procedures employed for the current study. Then, empirical results are reported.

2. Material Studied and Methods

2.1 Previous Research

Several literature materials about citation rates and factors that influence them guided this research. Citations play an important role in measuring the impact of institutions, departments, programs, and individuals (e.g., Kalaitzidakis et al., 2003; Kim, Morse, & Zingales, 2006). Individuals' citation rates are used to determine promotion decisions, salaries, and research awards (Hamermesh et al., 1982; Moore, Newman, & Turbull, 1998; Siow, 1991). Academic economists' salary adjustments are larger for a citation than for a publication (Hamermesh et al., 1982).

With respect to factors affecting citation rates, Ellison (2002) analyzed economics journals from 1970 to 1998 and determined that citation rates for second-tier and general interest journals had eroded. He also found a positive relationship between review time and citation rates. Laband and Tollison (2006) reviewed citation rates five years following publication for 73 journals in 1974 and 91 journals in 1996. Five years out, 26 percent of articles had no citations, and more than 85 percent of articles had fewer than 10 citations. They concluded that much of academic research is a wasted effort.

Although reader fatigue also may be expected to increase with the number of pages in an article, previous research finds that citations tend to increase with article length, which may reflect more content in longer papers (Hilmer & Lusk, 2009; Hudson, 2007; Laband & Tollison, 2006; Medoff, 2003). Both Hudson (2003) and Medoff (2003) found that the number of authors had no impact on citation rate, but Hudson (2007) found a lead author regional bias. Hudson (2007) also found a positive relationship between self-citations and non-self-citations attributable to an *advertising* effect and evidence of positive externalities from highly cited articles. Thus, accompanying articles from the particular volume also had higher citation rates. Several studies find that self-citations significantly increase either total citations or other non-self-citations (Hilmer & Lusk, 2009; Hudson, 2007; Laband & Tollison, 2006). Similarly, each of those previous several studies finds evidence that citation rates vary over time.

Laband and Tollison (2006) studied author order in the *American Economic Review* and *American Journal of Agricultural Economics*. They found that alphabetized two-author papers received more citations than non-alphabetized two-author papers in both journals and concluded that a preponderance of non-alphabetized papers in agricultural economics, compared to economics, may reflect the importance of nonmarket-based criteria to evaluate research in the former field. Both Hilmer and Lusk (2009) and Hudson (2007) found that the first article listed in each edition of a journal tends to garner more citations.

Hilmer and Lusk (2009) investigated citation rates in the *American Journal of Agricultural Economics* and the *Review of Agricultural Economics*. They considered whether internet technology affects citation rates by capturing citation rates following publication in 1991, 1993, 2001, 2003, and 2005, but they detected no such effect. They found that, on average, about 11 percent of articles in the *American Journal of Agricultural Economics* and nearly 50 percent of articles in the *Review of Agricultural Economics* had zero citations. Even so, the top-cited *Review of Agricultural Economics* article in their dataset still had more citations than 93 percent of the *American Journal of Agricultural Economics* articles analyzed. Tobit regression results indicated that lead article status, self-citations, and immediate citations within a year of publication had the largest positive impacts on non-self-citation rates. Article page length also had a significantly positive effect. Proceedings papers and comments/replies were cited less often, and no significant effect was ascertained for the number of authors or equations nor for dummy variables measuring whether an author was an AAEA fellow, at a top-tier school, or at a U.S. school or abroad. Hilmer and Lusk (2009) also examined whether author status as an AAEA fellow increased citation rates, but they found no effect.

Based on the previous research, citations have the potential to influence researcher careers, and for published journals, several factors tend to affect article citation rates. Those include page length, time since research paper publication, lead article status, and self-citations. Given this background, the current research builds on the literature and considers whether organizational structure characteristics support or inhibit a higher citation rate for journal articles published in the agricultural economics literature.

2.2 Research Design, Data Collection, and Analysis

The current study used data from a survey of *JARE* articles published between 1994 and 1998. Parcell et al. (2000) previously examined these same studies to determine best practices and consistency regarding methodologies, reporting practices, and clarity of published journal articles that had used regression procedures as a component of their methodologies. Of the 151 articles published in *JARE* during that period, 86 used regression analysis. Of those 86 articles, 68 also contained full information necessary for the current analysis. As a result, this study based its analysis on that subset of the 68 journal articles.

Questions posed in the Parcell et al. (2000) study were adapted from survey questions used by McCloskey and Ziliak (1996) to study articles employing regression analysis and published in the *American Economic Review* articles using regression analysis. The Parcell et al. study asked 25 questions for each journal article. Questions were stated such that the surveyor could respond with a "yes" to represent that the authors do this or report this, a "no" to represent that the authors do not do this or do not report this, and "not applicable" to represent that the question asked was not applicable to the article.

The survey data provided the base for our analysis, which was supplemented with annual citation data collected from Publish or Perish (Harzing, 2007). For each article, annual citation data were gathered for the time between a given article's publishing date, e.g., sometime between 1994 and 1998, and February 2010. Similarly, data on self-citations were collected and aggregated across years.   The remainder of the data for analysis in the current study was provided by Parcell et al. (2000).

Table 1 reports summary statistics for the 151 articles (2,294 pages) published in *JARE* from 1994 through 1998 and the subsample of 68 articles (896 pages) that complete the data for the current study. That is, the papers examined in the current study account for approximately 45 percent of the articles and 39 percent of the pages published in *JARE* during that period. Articles varied in length from eight pages to 20 pages. For the papers that used regression, the citation rate averaged 21.6 citations per article. Overall, about 14 percent of the reported citations were self-citations by the lead author.

Table 1. Summary statistics of surveyed articles that were published in the *Journal of Agricultural and Resources Economics* between 1994 and 1998 and used regression analysis

Variable	1994	1995	1996	1997	1998	All Years
Total articles published (n = 151)	32	26	30	27	36	151
Total pages published (n = 151)	463	410	418	426	578	2,294
Articles using regression (n = 86)	22	12	22	16	14	86
Current Study (n = 68)						
Percent of total articles published	62.5	34.6	53.3	48.1	27.8	45.0
Total pages reviewed	251	122	212	168	143	896
Percent of pages published	54.2	29.8	50.1	39.4	24.7	39.1
Shortest paper (pages)	8	11	10	8	9	8
Longest paper (pages)	17	18	16	17	20	22
Average number of citations	25.8	20.2	18.5	22	18.8	21.6
Average percent citations by the lead author	12.8	12.1	17.5	8.7	21.2	14.0

Note. In total, the *Journal of Agricultural and Resource Economics* published 151 articles between 1994 and 1998. Of those, 86 articles used regression analysis and 68 of the articles are used for the current analysis.

The distribution of citations and lead author self-citations is shown in Figure 1. Most journal articles received between 0.5 and 2.5 citations per year on average. A majority of those citations were not self-citations (Figure 1). Little correlation existed between number of citations and year published (< 0.14 in absolute value), first 100 pages (0.02), number of co-authors (0.07), or article length (0.18). As expected, number of co-authors and number of self-citations exhibited positive correlation (0.15).

Conceptually, the number of citations for a journal article may be modeled as dependent on a number of factors in a regression framework, as indicated by Equation 1:

$$Citations = f(number\ of\ co\text{-}authors,\ rank\ of\ lead\ author,\ self\text{-}citations\ by\ lead\ author,\ year\ published,$$
$$lead\ article,\ first\ 100\ pages,\ number\ of\ pages,\ type\ of\ data\ format,\ econometric\ model,$$
$$number\ of\ goodness\ of\ fit\ measures,\ sufficiently\ explained\ dependent\ variable,\ degrees\ of\ freedom\ reported,$$
$$statistical\ and\ economic\ significance\ both\ reported,\ simulation\ was\ used,\ number\ of\ equations,$$
$$number\ of\ exhibits,\ appendix\ present) \tag{1}$$

Table 2 summarizes explanatory variable descriptions and the expected relationship between each explanatory variable and the dependent variable. Many of the variables listed early in Equation 1 do not pertain to article organizational structure, but prior studies included them as variables that may impact citation rates. Several variables broadly relate to article organization, which may influence the chance for future citation. A binary variable was used to measure whether an article sufficiently explains its dependent variable and enables readers to easily discern the factor being analyzed. Degrees of freedom is a binary variable was used to indicate whether readers can assess statistical and economic significance. The power of the test is limited by degrees of freedom because it's based on sample size. Hence, reporting the number of observations is expected to be positively related to total citations. Whether the author reported economic significance in addition to statistical significance is represented by a binary variable. Describing economic significance is expected to have a positive statistical relationship with the total number of article citations. Whether or not simulation was used is also recorded as a binary variable.

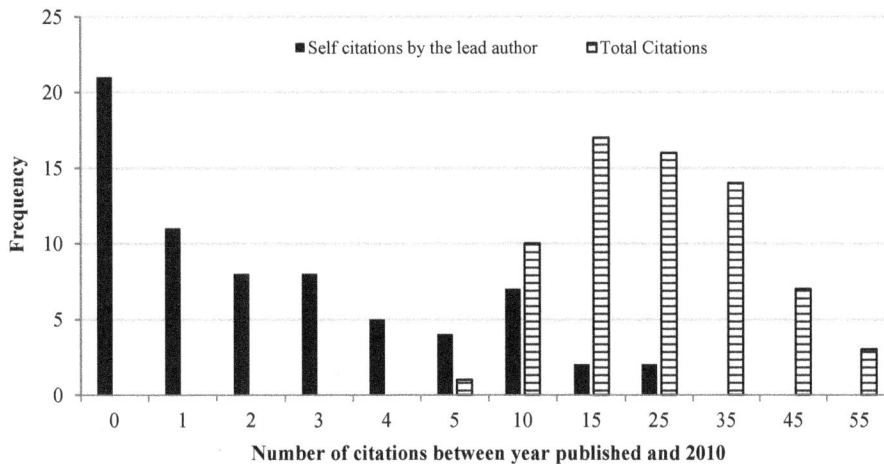

Figure 1. Distribution of citations and lead author self-citations for 68 journal articles published in the JARE between 1994 and 1998

Note. Of the journal articles studied for this analysis, most journal articles received between 0.5 and 2.5 citations per year on average. A majority of those citations were not self-citations. Articles with less than 5 citations and more than 55 citations were eliminated from the data to filter outliers.

Three variables refer to the level of detail provided in a journal article. A positive relationship is expected for total journal article citations and the number of equations and exhibits included, and a binary measure was used to indicate presence of an appendix. The number of goodness-of-fit measures shared (e.g., R^2, adjusted R^2, root mean squared error, log likelihood) is expected to have a positive effect on citation rates. A series of binary variables accounted for type of data format: time series, cross-sectional, or panel (*default* = panel). Additionally, binary variables were used to record the type of econometric model: ordinary least squares (OLS), limited dependent variable, or system (*default* = OLS).

Several other factors, not particularly linked to organizational structure, were assessed by the current study. Although the number of co-authors may be expected to have a positive effect on citations if greater collaboration generally leads to more valuable research or more self-citations and later *advertising* of the paper, previous research has found no such effect. Here, the rank of lead author is distinguished as non-academic, graduate student, assistant professor, associate professor, professor, and other (non-tenure track) academic (*default*). Among academicians, more seasoned, and perhaps better known, authors are expected to garner more citations, as experience and credentials likely translate into greater value as measured by impact on the profession.

A higher number of self-citations by the lead author is expected to positively affect the total number of citations. This variable is included to control for specific authors building a research portfolio on their prior research efforts, and it may also reflect *advertising* effects. The research accounted for publication year with a series of 0 or 1 binary variables (*default* = 1994) to capture fixed effects of some articles having been published longer than other articles.

Like suggested in the literature, article placement in a journal can influence citations. Because only three lead articles in our sample of papers both used regression procedures and qualified for the analysis, the analysis examined whether papers published in the first 100 pages (a 0 or 1 binary variable) had greater citation rates. Such an effect may reflect prestige or reader fatigue. For this research, number of pages is expected to be positively related to citation rates.

Table 2. Explanatory variable definitions and expected sign of impact

Variable	Definition	Expected sign
Number of authors		(+)
Rank of lead author	(binary variables: 1 = yes, 0 = no; default = other academic, e.g., non-tenure track)	
Graduate student		(-, ?)
Assistant		(-)
Associate		(?)
Professor		(+)
Non-academic		(?)
Self-citations	Article citations by lead author	(+)
Year article was published	(binary variables: 1 = yes, 0 = no; default = 1994)	
1995		(-)
1996		(-)
1997		(-)
1998		(-)
First 100 pages	(binary variable: 1 = yes, 0 = no)	(+)
Number of pages		(+)
Data format	(binary variables: 1 = yes, 0 = no ; default = time series)	
Cross-sectional		(?)
Panel		(?)
Regression technique	(binary variables: 1 = yes, 0 = no ; default = OLS)	
Limited dependent variable		(?)
System of equations		(?)
Number of goodness of fit measures		(+)
Sufficiently explained dependent variable		(+)
Degrees of freedom reported	(binary variable: 1 = yes, 0 = no)	(+)
Statistical and economic significance explained	(binary variable: 1 = yes, 0 = no)	(+)
Simulation used	(binary variable: 1 = yes, 0 = no)	(+)
Number of equations		(+)
Appendix present	(binary variable: 1 = yes, 0 = no)	(+)
Number of exhibits		(+)

Note. In most cases, organizational structure characteristics were expected to have a positive influence on citations. However, in some cases, there was uncertainty about the expected sign. Overall, variables expected to have a negative sign of impact were graduate student and assistant professor as lead author and year published.

Other variables were considered for inclusion in estimation of the model specified in equation 1. Author institutional affiliation was considered, but many of the articles included co-authors with multiple affiliation. Sample size would not allow for us to consider every affiliation. An external reviewer suggested two other variable including frequency of author publishing in the journal and then an acknowledgment affect, e.g., granting agency. We agree that frequency of author publishing is the same journal is interesting, but this variable could be highly correlated with self-citations of the lead author already accounted for in the model. As for an acknowledgement affect, most agricultural economists of the mid 1990s had research funded through experiment station Hatch Act funding from USDA. Grant funded research was less common at the time. Persons wishing to replicate this study with newer data could consider an acknowledgement affect.

To make the assessment about whether organizational structure variables and other factors influence citation rate, we considered several data analysis techniques. Presented models were chosen based on model appropriateness (i.e., normality of residuals) and to facilitate comparison to prior research on factors influencing citation rates. For instance, Laband and Tollison (2006) examined total citation rates as a function of self-citations and other variables using ordinary least squares (OLS) regression, and Hilmer and Lusk (2009) used Tobit models to assess similar effects on total-self citations. Given that self-citations are a component of total citations, we also considered models of total citations excluding self-citations as an explanatory variable. The *p*-values from Shapiro-Wilk tests applied to model residuals indicate that the null hypothesis of normality is rejected more often for Tobit models than for truncated OLS. The smaller sample size of 68 observations, of the 86 total available, reflects that outliers of less than or equal to four total citations and greater than or equal to 55 total citations are dropped to arrive at models for which the null hypothesis of normal residuals could not be rejected. Pseudo R^2 values indicate that little of the variation in citation rates is explained by the model, but no corresponding measure is available for truncated OLS. Fairly low R^2 values are also reported in prior citation rate studies (Hudson, 2007; Laband & Tollison, 2006).

3. Results and Discussion

The results indicate a lack of statistical significance for organizational structure variables influencing journal article citation rates (Table 3). Of those variables evaluated, only the use of panel data and the presence of an appendix had significant effects as assessed in the Tobit and truncated OLS regressions. For panel data, the variable was significant at the 10 percent level in the truncated OLS regression. Panel data's positive effect seems intuitive given that such data may allow researchers to better address some statistical issues than when only time-series or cross-sectional data were available. The negative effect of an appendix, however, is less easy to explain. It had statistical significance at the 5 percent level in the Tobit analysis. Perhaps, authors over use appendices as an easy fallback mechanism when they find it difficult to adequately explain complexities within the text. Such a possibility would be consistent with some degree of disorganization, which could confuse readers and subsequently lead to lighter citation of such papers.

Because most organizational structure characteristics were not significantly linked to influencing journal article citation rates, findings from the current analysis offer little evidence that a professional impact, measured as citations, may be expected to accrue from efforts to improve journal article organizational structure. Furthermore, at least for the journal analyzed, the editorial team could not expect to gain much in terms of impact factors, which are rooted in citations, from adopting a standardized organizational format.

Table 3. Marginal effects of Tobit and truncated OLS regressions

	Tobit			Truncated OLS		
	Total Citations		Total-Self Citations	Total Citations		Total-Self Citations
Number of Authors	-0.03	-0.03	-0.03	0.99	0.90	1.55
	(1.48)	(1.54)	(1.48)	(2.13)	(2.26)	(2.95)
Assistant	2.14	-0.64	2.14	-0.87	-4.41	0.43
	(4.31)	(4.29)	(4.31)	(6.30)	(6.44)	(8.64)
Associate	0.05	-1.91	0.05	-7.80	-10.20	-6.66
	(5.55)	(5.69)	(5.55)	(8.53)	(9.01)	(11.41)
Professor	-2.26	-2.43	-2.26	-8.08	-7.66	-7.51
	(4.51)	(4.68)	(4.51)	(6.88)	(7.06)	(9.09)
Grad Student	-7.03	-7.68	-7.03	-14.97*	-15.85*	-16.78*
	(5.71)	(5.92)	(5.71)	(8.93)	(9.46)	(13.40)
Non Academic	13.25**	10.36*	13.25**	12.61	9.57	16.15
	(5.67)	(5.74)	(5.67)	(7.91)	(8.08)	(10.50)
Self-Citations	0.85**	-	-0.15	1.02**	-	-0.54
	(0.37)		(0.37)	(0.52)		(0.78)
1995	-10.37**	-11.33**	-10.37**	-17.90***	-19.88***	-23.94***
	(4.67)	(4.82)	(4.67)	(6.92)	(7.34)	(9.74)
1996	-6.94*	-7.25*	-6.94*	-15.55***	-15.93**	-18.67**
	(3.72)	(3.86)	(3.72)	(6.04)	(6.32)	(8.48)
1997	-3.63	-4.24	-3.63	-7.94	-8.92	-10.39
	(4.19)	(4.34)	(4.19)	(5.72)	(6.00)	(7.56)
1998	-11.62**	-10.89**	-11.62**	-15.86**	-14.99**	-21.30**
	(4.75)	(4.91)	(4.75)	(7.07)	(7.32)	(10.27)
First 100 pages	-5.14	-3.86	-5.14	-9.74*	-8.49	-11.94
	(3.85)	(3.95)	(3.85)	(5.66)	(5.80)	(7.73)

Pages	0.87	1.15*	0.87	0.55	0.94	0.85
	(0.63)	(0.64)	(0.63)	(0.95)	(0.97)	(1.29)
Cross-Section	1.10	-1.82	1.10	1.64	-2.14	3.89
	(4.51)	(4.49)	(4.51)	(6.74)	(6.91)	(9.33)
Panel	4.69	4.29	4.69	11.81*	11.50*	15.80*
	(4.13)	(4.28)	(4.13)	(6.43)	(6.76)	(9.48)
LDV	1.19	-1.17	1.19	1.80	-1.59	1.08
	(4.36)	(4.39)	(4.36)	(6.21)	(6.32)	(8.46)
SYS	1.98	-1.35	1.98	3.50	-1.05	4.28
	(3.92)	(3.78)	(3.92)	(6.00)	(5.85)	(8.74)
Fit Statistics	1.71	0.50	1.71	2.36	1.00	2.51
	(2.49)	(2.53)	(2.49)	(3.45)	(3.59)	(4.83)
Explain Dep Var	-0.55	-2.12	-0.55	2.52	-0.68	0.64
	(4.98)	(5.12)	(4.98)	(7.64)	(7.76)	(10.88)
DOF	0.54	1.24	0.54	3.32	4.23	4.31
	(2.98)	(3.08)	(2.98)	(4.24)	(4.41)	(5.92)
Significance	-3.05	-3.85	-3.05	-4.95	-5.89	-6.68
	(3.65)	(3.77)	(3.65)	(5.21)	(5.51)	(7.13)
Simulation	2.55	2.86	2.55	5.03	5.25	5.54
	(3.71)	(3.84)	(3.71)	(5.59)	(5.80)	(7.67)
Equations	-0.17	-0.17	-0.17	-0.17	-0.14	-0.17
	(0.21)	(0.22)	(0.21)	(0.31)	(0.32)	(0.43)
Appendix	-10.82**	-11.13**	-10.82**	-7.32	-7.32	-10.72
	(4.76)	(4.93)	(4.76)	(7.92)	(8.41)	(11.23)
Exhibits	-0.85	-1.10	-0.85	-1.50	-1.93	-2.14
	(1.20)	(1.24)	(1.20)	(1.83)	(1.93)	(2.63)
R^2	0.06	0.05	0.05	-	-	-
Shapiro-Wilk Test of Normality of Residuals, P-value	0.042	0.177	0.048	0.240	0.052	0.237

Note. N = 68, *, **, *** denote statistical significance at 10%, 5%, 1% levels, respectively.

Of the organizational structure variables evaluated, only the use of panel data and the presence of an appendix had significant effects as assessed in the Tobit and truncated OLS regressions.

Consistent with expectations, the truncated OLS results provide some evidence that papers with graduate student lead authors garner fewer citations than those that have other (i.e., non-tenure track) academic researchers as lead authors (the default category) (Table 2). In comparison, the Tobit results suggested that articles with non-academic researchers as lead authors received relatively greater citation rates. Both Tobit and truncated OLS models indicated fewer citations for papers published in years following 1994 (the default year), which may partly reflect additional citations with the passage of time. Both models were also in agreement that self-citations have significantly positive effects on total citations, but this is not surprising given that the former is a component of the latter. In fact, given this point, it may be more appropriate to exclude self-citations from the model of total citations and used as the dependent variable total-self citations. For such a specification, the Tobit results seem more fitting than the truncated OLS results, as indicated by tests of normality of the residuals. In that case, the Tobit model indicates a positive effect of article length (i.e., pages) that is consistent with prior findings (Hilmer & Lusk, 2009; Hudson, 2007; Laband & Tollison, 2006). In contrast with prior findings of lead papers garnering greater citation rates (Hilmer & Lusk, 2009; Hudson, 2007), our truncated OLS results show some evidence of lower total citations for articles placed closer to the front of publications. Also contrary to the findings of Hilmer and Lusk (2009) and Hudson (2007), self-citations do not have a statistically positive impact on total-self citations in our results.

4. Conclusions

We had expected that characteristics of journal article organizational structure would influence citation rates. However, the results here suggest that journal article organizational structure has seemingly little overall impact on citation rates for the particular journal analyzed. To arrive at this conclusion, this research analyzed the effects of journal article organizational structure on subsequent citation rates. The analysis used journal article organizational data for 68 *Journal of Agricultural and Resource Economics* articles from 1994 to 1998 and citation rates for subsequent years through February 2010. Based on Tobit and truncated OLS regressions, panel data use and presence of an appendix were the two only organizational structure variables that had significant effects on journal article organizational structure. They had respective positive and negative effects. Other organizational structure variables tested in the regression and found to not have a statistically significant effect on citation rates included the number of exhibits, whether variables were sufficiently explained, number of degrees of freedom, whether economic significance was reported, number of equations, number of goodness-of-fit measures, and type of econometric model used.

In addition to evaluating journal article organizational structure's effect on citation rates, the research also considered whether other variables influence citation rates. Other variables included number of authors, rank of lead author, time since original publication, number of pages, and article placement in first 100 journal pages. Consistent with previous work, some evidence suggests that longer articles, presumably with greater content, are cited more frequently, and that publication date influences citation rates, which may partly reflect more citations with the passage of time. The lead author's status (e.g., graduate student, other academic, non-academic) may also influence citation rates.

This study has a few limitations. First, observations are limited to articles that used regression analysis, and as a result, it analyzes a relatively small sample size. Of the 151 total articles published during the sample period of 1994 through 1998, just 86 used regression analysis and only 68 were used for the current study. This study also did not account for school quality of the lead author or citations appearing in the year immediately following publication, but those variables have been shown to positively impact citation rates. Hence, future research could include these variables and utilize larger samples and different types of journals when revisiting the importance of journal article organizational structure for citation rates.

References

Ellison, G. (2002). The Slowdown in the Economic Publishing Process. *Journal of Political Economy, 110*(5), 947-93. http://dx.doi.org/10.1086/341868

Hamermesh, D. S., Johnson, G. E., & Weisbrod, B. A. (1982). Scholarship, Citations, and Salaries: Economic Rewards in Economics. *Southern Economic Journal, 49*(2), 472-81. http://dx.doi.org/10.2307/1058497

Harzing, A. W. (2007). *Publish or perish*. Retrieved from http://www.harzing.com/pop.htm

Hilmer, C. E., & Hilmer, M. J. (2005). How Do Journal Quality, Co-authorship, and Author Order Affect Agricultural Economists' Salaries? *American Journal of Agricultural Economics, 87*(2), 509-23. http://dx.doi.org/10.1111/j.1467-8276.2005.00738.x

Hilmer, C. E., & Lusk, J. L. (2009). Determinants of Citations to the Agricultural and Applied Economics Association Journals. *Review of Agricultural Economics, 31*(4), 677-694. http://dx.doi.org/10.1111/j.1467-9353.2009.01461.x

Hudson, J. (2007). Be Known by the Company You Keep: Citations – Quality or Chance. *Scientometrics, 71*(2), 231-8. http://dx.doi.org/10.1007/s11192-007-1671-6

Kalaitzidakis, P., Mamuneas, T. P., & Stengos, T. (2003). Rankings of Academic Journals and Institutions in Economics. *Journal of the European Economic Association, 1*(6), 1346-66. http://dx.doi.org/10.1162/154247603322752566

Kim, E., Morse, A., & Zingales, L. (2006). What Has Mattered to Economics since 1970. *Journal of Economic Perspectives, 20*(4), 189-202. http://dx.doi.org/10.1257/jep.20.4.189

Laband, D., & Tollison, R. (2006). Alphabetized Coauthorship. *Applied Economics, 38*(14), 1649-53. http://dx.doi.org/10.1080/00036840500427007

Ladd, G. W. (1991). Thoughts on Building an Academic Career. *Western Journal of Agricultural Economics, 16*(1), 1-10.

Lusk, J. L., & Hudson, M. D. (2009). Submission Patterns, Submission Policies, and Revealed Preferences for Agricultural Economics Journals. *Review of Agricultural Economics, 31*, 695-711. http://dx.doi.org/10.1111/j.1467-9353.2009.01462.x

McCloskey, D. N., & Ziliak, S. T. (1996). The Standard Error of Regression. *Journal of Economic Literature, 34,* 97-114.

Medoff, M. H. (2003). Collaboration and the Quality of Economics Research. *Labour Economics, 10*(5), 597-608. http://dx.doi.org/10.1016/s0927-5371(03)00072-1

Moore, W. J., Newman, R. J., & Turnball, G. K. (1998). Do Academic Salaries Decline with Seniority? *Journal of Labor Economics, 16*(2), 352-66. http://dx.doi.org/10.1086/209892

Parcell, J. L., Kastens, T. L., Dhuyvetter, K. C., & Schroeder, T. C. (2000). Agricultural Economists' Effectiveness in Reporting and Conveying Research Procedures and Results. *Agricultural and Research Economics Review, 29*(2), 173-82.

Perry, G. M. (1994). Ranking M.S. and Ph.D. Graduate Programs in Agricultural Economics. *Review of Agricultural Economics, 16*(2), 333-40. http://dx.doi.org/10.2307/1349473

Petrolia, D. R., & Hudson, D. (2013). Why Is the Journal of Agricultural & Applied Economics Not in the Major Citation Indices and Does It Really Matter? *Journal of Agricultural and Applied Economics, 45*(3), 381-388. http://dx.doi.org/10.1017/s1074070800004910

Siow, A. (1991). Are First Impressions Important in Academia? *The Journal of Human Resources, 26*(2), 236-55. http://dx.doi.org/10.2307/145922

Tomek, W. G. (1993). Confirmation and Replication in Empirical Econometrics: A Step toward Improved Scholarship. *American Journal of Agricultural Economics, 75,* 6-14. http://dx.doi.org/10.1093/ajae/75_special_issue.6

9

Community Analysis of Endophytic Bacteria from the Seeds of the Medicinal Plant *Panax notoginseng*

Shuai Liu[1,†], Dinghua Li[1,†], Xiuming Cui[1], Limei Chen[1] & Hongjuan Nian[1]

[1] Faculty of Life Science and Technology, Kunming University of Science and Technology, Kunming Key Laboratory of Sustainable Development and Utilization of Famous-Region Drug, Key Laboratory of *Panax notoginseng* Resources Sustainable Development and Utilization of State Administration of Traditional Chinese Medicine, Kunming, China

Correspondence: Hongjuan Nian, Faculty of Life Science and Technology, Kunming University of Science and Technology, Kunming, China. E-mail: hjnian@163.com

[†] *Shuai Liu and Dinghua Li contributed equally to this work.*

Abstract

Panax notoginseng is a traditional Chinese medicine. The roots of *P. notoginseng* can be used for treatment of diseases and raw materials in Chinese medicinal products. High yield and quality roots require cultivation in shade and humid conditions for 3 years. The long period cultivation makes *P. notoginseng* vulnerable to infect by pathogens. So control diseases are vital for the high yield and quality of *P. notoginseng*. The seed is the carrier systems of many probiotics and pathogens. To explore the indigenous bacterial community diversity, the endophytic bacteria from the seeds of the medicinal plant *P. notoginseng* were isolated and identified using traditional cultivation methods in combination with molecular technique. A total of 137 endophytic bacteria strains were isolated. The 16S rDNA of these strains was amplified and subjected to amplified ribosomal DNA restriction analysis (ARDRA) with restriction enzyme *Hae*III. All the isolated strains were grouped into 9 OTUs (Operational Taxonomic Units) on the basis of the similarity of the ARDRA band profiles. Each representative strain of 9 OTUs was selected for sequencing. γ-proteobacteria was the most dominant group among the isolates (98.5%), containing eight genera. *Pseudomonas* was the most dominant genus (58 of 135 isolates), whose isolates occurred in the seeds collected from all three places. The second dominant genus was *Enterobacter* (20.7%), followed by uncultured bacterium (14.8%) and *Stenotrophomonas* (10.4%). Among the six areas sampled, endophytic bacteria in the seeds collected from Panlong of Yanshan exhibited species diversity and contained the most isolates. These results suggest an abundant diversity of bacterial community within the seeds of *P. notoginseng*. These data provide insights into monitoring the seed health and disease outbreak during seeding.

Keywords: ARDRA, endophytic bacteria, diversity, OTUs, *Panax notoginseng*

1. Introduction

Endophytic bacteria ubiquitously inhabit a majority of plant species (Lodewyck et al., 2002). These organisms can be isolated from surface-disinfected plant tissues, including seeds, roots, stems and leaves and are not harmful to the hosts (Hallmann et al., 1998). Endophytic bacteria may promote plant growth and suppress plant diseases (Feng et al., 2006). Several studies have reviewed endophytic bacteria community structures and their potential biological functions (Senthilkumar et al., 2011; Sturz et al., 2000).

Panax notoginseng is a traditional Chinese medicine plant specifically grown in the Wenshan region of Yunnan Province. The root of *P. notoginseng* is often used to treat cardiovascular diseases, inflammation, different body pains, trauma, and internal and external bleeding due to injury (Sun et al., 2005). *P. notoginseng* is a 3-year-old plant. High quality roots require 3-year cultivation in the shade and humid conditions. The long period growth makes *P. notoginseng* vulnerable to be infected by pathogens. The root-rot disease is the most destructive one (Sun et al., 2004), which was caused by soil-borne fungal pathogens (including *Alternaria panax*, *Alternaria tenuis*, *Cylindrocarpon destructans*, *Cylindrocarpon didynum*, *Fusarium solani*, *Fusarium oxysporum*,

Phytophthora cactorum, Phoma herbarum and *Rhizoctonia solani*), bacterial pathogens (*Pseudomonas* sp. and *Ralstonia* sp.), and parasitic nematodes (such as *Ditylenchus* sp., *Rhabditis elegans* and *Meloidogyne* spp.) (Miao et al., 2006). Control diseases are vital for the high yield and quality of *P. notoginseng*.

The seeds of plants are the carrier systems of many probiotics and pathogens that play an important role in formation of rhizosphere microbial communities (Patkowska, 2001). However, studies concerning the interaction of the microbial community and plant seed genotype are lacking. Therefore, a better study of endophytic bacteria may help increase the current understanding of their function and potential role for the development of a more sustainable system for crop production. Few studies concerning endophytic bacterial community structures and their biological functions in *Panax* spp. have been reported, with the exception of reports on the endophytic bacterial community (Cho et al., 2007; Vendan et al., 2010). Ma et al. (2013) isolated 1000 endophytic bacterial strains from the root, stems, petioles, leaves and seeds of *P. notoginseng*, of which 104 strains exhibited antagonistic properties against at least one of three major pathogens (*F. oxysporum, Ralstonia* sp. and *Meloidogyne hapla*) related to the root-rot disease of *P. notoginseng*. Therefore, knowledge concerning the endophytic bacterial structures and species is vital for the growth of *Panax* plants, the monitoring of seed health and the control of seedling disease. Thus, the aims of the present study were to characterize the endophytic bacteria present in the seeds of the medicinal plant *P. notoginseng*, to obtain a better understanding of endophytic bacterial community structures and diversities and to identify potential biological control candidates against pathogens leading to seedling and root diseases.

2. Materials and Methods

2.1 Microorganisms and Culture Conditions

The bacterial strains were cultured on Luria-Bertani (LB) solid medium (peptone 10 g, yeast extract 5 g, NaCl 10 g, agar powder 15 g, and distilled water 1000 mL, pH 7.2) or in LB liquid broth at 37 °C.

2.2 Isolation and Purification of Endophytic Bacteria

The seeds were collected from 1-3-year-old healthy *P. notoginseng* plants, cultivated in Yanshan County, Maguan County, Guangnan County, Xichou County, Wenshan County and Qiubei County of the Wenshan region, Yunnan Province, China.

The samples were disinfected according to Li et al. (2010), with some modifications. The seeds were rinsed three times with sterile water, and subsequently, the moisture was absorbed using filter paper. The seeds were sterilized with 70% ethanol for 3 min, immersed in 2.6% sodium hypochlorite solution for 5 min, soaked in 70% ethanol for 30 s, and subsequently washed three times with sterile water. The water samples of the last rinse were inoculated onto LB agar plates as a negative control. Simultaneously, the surface sterilized seeds were pressed onto an LB agar plate to test the sterilization efficiency. The seeds that were not detected as contaminated by cultivable microorganisms were considered successfully surface disinfected and were subsequently used for the isolation of endophytic bacteria (Schulz et al., 1993).

The surface sterilized seeds were divided into three groups. Five seeds in each group were homogenized with 2 mL of sterile water. The homogenates were diluted 1000-fold, and 100-μL dilutions were then spread onto LB agar plates. Each sample was replicated three times. The plates were cultured at 37 °C for 72 h. The bacterial colonies with obvious morphological differences were purified. Colonies without distinct morphological differences were randomly selected.

2.3 DNA Extraction from Bacteria

The purified strains were inoculated in LB liquid medium and shaken at 200 rpm for 12 h at 37 °C. The cultures were centrifuged and subsequently collected. Genomic DNA was extracted using the Genomic DNA Purification Kit (TIANGEN Biotech, Beijing, China) according to the manufacturer's instructions.

2.4 Amplification of the Bacterial 16S rRNA Gene

The genomic DNA was used as the PCR template. A pair of primers, 799f and 1492r [13], were used to amplify the 16S rDNA. In total, a 20-μL PCR reaction mixture contained 1 μL of genomic DNA, 1 μL of each primer (10 μM), and 1 μL 2 × Es Taq MasterMix (Es Taq DNA Polymerase, 2 × Es Taq PCR buffer, 3 mM $MgCl_2$, and 400 μM dNTP) (ComWin Biotech, Beijing, China). After initial denaturing at 94 °C for 3 min, thermal cycling proceeded with, denaturing at 94 °C for 30 s, annealing at 50 °C for 30 s, and elongation at 72 °C for 45 s. At the end of 30 cycles, a final extension step was performed at 72 °C for 10 min.

2.5 Amplified Ribosomal DNA Restriction Analysis (ARDRA)

The specificity of the PCR product was examined on an agarose gel. The specific PCR product was used for digestion. The PCR product was completely digested with restriction enzyme HaeIII (NEB, Beijing, China), an enzyme that identifies four bases. The reaction mixtures contained 2 μL of 10 X incubation buffer, 2 μL of HaeIII (10 U/μL) and 16 μL of the PCR product. The digestion was performed according to the manufacturer's instructions. Reaction mixtures were incubated overnight at 37 °C. The total volume of each restriction-digested product was separated on a 1% (w/v) agarose gel and subsequently photographed. According to ARDRA patterns, the strains were grouped into Operational Taxonomic Units (OTUs) according to Sessitsch et al. (2002).

2.6 Sequencing and Phylogenetic Analysis

The PCR products of 16S rDNA from each ARDRA pattern were selected for sequencing at Sangon Biotech Co., Ltd (Shanghai, China). All obtained sequences were compared with the sequences in the GenBank database using the BLASTN search program. The most similar sequences were further aligned using Clustal W software (Tompson et al., 1994). All reference sequences were obtained from the National Center for Biotechnology Information (NCBI). Phylogenetic trees were constructed using the neighbor-joining method with two-parameter MEGA software (version6.0) (Saitou & Nei, 1987). Statistical significance levels of interior nodes were determined by bootstrap analysis (1,000 data resamplings) (Felsenstein, 1985).

2.7 Nucleotide Sequence Accession Numbers

The nucleotide sequences of the 16S rRNA gene for each analyzed strain have been deposited in GenBank and have been assigned accession numbers (Table 1).

3. Results

3.1 Isolation of Endophytic Bacteria

The surface sterilized seeds were divided into groups and then homogenized with sterile water. The homogenates were diluted and spread on LB agar plates. The plates were cultured at 37 °C for 72 h. Colonies with obvious morphological differences were purified. Colonies without morphological differences were randomly selected. A total of 137 colonies were isolated and purified.

3.2 16S rRNA Amplification and ARDRA Analysis

A single colony was inoculated in LB broth and cultured at 37 °C for 12 h. The culture sample was used for DNA extraction. A pair of PCR primers (799f and 1492r) was used to amplify the 16S rRNA gene of the 137 isolates. Genomic DNA was used for the templates. The fragment size of 16S rDNA is approximately 700bp (Figure 1). The specific PCR product was completely digested with the restriction enzyme HaeIII. Based on ARDRA patterns, the isolated bacterial strains were grouped into different OTUs (Table 1). An OTU was defined as a group of clones with an identical banding patterns obtained from digestion. According to the ARDRA patterns (Figure 2), 137 strains were grouped into 9 OTUs.

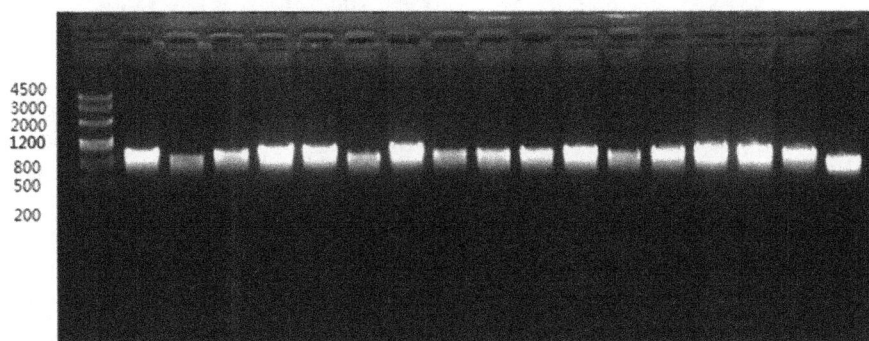

Figure 1. PCR amplification of the 16S rRNA gene

Note. The primers 799f and 1492r were used. Genomic DNA of endophytic bacteria was used as a template, and the annealing temperature of the PCR was 50 °C. The predicted product was approximately 700 bp.

Table 1. Number of phylotypes per OTU and their *Hae*III patterns

OTU Identification	Phylotypes ID	Amplified product (bp)	*Hae*III pattern
γ-proteobacter			
Closest to *Stenotrophomonas* sp.	WSDS21	689	335,204,150
	PL122	693	336,206,151
	PL232	700	335,211,154
	PL11	700	335,212,153
Closest to *Pseudomonas* sp.	WSDS121	699	320,179,119,81
	SLY22	703	321,179,122,81
	SLY13	694	318,178,118,80
	PL12	701	322,178,120,81
	PL133	698	321,180,118,79
Closest to Uncultured bacterium	YNZJ312	703	331,197,148,27
	YNZJ323	699	328,201,141,29
	SLY321	694	327,198,141,28
	DYPE121	704	332,196,145,31
Closest to *Pectobacterium carotovorum* subsp.	SLY33	699	446,244,9
Closest to *Yokenella regensburgei* strain	MGJ121	682	357,201,79,33,12
Closest to *Citrobacter* sp.	PL212	703	501,118,84
Closest to *Kluyvera ascorbata* strain	MGJ136	700	357,217,126
Closest to *Enterobacter* sp.	PL231	703	301,212,179,11
	SLY32	703	302,211,179,11
Firmicutes			
Closest to *Paenibacillus* sp.	PL222	700	441,177,52,30

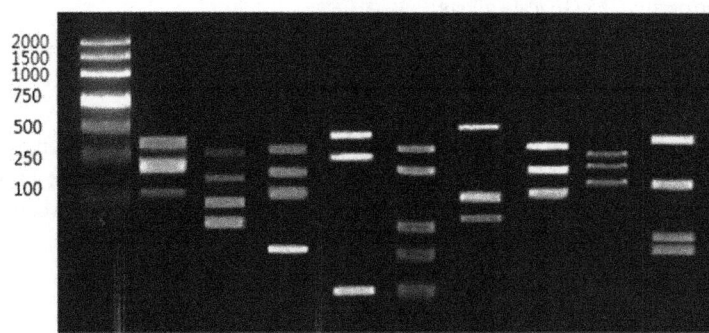

Figure 2. ARDRA analysis patterns of the 16S rDNA.

Note. The specific PCR products were completely digested using the restriction enzyme *Hae* III. The digested products were separated by electrophoresis. Strains with the same digestion patterns were grouped into OTUs.

3.3 Diversity and Distribution Analysis of Endophytic Bacteria

Representative isolates of specific groups were selected for sequencing and compared with the sequences in GenBank using the BLASTN search program. All isolates showed high similarities (≥ 99%) with their closest related species. The details of representative strains are listed in Table 2. Sequence analysis revealed the bacterial diversity of *P. notoginseng* seeds. These strains contain 2 Gram-positive species in one genus and 135 Gram-negative species in eight genera, indicating the complexity of endophytic population present in the seeds of *P. notoginseng* plants. The relationship between the isolates and the reference species is shown in the phylogenetic tree (Figure 3). γ-proteobacteria was the most dominant group among the isolates (135 of 137 isolates) and comprised eight genera, with *Pseudomonas* being dominant (58 of 135 isolates). The second dominant genus was *Enterobacter* (20.4%), followed by uncultured bacterium (14.6%) and *Stenotrophomonas* (10.2%).

Table 2. Similarity of the16S rDNA sequences of partial endophytic bacterial strains from *P. notoginseng* seeds

Group	Isolates (accession number)	No. of isolates	Closest relative (accession number)[a]	Similarity (%)
γ-proteobacteria	WSDS21 (KX688530)	3	*Stenotrophomonas maltophilia* strain G10b (KC136828.1)	99
	PL11 (KX688526)	4	*Stenotrophomonas* sp. REp-tet_144 (JX899643.1)	99
	PL232 (KX688525)	5	Uncultured *Stenotrophomonas* sp. (LC002923.1)	99
	PL122 (KX688520)	2	*Stenotrophomonas maltophilia* strain ZJB-14120 (KM655831.1)	99
	WSDS121 (KX688529)	3	*Pseudomonas* sp. Tibet-YD5003-3 (KF805078.1)	99
	SLY22 (KX688532)	4	*Pseudomonas* sp. BS29 (KR063209.1)	99
	SLY13 (KX688533)	1	*Pseudomonas* sp. +Y33 (JX113247.1)	99
	PL12 (KX688521)	26	*Pseudomonas beteli* strain RRLJ SMAR (DQ299947.1)	99
	PL133 (KX688522)	24	*Pseudomonas* sp. PW49 (KT726998.1)	99
	YNZJ312 (KX688537)	8	Uncultured bacterium clone Y1-5 (JF766465.1)	99
	YNZJ323 (KX688538)	9	Uncultured bacterium clone M1_209_H3 (JN683972.1)	99
	SLY321 (KX688535)	2	Uncultured bacterium clone 3Y-35 (EU786145.1)	99
	DYPE121 (KX688536)	1	Uncultured bacterium clone c93 (KC954365.1)	99
	SLY33 (KX688534)	1	*Pectobacterium carotovorum* subsp. carotovorum strain RN24 (KC790284.1)	99
	MGJ121 (KX688527)	2	*Yokenella regensburgei* strain NvH01 (KJ397957.1)	100
	PL212 (KX688523)	7	*Citrobacter* sp. F41 (FJ405282.1)	99
	MGJ136 (KX688528)	5	*Kluyvera ascorbata* strain IHB B 7177 (KJ767338.1)	99
	PL231 (KX688524)	24	*Enterobacter* sp. HT-Z52-B2 (KJ516915.1)	100
	SLY32 (KX688531)	4	*Enterobacter aerogenes* strain IEY (GQ165811.1)	100
Firmicutes	PL222 (KX688519)	2	*Paenibacillus* sp. MC5-3 (FJ932657.1)	99

Note. [a] Closest relative species and its accession number in the 16S rDNA sequence database.

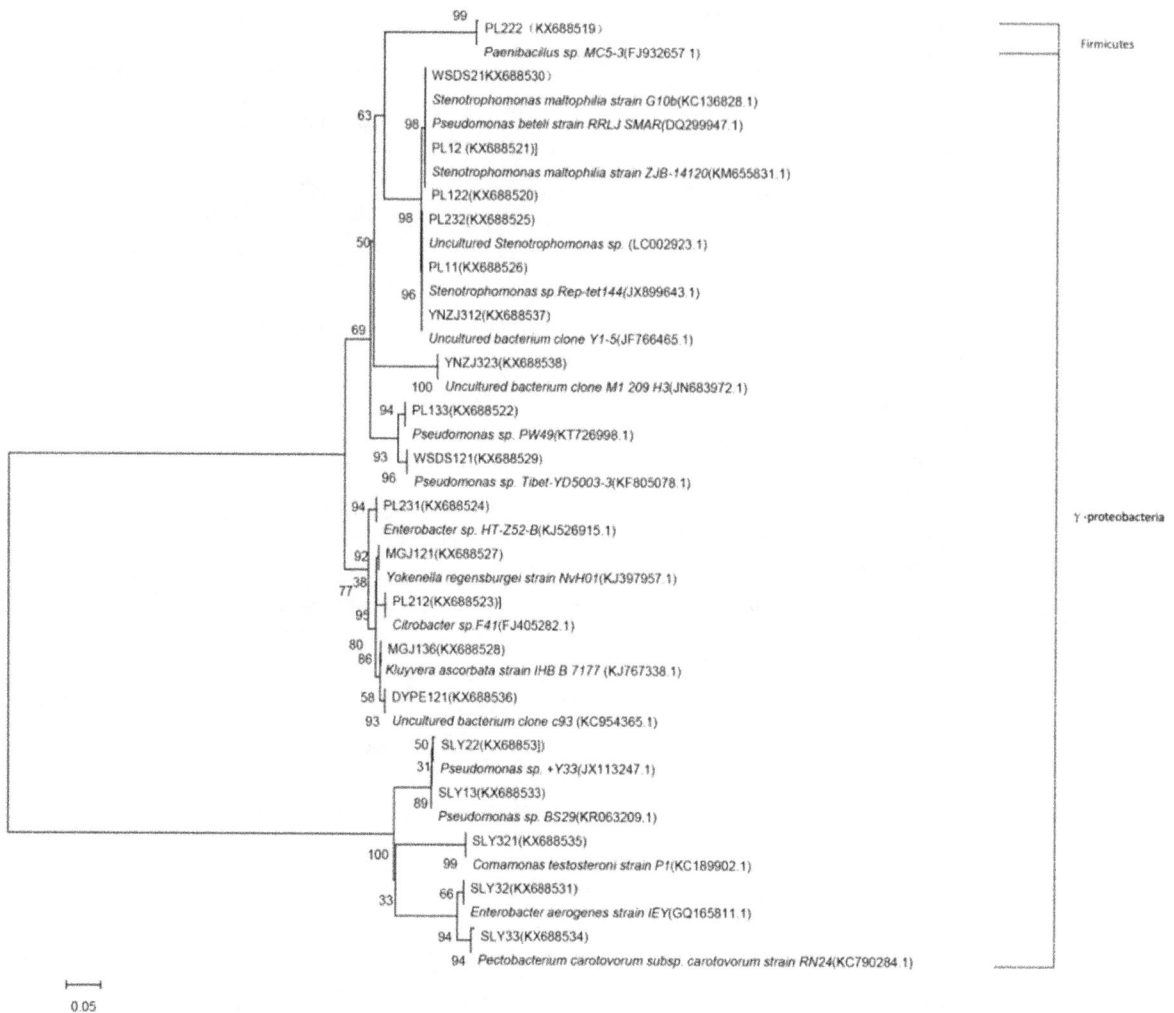

Figure 3. Phylogenetic tree based on partial 16S rDNA sequences of the endophytic bacteria of the *P. notoginseng* seeds using the neighbor-joining method

Note. The numbers in round brackets indicate the accession number in GenBank. The numbers at branch points indicate the bootstrap threshold (samples drawn 1000 times). Scale bar, 0.05 substitutions per base position.

The diversity of the endophytic bacteria of *P. notoginseng* plants was assessed using the seeds collected from the following six growing areas of the Wenshan region: Panlong of Yanshan (PL), Wenshandongshan of Wenshan (WSDS), Zhujie of Guangnan (ZJ), Hanqing of Maguan (MG), Shuanglongying of Qiubei (SLY), and Xichou (DYPE). The *Pseudomonas* genus showed a predominant existence and wide distribution in seeds collected from three places, which included Panlong of Yanshan, Wenshandongshan of Wenshan and Shuanglongying of Qiubei. Both the second and the third dominant genera, *Enterobacter* and *Stenotrophomonas*, were isolated from Panlong of Yanshan and Shuanglongying of Qiubei. The most endophytic bacteria isolates (94 of 137 isolates) originated from Panlong of Yanshan. With the exception of Panlong, the highest number of species (5 of 9 species) occurred in the seeds collected from Shuanglongying of Qiubei.

4. Discussion

Molecular approaches based on 16S rRNA gene analysis have been successfully used for bacterial community analysis. The ARDRA technique has been widely applied in genetic diversity studies of rhizobacteria (Dellagnezze et al., 2016; Mehri et al., 2011; Nievas et al., 2012; Santoro et al., 2016; Sanyal et al., 2016). In the present study, ARDRA analysis was applied to group the endophytic bacterial strains into different OTUs. The sequencing of 16S rDNA proved that each representative strain from each OTU represents a genus. The results suggest that this method is accurate and effective for the isolation and identification of the endophytic bacteria from *P. notoginseng* seeds, provides a preliminary screening before sequencing and avoids several rounds of sequencing.

Endophytes have been considered as abundant sources for probiotics. Ratnaweera et al. (2013) isolated eight endophytic fungi from *Opuntia dillenii*, seven of which exhibited antimicrobial activity. The most biologically active species is *Fusarium* sp., and the second most active is *Aspergillus niger*. Chadha et al. (2015) suggested that plant endophytic fungi can protect plants against various pathogen and pests and help plants survive under harsh biological or abiotic stresses. Tantirungkij et al. (2015) reported that rice leaves harbor several new yeast strains. Khan et al. (2015) isolated two fungal strains, *Fusarium tricinctum* RSF-4L and *Alternaria alternata* RSF-6L, which promote the growth of the host by phytohormone secretion. Ma et al. (2013) isolated endophytic bacteria from *P. notoginseng*, and 104 strains showed antagonism against *Fusarium oxysporum*, *Ralstonia* sp. and *Meloidogyne hapla*, which are three major pathogens associated with the root-rot disease complex of *P. notoginseng*. Shahzad et al. (2016) isolated an endophytic *Bacillus amyloliquefaciens* with the potential to produce gibberellins (GAs) and that plays a role in improving host-plant physiology.

In the present study, the bacterial endophytes isolated from the seeds of *P. notoginseng* belonged to four bacterial groups, including γ-proteobacteria and *Firmicutes*, *Enterobacter*, uncultured bacteria and *Firmicutes*. These results suggest that *P. notoginseng* seeds carry abundant microbial resources. Among the isolated bacteria, the dominant bacteria was *Pseudomonas* sp., followed by *Enterobacter* sp., uncultured bacteria clones, and *Stenotrophomonas* sp. To our knowledge, this report is the first comprehensive study on the isolation of endophytic bacteria from *P. notoginseng* seeds.

Strains isolated from the seeds may have biological activity against pathogens or perhaps are pathogens leading to diseases that occur during seed germination, seedling growth or plant development. Some *Pseudomonas* strains serve as plant growth-promoting rhizobacteria (PGPR) or biological control agents against plant pathogens (de Bruijn et al., 2007; Raaijmakers et al. 2010). *Pseudomonas* sp. is also one of bacterial pathogens that cause *P. notoginseng* root-rot disease (Miao et al., 2006). In the present study, the dominant endophytic bacteria were *Pseudomonas* sp. Whether the subsequent harboring of these *Pseudomonas* sp. strains leads to disease or plays a beneficial role on plants requires further research. *Stenotrophomonas maltophilia* has been reported to promote plant growth due to its production of phytohormones (Park et al., 2005; Naz et al., 2009) and to be a biological control agent due to its production of antibacterial compounds and secretion of fungicidal metabolites (Messiha et al., 2007; Taghavi et al., 2009). *Pectobacterium carotovorum* is a plant pathogen with a diverse host range and that causes bacterial soft rot (Mansfield et al., 2012). *Enterobacter aerogenes* is a nosocomial and pathogenic bacterium that causes opportunistic human infections. Whether these pathogens cause diseases on *P. notoginseng* remains unknown. Therefore, the isolation of these endophytic bacterial strains not only helps to further current understanding of the outbreak of seedling or plant diseases, but also provides good candidates for biological control against soil-borne root diseases.

Acknowledgements

This work was supported by the National Natural Science Foundation of China (31560246 and 31160020) and China Scholarship Council (201508535046).

References

Chadha, N., Mishra, M., Rajpal, K., Bajaj, R., Choudhary, D. K., & Varma, A. (2015). An ecological role of fungal endophytes to ameliorate plants under biotic stress. *Arch Microbiol, 197*(7), 869-81. https://doi.org/10.1007/s00203-015-1130-3

Chelius, M. K., & Triplett, E. W. (2001). The diversity of archaea and bacteria in association with the roots of *Zea mays* L. *Microb Ecol, 41*, 252-263. https://doi.org/10.1007/s002480000087

Cho, K. M., Hong, S. Y., Lee, S. M., Kim, Y. H., Kahng, G. G., Lim, Y. P., ... Yun, H. D. (2007). Endophytic bacterial communities in ginseng and their antifungal activity against pathogens. *Microb Ecol, 54*, 341-351. https://doi.org/10.1007/s00248-007-9208-3

de Bruijn, I., de Kock, M. J. D., Yang, M., de Waard, P., van Beek, T. A., & Raaijmakers, J. M. (2007). Genome-based discovery, structure prediction and functional analysis of cyclic lipopeptide antibiotics in Pseudomonas species. *Mol Microbiol, 63*, 417-428. https://doi.org/10.1111/j.1365-2958.2006.05525.x

Dellagnezze, B. M., Vasconcellos, S. P., Melo, I. S., Santos Neto, E. V., & Oliveira, V. M. (2016). Evaluation of bacterial diversity recovered from petroleum samples using different physical matrices. *Braz J Microbiol, 47*(3), 712-23. https://doi.org/10.1016/j.bjm.2016.04.004

Felsenstein, J. (1985). Confidence limits on phylogenies: An approach using the bootstrap. *Evolution, 39*, 783-791. https://doi.org/10.2307/2408678

Feng, Y., Shen, D., & Song, W. (2006). Rice endophyte *Pantoea agglomerans* YS19 promotes host plant growth and affects allocations of host photosynthates. *J Appl Microbiol, 100*, 938-945. https://doi.org/10.1111/j.1365-2672.2006.02843.x

Hallmann, J., Quadt-Hallmann, A., Rodrguez-Kabana, R., & Kloepper, J. W. (1998). Interactions between *Meloidogyne incognita* and endophytic bacteria in cotton and cucumber. *Soil Biol Biochem, 30*, 925-937. https://doi.org/10.1016/S0038-0717(97)00183-1

Khan, A. R., Ullah, I., Waqas, M., Shahzad, R., Hong, S. J., Park, G. S., ... Shin, J. H. (2015). Plant growth-promoting potential of endophytic fungi isolated from *Solanum nigrum* leaves. *World J Microbiol Biotechnol, 31*(9), 1461-6. https://doi.org/10.1007/s11274-015-1888-0

Li, Y. H., Zhu, J. N., Zhai, Z. H., & Zhang, Q. A. (2010). Endophytic bacterial diversity in roots of *Phragmites australis* in constructed Beijing Cuihu Wetland (China). *FEMS Microbiol Lett, 309*, 84-93. https://doi.org/10.1111/j.1574-6968.2010.02015.x

Lodewyck, C., Vangronsveld, J., Porteous, F., Moore, E. R. B., Taghavi, S., Mezgeay, M., & van der Lelie, D. (2002). Endophytic bacteria and their potential application. *Crit Rev Plant Sci, 86*(6), 583-606. https://doi.org/10.1080/0735-260291044377

Ma, L., Cao, Y. H., Cheng, M. H., Huang, Y., Mo, M. H., Wang, Y., Yang, J. Z., & Yang, F. X. (2013). Phylogenetic diversity of bacterial endophytes of *Panax notoginseng* with antagonistic characteristics towards pathogens of root-rot disease complex. *Antonie Van Leeuwenhoek, 103*(2), 299-312. https://doi.org/10.1007/s10482-012-9810-3

Mansfield, J., Genin, S., Magori, S., Citovsky, V., Sriariyanum, M., Ronald, P., ... Foster, G. D. (2012). Top 10 plant pathogenic bacteria in molecular plant pathology. *Mol Plant Pathol, 13*, 614-629. https://doi.org/10.1111/j.1364-3703.2012.00804.x

Mehri, I., Turki, Y., Chair, M., Chérif, H., Hassen, A., Meyer, J. M., & Gtari, M. (2011). Genetic and functional heterogeneities among fluorescent *Pseudomonas* isolated from environmental samples. *J Gen Appl Microbiol, 57*, 101-114. https://doi.org/10.2323/jgam.57.101

Messiha, N. A. S., Van Diepeningen, A. D., Farag, N. S., Abdallah, S. A., Janse, J. D., & Van Bruggen, A. H. C. (2007). *Stenotrophomonas maltophilia*: a new potential biocontrol agent of *Ralstonia solanacearum*, causal agent of potato brown rot. *Eur J Plant Pathol, 118*, 211-225. https://doi.org/10.1007/s10658-007-9136-6

Miao, Z. Q., Li, S. D., Liu, X. Z., Chen, Y. J., Li, Y. H., Wang, Y., ... Zhang, K. Q. (2006). The causal microorganisms of *Panax notoginseng* root rot disease. *Sci Agric Sin, 39*, 1371-1378.

Naz, I., Bano, A., & Hassan, T. U. (2009). Isolation of phytohormones producing plant growth promoting rhizobacteria from weeds growing in Khewra salt range, Pakistan, and their implication in providing salt tolerance to *Glycine max* L. *Afr J Biotechnol, 8*, 5762-5768. https://doi.org/10.5897/AJB09.1176

Nievas, F., Bogino, P., Nocelli, N., & Giordano, W. (2012). Genotypic analysis of isolated peanut-nodulating rhizobial strains reveals differences among populations obtained from soils with different cropping histories. *Appl Soil Ecol, 53*, 74-82. https://doi.org/10.1016/j.apsoil.2011.11.010

Park, M., Kim, C., Yang, J., Lee, H., Shin, W., Kim, S., & Sa, T. (2005). Isolation and characterization of diazotrophic growth promoting bacteria from rhizosphere of agricultural crops of Korea. *Microbiol Res, 160*(2), 127-33. https://doi.org/10.1016/j.micres.2004.10.003

Patkowska, E. (2001). Formation of bacterial and fungal communities in the rhizosphere of soybean [*Glycine max*. [L.] Merrill] and their antagonism towards phytopathogens. *J Plant Prot Res, 41*(2), 181-191.

Raaijmakers, J. M., de Bruijn, I., Nybroe, O., & Ongena, M. (2010). Natural functions of lipopeptides from Bacillus and Pseudomonas: More than surfactants and antibiotics. *FEMS Microbiol Rev, 34*, 1037-1062. https://doi.org/10.1111/j.1574-6976.2010.00221.x

Ratnaweera, P. B., de Silva, E. D., Williams, D. E., & Andersen, R. J. (2015) Antimicrobial activities of endophytic fungi obtained from the arid zone invasive plant *Opuntia dillenii* and the isolation of equisetin, from endophytic *Fusarium* sp. *BMC Complement Altern Med, 15*, 220. https://doi.org/10.1186/s12906-015-0722-4

Saitou, N., & Nei, M. (1987). The neighbor-joining method: A new method for reconstructing phylogenetic trees. *Mol Biol Evol, 4*, 406-425.

Santoro, M. V., Bogino, P. C., Nocelli, N., Cappellari Ldel, R., Giordano, W. F., & Banchio, E. (2016). Analysis of plant growth-promoting effects of fluorescent *Pseudomonas* strains isolated from *Mentha piperita* rhizosphere and effects of their volatile organic compounds on essential oil composition. *Front Microbiol, 7*, 1085. https://doi.org/10.3389/fmicb.2016.01085

Sanyal, S. K., Mou, T. J., Chakrabarty, R. P., Hoque, S., Hossain, M. A., & Sultana, M. (2016). Diversity of arsenite oxidase gene and arsenotrophic bacteria in arsenic affected Bangladesh soils. *AMB Express, 6*, 21. https://doi.org/10.1186/s13568-016-0193-0

Schulz, B., Wanke, U., Draeger, S., & Aust, H. J. (1993). Endophytes from herbaceous plants and shrubs: effectiveness of surface sterilization methods. *Mycol Res, 97*, 1447-1450. https://doi.org/10.1016/S0953-7562(09)80215-3

Senthilkumar, M., Anandham, R., Madhaiyan, M., Venkateswaran, V., & Tongmin, Sa. (2011). Endophytic bacteria: perspectives and applications in agricultural crop production. In D. K. Maheshwari (Ed.), *Bacteria in agrobiology: Crop ecosystems*. Springer, Berlin. https://doi.org/10.1007/978-3-642-18357-7_3

Sessitsch, A., Reiter, B., Pfeifer, U., & Wilhelm, E. (2002). Cultivation independent population analysis of bacterial endophytes in three potato varieties based on eubacterial and Actinomycetes-specific PCR of 16S rDNA genes. *FEMS Microbiol Ecol, 39*, 23-32. https://doi.org/10.1111/j.1574-6941.2002.tb00903.x

Shahzad, R., Waqas, M., Khan, A. L., Asaf, S., Khan, M. A., Kang, S. M., ... Lee, I. J. (2016). Seed-borne endophytic *Bacillus amyloliquefaciens* RWL-1 produces gibberellins and regulates endogenous phytohormones of *Oryza sativa*. *Plant Physiol Biochem, 106*, 236-43. https://doi.org/10.1016/j.plaphy.2016.05.006

Sturz, A. V., Christie, B. R., & Nowak, J. (2000). Bacterial endophytes: Potential endophytes: Potential role in developing sustainable systems of crop production. *Crit Rev Plant Sci, 19*, 1-30. https://doi.org/10.1016/S0735-2689(01)80001-0

Sun, H. X., Qin, F., & Ye, Y. P. (2005). Relationship between haemolytic and adjuvant activity and structure of protopanaxadiol-type saponins from the roots of *Panax notoginseng*. *Vaccine, 23*, 533-5542. https://doi.org/10.1016/j.vaccine.2005.07.036

Sun, J. H., Ma, N., Chen, Z. J., Wang, C. L., & Cui, X. M. (2004). Effects of root rot on saponin content in *Panax notoginseng*. *J Chin Medicinal Mater, 27*, 79-80.

Taghavi, S., Garafola, C., Monchy, S., Newman, L., Hoffman, A., Weyens, N., ... van der Lelie, D. (2009). Genome survey and characterization of endophytic bacteria exhibiting a beneficial effect on growth and development of poplar trees. *Appl Environ Microbiol, 75*, 748-757. https://doi.org/10.1128/AEM.02239-08

Tantirungkij, M., Nasanit, R., & Limtong, S. (2015). Assessment of endophytic yeast diversity in rice leaves by a culture-independent approach. *Antonie Van Leeuwenhoek, 108*(3), 633-47. https://doi.org/10.1007/s10482-015-0519-y

Tompson, J. D., Higgins, D. G., & Gibson, T. J. (1994). CLUSTAL W: Improving the sensitivity of progressive multiple sequence alignment through sequence weighting, position-specific gap penalties and weight matrix choice. *Nucleic Acids Res, 22*, 4673-4680. https://doi.org/10.1093/nar/22.22.4673

Vendan, R. T., Yu, Y. J., Lee, S. H., & Rhee, Y. H. (2010). Diversity of endophytic bacteria in ginseng and their potential for plant growth promotion. *J Microbiol, 48*, 559-565. https://doi.org/10.1007/s12275-010-0082-1

10

Evaluation of the Efficiency of Duckweeds, *Lemna* sp. and *Spirodela* sp., in the Treatment of Tilapia Effluents

Itzel Galaviz-Villa[1], Cinthya Sosa-Villalobos[1], Alicia García-Sánchez[1], Ma. Refugio Castañeda-Chavez[1], Fabiola Lango-Reynoso[1] & Isabel Amaro-Espejo[1]

[1] Division of Graduate and Research Studies, Technological Institute of Boca del Río, Veracruz, México

Correspondence: Cinthya Sosa-Villalobos, Division of Graduate and Research Studies, Technological Institute of Boca del Río, Km. 12 Carretera Veracruz-Córdoba. C.P. 94290, Boca del Río, Veracruz, México. E-mail: ca.sosavi@gmail.com

Abstract

Farming aquatic plants can be used as an alternative in the treatment of effluents from aquaculture production units and in turn, in the production of biomass plant for feeding terrestrial and aquatic organisms. This research aims to evaluate the efficiency of duckweeds *Spirodela* sp. and *Lemna* sp. in the treatment of tilapia effluents (*Oreochromis niloticus*). The experiment was performed in triplicate and was conducted under natural environmental conditions within the facilities of the Laboratory of Applied Aquaculture Research (LAAR) of the Technological Institute of Boca del Río (ITBOCA). Each treatment contained 230 liters of effluent in each tube, with a water column of 40 cm and a seeding density of 400 g/m^2 of vegetative biomass. The evaluation of the efficiency of *Spirodela* sp. and *Lemna* sp., in the removal of dissolved nutrients, was performed in 7 monitoring times; 12, 24, 36, 48, 72, 96 and 120 h. The results showed an efficiency in nutrient removal at 120 h of 75, 74 and 66% of N-NH$_3$; 96, 92 and 75% N-NO$_2$; 93, 88 and 75% N-NO$_3$; 75, 72 and 64% N-NTK; 73, 60 and 58% of N-org., and 73, 63 and 68% of P. On the other hand, the removal of TSS and BOD5, during the first 24 h, was 83, 54, 58% and 65, 59, 33%, in the treatments. The efficiency in nutrient removal of both duckweeds, showed that both plants can be used in the treatment of effluents, being a sustainable and economical alternative for the aquaculture industry.

Keywords: phytoremediation, aquatic plants, aquaculture effluents

1. Introduction

One of the main constraints on aquaculture production is the high concentration of nutrients that are generated as a result of the excretions of fish, the provided food and other inputs used to control crop (Tacon & Foster, 2003). Currently, the development model most used in Aquaculture Production Units (APU) is the intensive system, which makes the activity increasingly dependent on inputs such as; balanced food, electricity, hormones, antibiotics, etc. (Espinosa & Bermudez, 2012).

In the state of Veracruz, most of the aquaculture production units of tilapia discharge their effluents directly, without previous treatment into bodies of surface water becoming a source of pollution for the aquatic environment (Palomarez-García, 2010). These effluents include uneaten feed, metabolic excretions, feces, dead fish and organic and inorganic solid waste. If the flow of these compounds, into the environment exceeds the assimilative capacity of ecosystems; it could cause severe impacts, both in the water column and in benthos, such as; eutrophication, oxygen depletion and alteration of local biodiversity (IUCN, 2007). The degree of environmental impact is straightly related to the production system used. These systems are: extensive, semi-intensive and intensive. As the system is intensified, a greater number of inputs and raw materials are used (Flores, González, & Prado, 2007).

The environmental impact of aquaculture is an issue of global dimensions; freshwater aquaculture raises nutrient loading of river systems, which probably will intensify in the future. Much of the environmental impact of aquaculture is derived from the sum of individual farms. Although, environmental impact assessment and licensing and certification systems are required for individual intensive and for large-scale farms. There have not been approaches based on mitigation, nor management measures that collectively cover the overall impact of small farms. Some farms have an impact that affects their own culture systems, such as; generating hypoxia, fish kills or stress, or by creating conditions that favor the spread of disease (FAO, 2014).

Production of aquatic macrophytes in aquaculture effluents is considered an alternative for reducing large volumes of nutrients discharged into water and soil of national assets. These macrophytes are characterized by being highly productive and present an accelerated growth factor that has caused the research being directed toward its control, with emphasis on its eradication; however, in many developing countries, this vegetation is used as feed for farm animals. Due to its high productivity it generates excellent crops. Besides, they do not require most of the agricultural work, or the purchase of inputs such as seeds and fertilizers (Pacheco, 2009).

Ecosystems dominated by aquatic macrophytes are considered as the most productive in the world. Aquatic plants have the ability to assimilate nutrients and create favorable conditions for microbial decomposition of organic matter. For this reason, they are known as self-purifying for aquatic environments and are used in the treatment of wastewater (Brix & Schierup, 1989; Pardo, 2006; Miranda & Quiróz, 2013). Macrophytes of the *Lemnaceae* family have shown great ability in the assimilation of nutrients. Under optimal growth conditions, such as nutrient availability, light and optimal water temperature, can double its biomass every two or three days. And achieve protein content between 15% and 45%, because they assimilate large amounts of nitrogen mainly in ammonium form and phosphorus in orthophosphate form. When used for the treatment of wastewater, they generate high quality biomass, because the ammonium is converted directly into protein. Among the potential uses of aquatic plants, we can refer to its operation as feed for cattle, sheep and pigs. They are used through the ensiling and drying process, then mixed with other feed; in feed for aquaculture production, in obtaining fertilizers, pulp or paper pulp production, for biological purification of wastewater and in the production of biofuels (Oron, 1994; Lallana, 1997; Zetina, 2010).

According to Troell et al. (2005) comprehensive methods of biotransformation, besides bringing ecological and social benefits, allow additional production of food without other input costs. For this, it is necessary to perform carefully planned scientific studies to prove that aquaculture is a production alternative with great potential for sustainable growth (FAO, 2003; Pardo, 2006). This research project aims to evaluate nutrient removal from tilapia effluents culture using aquatic macrophytes, *Spirodela* sp. and *Lemna* sp., as an alternative in effluents treatment and as protein production for supplementary feeding of their crops.

2. Methods

The experiment was conducted under natural conditions within the premises of the Laboratory of Applied Aquaculture Research (LAAR) of the Technological Institute of Boca del Río (ITBOCA).

2.1 Collection and Identification of Macrophytes

The collection of aquatic macrophytes *Spirodela* sp. and *Lemna* sp. was performed manually in wetlands of Medellín de Bravo, Ver., and 30 × 30 cm monofilament nylon net was used. The samples were stored in plastic bags with airtight seal and transported in a cooler at room temperature, to the Research and Aquatic Resources Laboratory (RARL) of the Technological Institute of Boca del Río. The identification of species and genus was performed with an ITALY OPTICS® stereomicroscope, with the taxonomic keys described by (Rzedowski et al., 2005; Landolt & Schmidt-Mumm, 2009; Mora-Olivo, 2013).

2.1.1 Acclimation and Cultivation of Macrophytes

Macrophytes *Spirodela* sp. and *Lemna* sp. were washed with drinking water and rinsed with distilled water, allowed to drain for one hour and then weighed on a CQT 202 Core balance from Adam Equipment; 200 g of each gender were planted in tubs of 243 × 60.5 × 29 cm. The tubs were filled with effluent from (*O. Niloticus*) tilapia culture in juvenile stage (30 organisms) at a level of depth of 25 cm. The water was replaced every fifth day and 30% of the biomass was harvested each week. Acclimation and culture were for 30 days. Fish were fed three times a day with balanced food, Silver Cup® 3.5 mm (32% protein).

2.2 Experimental Unit

In a geomembrane circular pond of 0.8 × 1.5 m in diameter with a capacity of 4.95 m^3, a semi-intensive culture of tilapia (*O. Niloticus*) in growing stage (120 organisms, 30 org/m^3) was maintained for 25 days. The organisms were fed three times a day, with a commercial diet of (5.5 mm Silver Cup®) 32% protein.

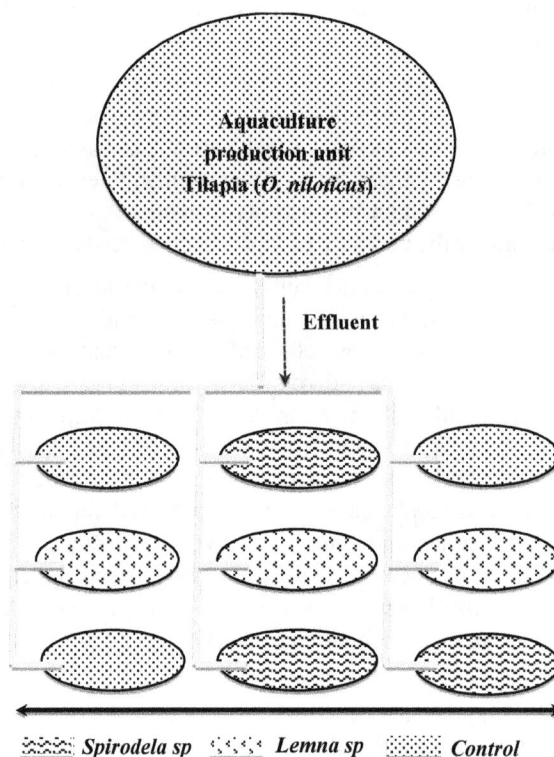

Figure 1. Experimental unit

The assessment of macrophytes for wastewater treatment consisted of three treatments in triplicate, with a control, *Spirodela* sp. and *Lemna* sp. Each treatment contained 230 liters of effluent in each tub (Figure 1) with a water sheet of 40 cm and a seeding density of 400 g/m² of vegetative biomass.

2.3 Laboratory Analysis

2.3.1 Physico-Chemical Analysis of the Effluent

Effluent temperature and pH measuring were performance *in situ* directly in each of the tubs (treatment and control) of the experimental system; at time 0 (effluent), 12, 24, 36, 48, 72, 96 and 120 hours. An amount of 1.5 L of the effluent was collected in plastic containers and taken to the laboratory for the different physico-chemical determinations. For determination of total phosphorus, 500 ml of effluent were filtered through Whatman nitrocellulose membrane filters (45 mm, 0.45 μm). The remaining 1000 ml were fixed with H_2SO_4 to get a pH less than 2. Filters and the acidified effluent were stored in refrigeration until their analysis.

Table 1. Average values of the physico-chemical characterization in the effluent

Parameter	Effluent
pH	7.4±0.00
T (°C)	29.0±0.00
NH_3-N (mg/L)	4.14±0.11
NO_2-N(mg/L)	1.0±0.00
NO_3-N (mg/L)	10±0.00
NOrg. (mg/L)	0.93±0.17
TKN (mg/L)	5.08±0.17
Total Phosphorus (mg/L)	0.83±0.02

The average values of the physico-chemical characterization in the effluent are shown in Table 1. Procedures corresponding to standardized methods for analysis of drinking water and wastewater were followed (APHA, 1995).

2.3.2 Chemical Composition of Duckweeds

Physico-chemical analyses of macrophytes were performed, in order to know the volume and composition of biomass produced. Macrophytes were harvested, weighed and dried in an oven at 60 °C. Determining the chemical composition of duckweeds was performed at the end of the experiment, according to official methods of analysis of AOAC (1995).

2.4 Assessment on Removal Efficiency

The relation proposed by Paniagua-Michel and García (2003) (Equation 1) was used to evaluate the removal efficiency (RE), expressed as a percentage, in the different treatments.

$$\%RE = [(\text{Influent Concentration} - \text{Effluent Concentration})/\text{Influent Concentration}] \times 100 \qquad (1)$$

2.5 Statistical Analysis

The program Stadistic version 7.0 was used for analysis of variance ($p < 0.05$) and to determine significant differences between treatment and monitoring times. The test a posteriori Tukey ($p < 0.05$) was applied for multiple comparisons of means.

3. Results

RE percentages of ammonia nitrogen (NH_3-N), nitrites (NO_3-N), nitrates (NO_2-N) and total phosphorus (TP), are shown in Figure 2. A maximum value of 75% removal of NH_3-N with *Spirodela polyrrhiza* at 120 hours was observed. The removal of nitrites and nitrates was higher at 120 h, 93 and 96%, respectively with *Spirodela polyrrhiza*. The phosphorus was removed with a maximum value of 73% both at 72 h, and at 120 hours with *Spirodela polyrrhiza*.

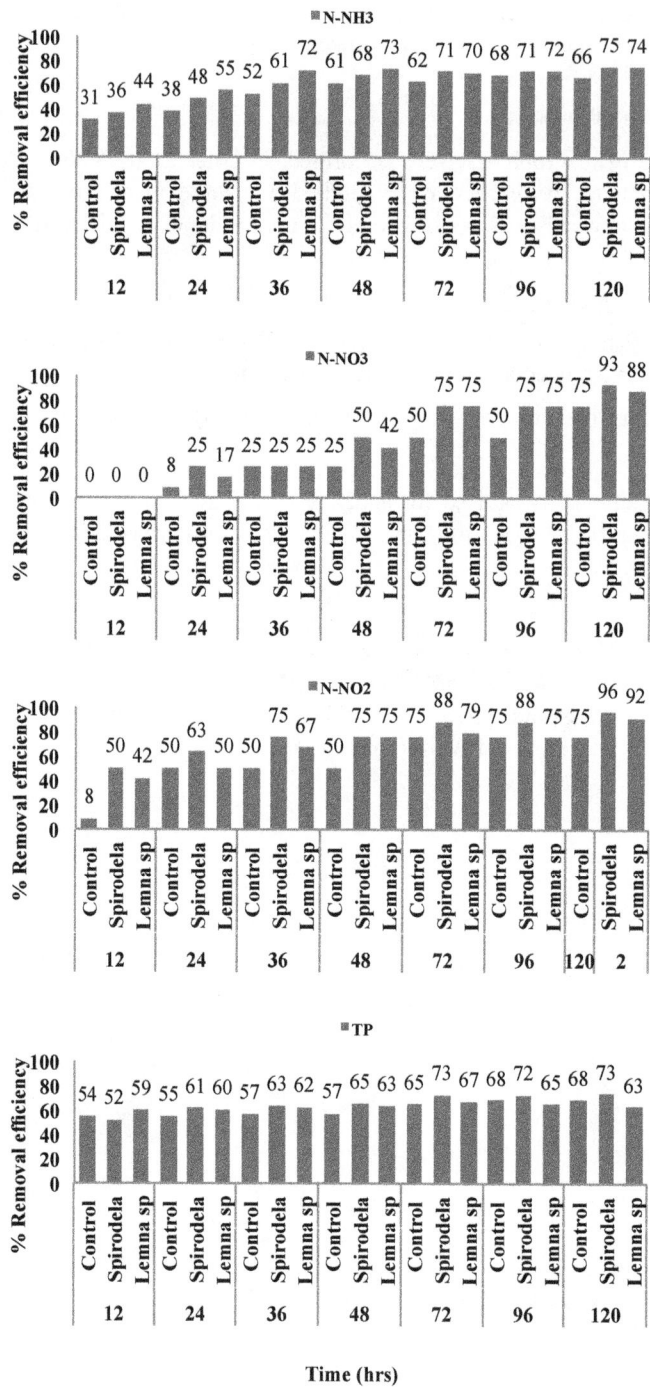

Figure 2. Nutrient removal efficiency

Some differences between the times of monitoring and treatments are observed in Figure 3, when statistical analysis is applied.

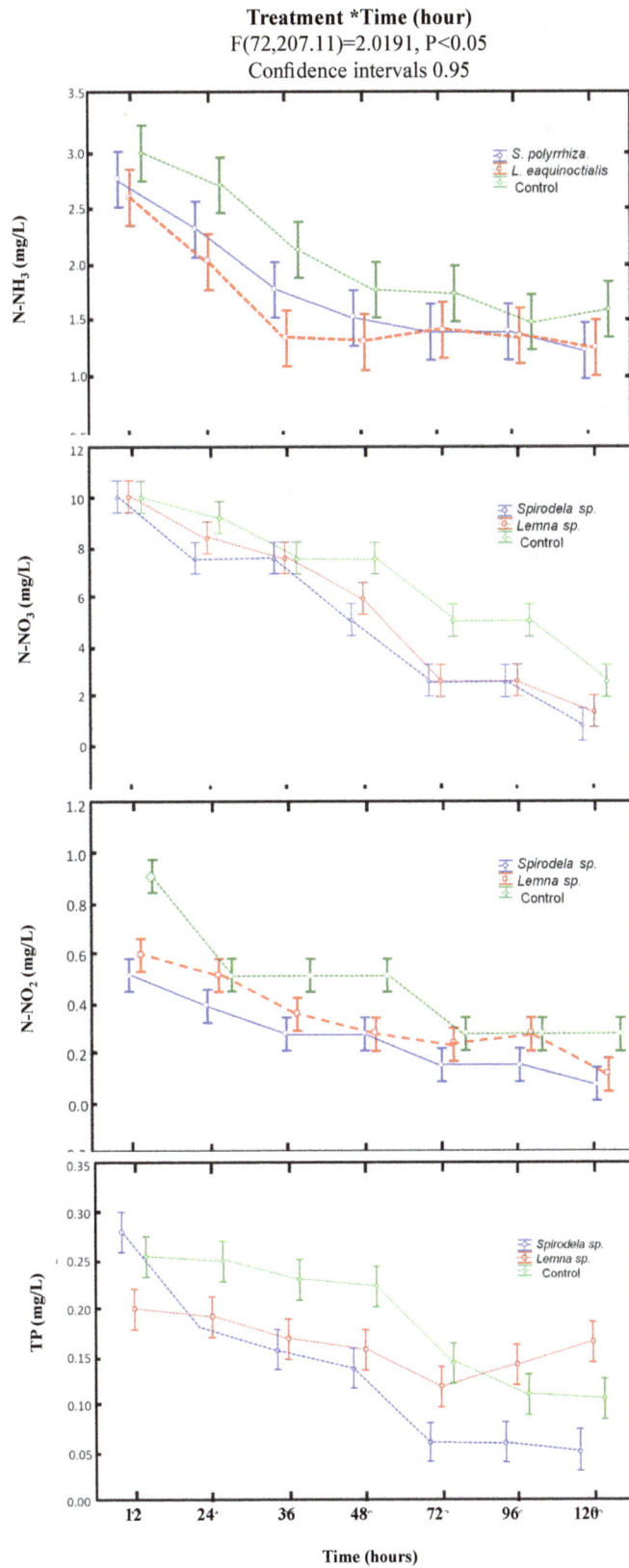

Figure 3. Analysis of variance (p < 0.05) of N-NH$_3$, N-NO$_2$, N-NO$_3$, Total Phosphorus and control vs time

The analysis of variance showed that there is no significant (Tukey, $P < 0.05$), with respect to monitoring times and concentrations of $N\text{-}NH_3$, $N\text{-}NO_2$ and $N\text{-}NO_3$. Statistical analysis confirmed that the % removal efficiency of $N\text{-}NH_3$ was the treatment of *Spirodela polyrrhiza*. In the case of total phosphorus, there are significant differences between treatments and control.

Biochemical Oxygen Demand (BOD_5) decreased in the first 24 hours, 65%, 59%, and 33% in treatments with *Spirodela* sp., *Lemna* sp. and control respectively. The percentage of removal of Total Suspended Solids (TSS), which decreased at 120 h, was 83%, 54% and 58%, in treatments with *Spirodela* sp., *Lemna* sp. and control respectively. The average pH value in the effluent was 7.4 and 7.3±0.12 in treatments with *Spirodela* sp. and *Lemna* sp. The control treatment showed the highest variation in pH of 7.8±0.58. The control treatment showed a decrease in nutrient concentration of 66% $NH_3\text{-}N$, 88% P, and 58% TSS, after five days of treatment.

The chemical composition of duckweeds, after 120 hours of treatment, is summarized in Table 2.

Table 2. Chemical composition of duckweeds

SP	% DM	% CP	% A	% CF
Spirodela sp.	92.13	33.91	15.07	12.46
Lemna sp	95.12	30.10	10.83	9.45

Note. SP: Species; DM: Dry matter; CP: Crude protein; A: Ash; CF: Crude fiber.

DM content was 92 to 95%, in macrophytes *Spirodela* sp. and *Lemna* sp., the content of CP was 34 and 30% with *Lemna* and *Spirodela* sp., in A the percentage ranged from 10.83 to 15, and in the content of CF, values ranged from 9.45 to 12.45 with *Lemna* sp. and *Spirodela* sp., respectively.

4. Discussion

Waha et al. (2005) indicated that the capacity of absorption of nutrients by aquatic plants is related to the growth rate, the established population and the composition of plant tissue.

Al-Qutob and Nashashibi (2012) mentioned that duckweeds from the effluent prefer ammoniacal nitrogen as a nitrogen source and preferably will remove ammonia, even in the presence of high concentrations of nitrate. A greater efficiency than that reported by Obek and Hazar (2002) was obtained, where a removal of 90% of phosphorus was gotten with *Lemna minor*, which was cultivated in domestic wastewater during 6 days of treatment. The macrophyte *Spirodela* polyrrhiza, has been the most effective in removing nutrients. According to Brix and Schierup (1989), Oron (1994), Pardo et al. (2006), Luchini and Huidobro (2008), and Miranda (2013), aquatic macrophytes belonging to the family *Lemnaceae* have shown a great capacity in the assimilation of nutrients in the treatment of waste water, as can be confirmed in this study.

Mohedano (2004), worked with effluents from tilapia (*O. niloticus*) using *Lemna valdiviana* treatments and obtained removal percentages for NH3-N of 94.44%, phosphorus 96.30%, TSS 93.37%, and NO_3 91.72%, after 13 days of treatment. However, in this study similar removal rates were obtained in less time (five days).

The efficiency that these macrophytes have when assimilating nutrients is explained by Alaerts, Mahbubar, and Kelderman (1996), they mention that bacteria and algae are responsible for nitrification and denitrification of compounds dissolved in water; where duckweed (*Lemna* sp.) may be responsible for the removal of 16 to 47% nitrogen and 9 to 61% phosphorus. Mohedano (2004) also reported high levels of removal in control treatment of 70.36% $NH_3\text{-}N$, 63% P, and 70.33% TSS, but twice as long (13 days) in effluents from tilapia.

It is noteworthy that after the third day of acclimation of *Lemna* sp., yellow and white fronds, and necrosis began to be observed. Mohedano (2010) reported that the rapid mortality of macrophytes may be due to high levels of toxicity of $NH_3\text{-}N$, wind, algae pollution, diseases, senescence, high density, lack of light, and CO_2. Tolerances of $NH_3\text{-}N$ for tilapia culture are in the range of 0.6 to 2.0 mg/L. The level of $NH_3\text{-}N$ of the effluent at the beginning of the experiment was 4.14 mg/L. In this regard Abdalla, McNabb, and Batterson (1996) mentioned that tilapia (*O. niloticus*) can tolerate levels from 1.1 to 4.1 mg/L of $NH_3\text{-}N$ for long periods of time (up to 96 hours), showing a 50% mortality; however, during this experiment there was a mortality of 3.3%, the sensitivity of this species varies according to the size of the fish and the water temperature. Chen, Jin, Zhang, Fang, Xiao, and Zhao (2012) reported that the duckweed takes phosphorus and nitrogen from water. The nitrogen is first converted into ammonia and then taken up by the roots. For this reason, they have been used in diets for ducks, fish, and shrimp (Ponce, Febrero, González, Romero, & Estrada, 2005). Statistical analysis applied, allowed us to observe that there was no significant difference between monitoring and treatment times, however in some cases it was noted that the control treatment had significant differences.

5. Conclusions

The efficiency of aquatic macrophytes *Spirodela* sp. and *Lemna* sp. were evaluated, which accomplished high nutrient removals. According to the results, the effluent from ponds of tilapia (*O. Niloticus*), can be considered an appropriate remedy to be implemented for aquatic macrophytes crops, in order to reduce nutrient concentrations that pollute the bodies of receiving water, which in turn generate biomass that can be used for feeding farmed organisms, reducing production costs. The proximate analysis performed in both macrophytes showed a high protein content and a suitable level of crude fiber, these results allow their use as aquaculture diets.

References

Abdalla, A., McNabb, C., & Batterson, T. (1996). Ammonia dynamics in fertilized fishponds stocked with Nile Tilapia. *The Progressive Fish-Culturist, 58*(3), 117-123. http://dx.doi.org/10.1577/1548-8640(1996)058<01 17:ADIFFP>2.3.CO;2

Alaerts, G., Mahbubar, R., & Kelderman, P. (1996). Performance analysis of a full-scale duckweed-covered sewage lagoon. *Water Research, 30*(4), 843-852. http://dx.doi.org/10.1016/0043-1354(95)00234-0

Al-Qutob, M. A., & Nashashibi, T. S. (2012). Duckweed Lemna minor (*Liliopsida, Lemnacea*) as a natural biofilter in brackish and fresh closed recirculating systems. *AACL BIOFLUX, 5*(5), 380-392. Retrieved from https://www.researchgate.net/publication/232715627_Al-Qutob_M_A_Nashashibi_T_S_2012_Duckweed_ Lemna_minor_Liliopsida_Lemnaceae_as_a_natural_biofilter_in_brackish_and_fresh_closed_recirculating_ systems_AACL_Bioflux_55380-392

AOAC (International Association of Official Analytical Chemists). (1995). *Official methods of analysis of AOAC International*. AOAC, Washington, D.C.

Brix, H., & Schierup, H. (1989). The Use of Aquatic Macrophytes in Water-pollution Control. *Ambio, 18*(2), 100-107. Retrieved from http://mit.biology.au.dk/~biohbn/hansbrix/pdf_files/Ambio_1989_100-107.pdf

Chen, Q., Jin, Y., Zhang, G., Fang, Y., Xiao, Y., & Zhao, H. (2012). Improving production of bioethanol from Duckweed (*Landoltia punctate*) by pectinase pretreatment. *Energies, 5*(8), 3019-3022. http://dx.doi.org/ 10.3390/en5083019

de Waha Baillonville, T., Diara, H. F., Watanabe, I., Berthet, C., & Van, H. (1991). Assessment and attempt to explain the high performance of Azolla in subdesertic tropics. *Plant Soil, 137*(1), 145-1149. http://dx.doi.org/10.1007/BF02187446

Eaton, A. D., Franson, M. A., Greenberg, A. E., & Clesceri, L. S. (Eds.). (1995). *Standard Methods for the Examination of Water and Wastewater*. American Public Health Association (APHA), the American Water Works Association, and the Water Environment Federation, USA.

Espinosa, P., & Bermúdez, M. (2012). *La acuicultura y su impacto al medio ambiente* (Vol. 2, pp. 221-232). Retrieved from http://www.ciad.mx/archivos/revista-dr/RES_ESP2/RES_Especial_2_10_Bermudez.pdf

Flores, J., González, E., & Prado, P. (2007). Puntos críticos en la evaluación de impacto ambiental de la Camaronicultura en el Pacífico de Nicaragua, durante su proceso productivo: Producción de larvas, operación y abandono de Granjas. *Universitas, 1*(1), 33-38. http://dx.doi.org/10.5377/universitas.v1i1.1631

Food and Agriculture Organization of the United Nations. (2003). *Acuicultura sostenible para el futuro*. FAO. Retrieved March 23, 2016, from http://www.fao.org/spanish/newsroom/news/2003/21619-es.html

Food and Agriculture Organization of the United Nations. (2014). *El estado mundial de la pesca y la acuicultura*. Departamento de Pesca y Acuicultura de la FAO. Retrieved March 23, 2016, from http://www.fao.org/ fishery/sofia/es

González, R., Fonseca, E., Rico, R., Romero, O., & Ponce, J. (2013). Utilización de Lemna trinervis en la alimentación de la tilapia aurea. *Revista Granma Ciencia, 17*(2), 9.

Lallana, V. (1997). Las plantas acuáticas del rio Paraná. *Su importancia en el ecosistema* (Artículo Técnico de divulgación AT-01). Panamá.

Landolt, E., & Schmidt, U. (2009). Flora de Colombia. *Lemnaceae* (Vol. 24, p. 54). Instituto de Ciencias Naturales, Universidad Nacional de Colombia, Bogotá D. C. Colombia.

Luchini, L., & Huidobro, S. (2008). *Perspectivas en acuicultura: Nivel mundial, regional y local* (pp. 87-90). Dirección de Acuicultura-Subsecretaría de Pesca y Acuicultura. Secretaría de Agricultura, Ganadera, Pesca y Alimentos. http://produccionbovina.com.ar/produccion_peces/piscicultura/113-perspectivas.pdf

Miranda, M., & Quiroz, A. (2013). *Efecto del fotoperíodo en la remoción de plomo por Lemna Gibba L. (Lemnaceae)*. Polibotánica. Retrieved March 23, 2016, from http://www.redalyc.org/articulo.oa?id=62 127866010

Mohedano, A. (2010). *Uso de macrófitas lemnáceas (Landoltia punctata) no polimento e valorização do efluente de suinocultura e na fixação de carbono* (Doctoral dissertation). Retrieve from https://repositorio.ufsc.br/ xmlui/handle/123456789/94053

Mohedano, R. (2004). *Tratamiento de efluentes y producción de alimento, en cultivos de tilapia (Oreochromis niloticus), a través de macrófitas acuáticas Lemna valdiviana (Lemnaceae). Una contribución para la sustentabilidad de la acuacultura* (Dissertation). Universidad Federal de Santa Catartina, Brasil

Mora, A., Villaseñor, J., & Martínez, M. (2013). Las plantas vasculares acuáticas estrictas y su conservación en México. *Acta Botánica Mexicana, 103*, 27-63. Retrieved from http://www.redalyc.org/articulo.oa?id=57425 775001

Öbek, E., & Hasar, H. (2001). Role of duckweed (*Lemna minor* L.) harvesting in biological phosphate removal from secondary treatment effluents. *Fresenius Environmental Bulletin, 11*(1) 27-29.

Oron, G. (1994). Duckweed culture for wastewater renovation and biomass production. *Agricultural Water Management, 26*, 27-40. http://dx.doi.org/10.1016/0378-3774(94)90022-1

Pacheco, R. (2009). *Rendimiento de la Lemna sp. Con el empleo de niveles de fertilización orgánica e inorgánica* (Dissertation). Universidad de Granma, Facultad de Medicina Veterinaria. Bayamo, Provincia Granma, Cuba.

Palomarez, J. (2010). *Valoración de la calidad de los influentes y efluentes de las granjas acuícolas de la cuenca baja del río Jamapa, Veracruz* (Dissertation). Colegio de Postgraduados campus Veracruz. Retrieved from http://hdl.handle.net/10521/436

Paniagua, J., & Garcia, O. (2003). Ex-situ bioremediation of shrimp culture effluent using constructed microbial mats. *Aquacultural Engineering, 28*, 131-139. http://dx.doi.org/10.1016/S0144-8609(03)00011-6

Pardo, S., Suárez, H., & Soriano, E. (2006). Tratamiento de efluentes: Una vía para la acuicultura responsable. *Revista MVZ Córdoba, 11*(1)**,** 20-29. Retrieved from http://www.redalyc.org/articulo.oa?id=69309903

Ponce, J., Febrero, S., González, R., Romero, O., & Estrada, O. (2005). Perspectivas de la *Lemna* sp. para la alimentación de peces. *Revista Electrónica de Veterinaria, 6*(3), 1-6. Retrieved from http://agris.fao.org/agris-search/search.do?recordID=DJ2012035032

Rzedowski, G., & Rzedowski, J. (2005). *Flora fanerogámica del Valle de México*. Instituto de Ecología, A.C. y Comisión Nacional para el Conocimiento y Uso de la Biodiversidad, Pátzcuaro, Michoacán. Retrieved from http://www.scielo.org.mx/scielo.php?pid=S1870-34532007000200020&script=sci_arttext

Tacon, A., & Forster, I. (2003). Aquafeeds and the environment: Policy implications. *Aquaculture, 226*(1-4), 181-189. http://dx.doi.org/10.1016/S0044-8486(03)00476-9

Troell, M., Neori, A., Chopin, T., & Buschmann, A. (2005). Biological wastewater treatment in aquaculture–More than just bacteria. *World Aquaculture, 36*, 27-29.

UICN (Unión Mundial para la Naturaleza). (2007). *Guía para el Desarrollo Sostenible de la Acuicultura Mediterránea*. Interacciones entre la Acuicultura y el Medio Ambiente. Retrieved from http://www.apromar.es/noticias/general/UICN-SGPM-FEAP%20Guia-1.pdf

Zetina, C., Reta, C., Ortega, M., Ortega, E., Sánchez, M., Herrera, J., & Becerril, M. (2010). Utilization of the duckweed (*Lemnaceae*) in the production of tilapia (*Oreochromis* spp.). *Archivos de Zootecnia, 59*(R), 133-155.

11

Using an "Index of Merit" to Evaluate Winterhardy Pea Lines

Azize Homer[1] & Robin W. Groose[1]

[1] Department of Plant Sciences, University of Wyoming, Laramie, WY, USA

Correspondence: Azize Homer, Department of Plant Sciences, University of Wyoming, Laramie, WY, USA.
E-mail: ademirbas@hotmail.com

Abstract

Winter feed pea (*Pisum sativum* ssp. arvense) might serve as a partial or complete replacement for fallow in the winter wheat-summer fallow (WW-SF) system with potential to integrate cereal and livestock production in the Central Great Plains (CGP). The objective of this study was to evaluate advanced winter pea lines bred in the Wyoming environment in comparison with existing winter feed pea cultivars that were bred elsewhere. Six elite lines, one a blend of two lines, and three check cultivars were compared for overall merit, based on yield for forage and seed, and in two different production systems, dryland and irrigated, and at two locations (Lingle WY and Laramie WY) during the 2010-2011 and 2011-2012 growing seasons. Indices of merit, calculated in two ways: a mean-adjusted index and a standardized index, were used to simultaneously evaluate lines/cultivars for forage and seed yield. Based on the results from both indices, five Wyoming-bred elite lines (one a blend of two lines) ranked in the top five lines of 10 lines/cultivars tested. Importantly, three Wyoming-bred lines (Wyo#11, Wyo#11 +Wyo#13, and Wyo#13) all ranked significantly higher for overall merit than any existing winter feed pea cultivar tested in this study: 'Common', 'Specter' and 'Windham'. Because four measures of merit in the both indices are positively correlated no serious compromises or "trade-offs" are manifested among these four traits. This research shows that winter pea has potential value for forage and seed yield, mostly depending on growing season precipitation in the CGP.

Keywords: index of merit, selection, winter pea, breeding, yield, dryland farming

1. Introduction

Plant breeders usually consider selecting multiple traits in their crop improvement programs. New cultivars of a crop should perform similar to or better than existing cultivars to be acceptable by growers. Breeders may use one of the three methods of tandem selection, independent culling, and index selection to simultaneously select several traits (Hazel & Lush, 1943; Luby & Shaw, 2008; Acquaah, 2012).

An "index of merit" is essentially a "selection index", as originally defined by Smith (1936) which has been used widely by breeders, and which can take many forms (Simmonds & Smartt, 1999; Sleper & Poehlman, 2006; Acquaah, 2012). Hazel and Lush (1943) showed that selection for a total score or index of net desirability is much more effective than selection for one trait at a time.

Brown and Caligari (2008) state that "in almost all studies carried out it has been shown that index selection is more effective in identifying genotypes that are 'superior' for many different traits" in contrast to alternative methods such as tandem selection or independent culling. Acquaah (2012) notes that it is often the case that using the concept of "selection on total merit, the breeder would make certain compromises, selecting individuals [or in this study lines/cultivars] that may not have been selected if the choice was based on a single trait." Acquaah (2012) goes on to say that "An index by itself is meaningless, unless it is used in comparing several individuals [or in this study lines/cultivars] on a relative basis." A classic selection index takes the form

$$I = b_1x_1 + b_2x_2 + b_3x_3 + ... + b_nx_n \qquad (1)$$

Where, x_1, x_2, x_3, to x_n are the phenotypic performance for each line/cultivar for n traits of interest, and where b_1, b_2, b_3, to b_n are relative weights attached to the respective traits. Weights are often the respective relative economic importance of each trait. Acquaah (2012) refers to this as a "basic index" which is an additive index that may be used in cultivar assessment in registration trials. Brown and Caligara (2008) note that multiplicative selection indices are also possible, but they provide no examples of the use of such indices in plant breeding.

Jost et al. (2012) evaluated the effectiveness of selection indices (classic, base, parameters and weight free, based on desired gains, multiplicative, and rank sum) for the identification of inbred common bean lines with higher grain yield, desirable morphological and phenological traits, and better nutritional quality traits. They concluded that the classical, base and multiplicative indices provided superior genetic progress in the selection of inbred common bean lines. In this study, we consider only additive selection indices, which can be directly analyzed without transformation as linear additive models (LAM) of merit for yield via analysis of variance (ANOVA).

Sometimes, when different traits are measured with different units, or when the variance for different traits varies among lines/cultivars (i.e., heteroskedasticity), a selection index may be based on standardized measures of components of merit (Acquaah, 2012). Sleper and Poehlman (2006), in their discussion of "selection index", emphasize that the procedure often necessitates making "personal judgments" on the value to assign to each trait.

Simmonds and Smartt (1999) note that detailed economic analysis is seldom performed and that plant breeders often use an "intuitive selection index" over the course of a breeding program. These authors also discuss how "Index equations may be constructed on a purely economic basis or on a genetical one or on both". They note that a "genetic selection index" often starts from the assumption that characters are equally important and need no economic weighting (or, in other words, are weighted equally). That is the approach taken in this study.

The "intuitive selection index" of Simmonds and Smartt (1999) is related to the "concept of general worth" discussed by Acquaah (2012), where "a number of traits, which considered together, define the overall desirability of the cultivar" and where "yield of the economic product is almost universally the top priority." In this study, we focus on yield of forage and seed, and in dryland and irrigated systems.

Although all four yield traits, forage dryland (FD), forage irrigated (FI), seed dryland (SD), and seed irrigated (SI), were expressed in the same units, i.e., kg ha^{-1}, these can be considered distinctly different measures of merit (and specifically, yield) involving forage vs. seed, and dryland vs. irrigated, even if the underlying genetic basis for these traits is largely the same. Also, forage and seed yields were measured at different times, i.e., early summer for forage and mid-summer (at maturity) for seed.

Index selection may be used to test top ranking varieties in multiple environments before and after registration in order to assess their value for cultivation and use (Przystalski et al., 2008). The objective of this study was to identify the best lines/cultivars for dual use (forage vs. seed) in different production systems (dryland vs. irrigated, and where irrigated test may indicate potential maximal production in good years on dryland, when moisture is not limiting).

2. Materials and Methods

Diverse *arvense* genotypes were hybridized in the greenhouse, and natural and artificial selection began in an F_2 spaced-plant nursery. Selection continued among single plants within superior segregating families, and finally among bulked progenies of advanced lines, integrating elements of both pedigree selection and the bulk breeding method. As breeding populations were advanced from the F_2 through F_9 generations, the number of lines retained was reduced, as seed of elite, advanced lines was increased.

In the 2010-2011, and 2011-2012 winter annual growing seasons, seven Wyoming advanced breeding lines which were selected in the dryland WW-SF environment were evaluated together with three U.S. winter pea cultivars ('Common', 'Specter', and 'Windham', all from the Pacific Northwest) in Lingle (42°15′N, 104°20′W, elevation 1272 m), and Laramie, WY (41°18′N, 105°35′W, elevation 2184 m), under dryland and irrigated conditions in RCBD experiments. "Index of merit", related to "selection indices," was used to simultaneously evaluate the lines/cultivars for several traits including forage and seed yield.

The index of merit was calculated in two different ways; a "mean-adjusted index of merit" and "a standardized index of merit". In both cases, the indices of merit were based on relative measures of forage dryland (FD), forage irrigated (FI), seed dryland (SD), and seed irrigated (SI)), for the 10 lines/cultivars, and take this form,

$$I = .25x_{FD} + .25x_{FI} + .25x_{SD} + .25x_{SI} \qquad (2)$$

Where, the four measures of relative yield were weighted equally and the relative phenotypic values were calculated from data summarized in Table 1, where means are for 10 lines/cultivars for each of the four traits.

For the "mean-adjusted index of merit," for each line/cultivar, and for each trait i, x_i is a relative deviation from the mean for all 10 lines/cultivars, and presented as a percentage. Thus for a given line/cultivar, for each trait the phenotypic value takes this form,

$$x_i = [(\text{value for trait for line} - \text{mean for all lines})/\text{mean for all lines}] \times 100\% \qquad (3)$$

Calculated this way, the individual x_i values are positive and negative percent deviations from the overall mean for each trait, and the mean of x_i values is zero.

For the "standardized index of merit", for each line/cultivar, and for each trait i, x_i is a standardized deviation from the mean for all 10 lines/cultivars. Here the phenotypic value takes this form,

$$x_i = \text{(value for trait for line – mean for all lines)/standard deviation for all lines} \tag{4}$$

Calculated this way, the individual x_i values are positive and negative dimensionless deviations from the overall mean for each trait, and the mean of x_i values is 0, with a variance of 1, and is related to the Z-distribution in statistics, with deviations, x_i, normally distributed and where $\bar{x} = 0$ and $\sigma_x^2 = 1$ (Snedacor & Cochran, 1967; Acquaah, 2012).

Table 1. Four individual measures of merit for yield of winter peas for forage and seed under dryland and irrigated production systems

Lines/Cultivars	Forage DM yield (kg ha^{-1})		Seed yield (kg ha^{-1})	
	dryland	irrigated	dryland[†]	irrigated
Wyo #11	693.7 a [‡]	2739.4 c	953.8 a	2248.8 b
Wyo#11+Wyo#13	588.1 c	2858.9 b	849.3 ab	2518.3 a
Wyo #13	561.6 c	2987.8 a	982.3 a	1853.3 d
Wyo #8	479.8 e	2501.6 d	819.2 ab	2517.8 a
Wyo #6	639.8 b	1983.3 h	764.3 abc	2049.3 c
Common	530.4 d	2146.6 f	708.3 bc	1863.6 d
Wyo #12	438.8 f	2215.1 e	538.2 d	1854.6 d
Specter	380.5 g	2257.8 e	624.0 c	1721.5 e
Windham	376.9 g	1759.7 i	780.5 abc	1621.3 f
Wyo #10	375.9 g	2095.7 g	622.0 c	1381.5 g
Mean	506.6	2354.6	764.2	1963.0

Note. [†]Results from only 2010-2011 growing season combined over two locations; [‡] Values followed by the same letter in a column are not significantly different (p = 0.05) based on LSD.

There is no *a priori* knowledge of how locally bred and adapted winter pea cultivars might be adopted by the CGP producers. It cannot be predicted to what extent they might be grown for forage vs. seed, or in dryland vs. irrigated production systems. Therefore, FD, FI, SD, and SI were weighted, in the indices of merit equally, $b_{FD} = b_{Fi} = b_{SD} = b_{SI} = 0.25$ (as per Acquaah, 2012, where relative economic values of different measures of phenotypes were not known, and were therefore weighted equally).

For both indices of merit, summary data for the 10 lines/cultivars, and for the four yield traits, were analyzed with one-way ANOVA, randomized complete block design (RCBD), where main effects were lines and traits (FD, FI, SD, and SI). Here, traits may also be considered blocks.

3. Results and Discussion

Means for all four traits, forage dryland (FD), forage irrigated (FI), seed dryland (SD), and seed irrigated (SI) for the 10 lines/cultivars, were summarized in Table 1. Measures of these traits were based on mean performance in randomized, replicated trials of lines/cultivars over two years and at two locations, except for SD, which was measured only in the first year because severe drought at both locations prevented seed production at both locations in the second year. For each of the four traits, 160 plots were established, 640 plots overall. Means are least-square means as determined by ANOVA.

Although all yield data in Table 1 were presented in the same units, kg ha^{-1}, the traits represent different plant products harvested at different times (forage vs. seed) and under different production systems (dryland vs. irrigated). Standard deviations for the four traits were highly correlated with means, r = .9944 (Prob = .0028, n = 4), a classic indication of heteroscedasticity, and a strong indication to adjust data relative to means ("mean-adjusted index of merit") or standard deviations ("standardized index of merit"), as suggested by Acquaah (2012).

3.1 Mean-Adjusted Index of Merit

For the "mean-adjusted index of merit", deviations from means for the four yield traits (FD, FI, SD, and SI) for the 10 lines/cultivars are presented in Table 2 in units of plus or minus 1000 kg ha^{-1}.

Table 2. Deviations from means for the four yield traits for 10 winter pea line/cultivars

Lines/Cultivars	Forage		Seed	
	Dryland	Irrigated	Dryland	Irrigated
Wyo #11	187.1	384.8	189.6	285.8
Wyo11+Wyo13	81.5	504.3	85.1	555.3
Wyo #13	55.1	633.2	218.1	109.7
Wyo #8	-26.8	147.0	55.0	554.9
Wyo #6	133.2	-371.3	0.1	86.3
Common	23.9	-208.0	-55.9	-99.4
Wyo #12	-67.7	-139.5	-226.0	-108.3
Specter	-126.0	-96.7	-140.2	-241.5
Windham	-129.7	-594.9	16.4	-341.7
Wyo #10	-130.6	-258.9	-142.2	-581.5
Mean	0.0	0.0	0.0	0.0
Std Dev.	114.5	402.2	144.6	373.0

The ANOVA for the "mean-adjusted index of merit" showed that nearly 70% of variation is due to lines/cultivars (Table 3), indicating that; overall, the primary source of the variation observed in this study is genetic. For the "mean-adjusted index of merit," relative deviations from means for the four yield traits (FD, FI, SD, and SI) for the 10 lines/cultivars were presented as an overall percentage mean deviation for each line/cultivar, together with mean separations (Figure 1).

Overall means for "mean-adjusted index of merit" of line/cultivars ranged from 23.16% (line Wyo #11, rank 1) down to -21.25% (line Wyo #10, rank 10; Figure 1). All check cultivars of winter pea ('Common', 'Specter', and 'Windham') ranked in the bottom five of ten lines/cultivars. Five Wyoming-bred elite lines (one a blend of two lines) ranked in the top five lines of 10 lines/cultivars tested. Importantly, three Wyoming-bred lines (Wyo #11, Wyo #11 + #13, and Wyo #13) all ranked significantly higher (p = 0.05) for overall merit than any existing winter feed pea cultivar tested in this study: 'Common', 'Specter' and 'Windham'. Five of the seven Wyoming bred and selected lines out-performed both of the Palouse-bred cultivars ('Specter' and 'Windham').

Table 3. Sums of squares (SS), mean squares (MS) and percent contribution to total sums of squares (%TSS) of analysis of variance for indices of merit

Source	Mean-adjusted index of merit					Standardized Index of Merit				
	df	SS	MS	Pr > F	%TSS	df	SS	MS	Pr > F	%TSS
Line	9	9532	1059	< .0001	69.58	9	25.22	2.80	< .0001	70.07
Error	30	4167	139		30.42	30	10.77	0.40		29.93
Total	39	13699				39	35.99			

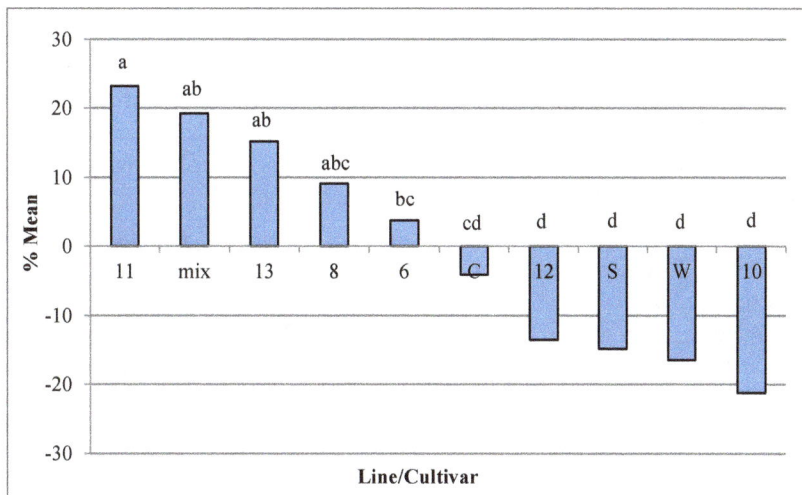

Figure 1. Mean-Adjusted Index of Merit for Wyoming-bred winter pea lines (numbered and "mix") in comparison with check cultivars 'Common' (C), 'Specter' (S), and 'Windham' (W)

Note. Lines with same letters are not significantly different at p = 0.05 based on LSD test.

3.2 Standardized Adjusted Index of Merit

For the "standardized index of merit," deviations from means for the four yield traits (FD, FI, SD, and SI) for the 10 lines/cultivars were the same as for the "mean-adjusted index of merit" as presented in Table 2 in units of plus or minus 1000 kg ha^{-1}.

As in "mean-adjusted index of merit," ANOVA for the "standardized index of merit" showed that 70% of variation is due to lines/cultivars (Table 3), indicating that, overall, the primary source of variation in these studies is genetic. For the "standardized index of merit," relative deviations from means for the four yield traits (FD, FI, SD, and SI) for the 10 lines/cultivars are presented graphically in Figure 2 together with mean separations.

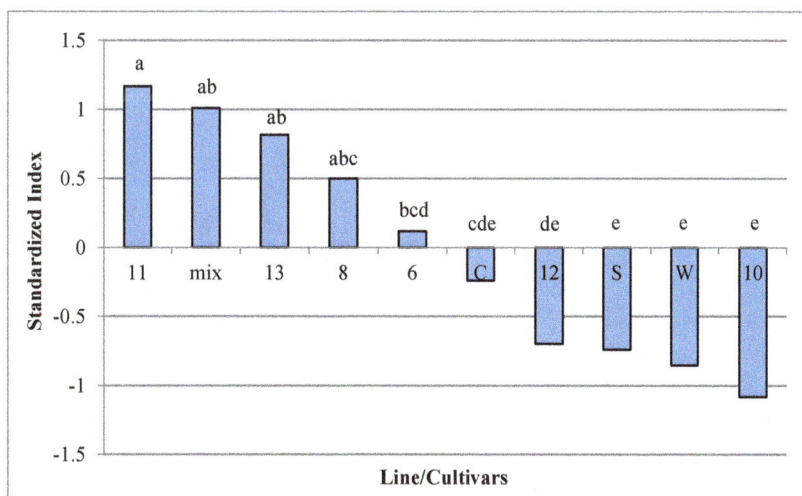

Figure 2. Standardized Index of Merit for Wyoming-bred winter pea lines (numbered and "mix") in comparison with check cultivars 'Common' (C), 'Specter' (S), and 'Windham' (W)

Note. Lines with same letters are not significantly different at p = 0.05 based on LSD test.

Here the means separation is slightly more refined than for the "mean-adjusted index of merit", with five, rather than four group means. Ranks remain the same. Again, all check cultivars of winter pea ('Common', 'Specter', and 'Windham') ranked in the bottom five of ten. Here, four Wyoming-bred lines (Wyo #11, Wyo Mix of #11 and #13,

Wyo #13, and Wyo #8) ranked significantly higher for overall merit than any existing winter feed pea cultivar tested in this study: 'Common', 'Specter' and 'Windham'.

In both "indices of merit", the two Wyoming lines that ranked in the bottom five of ten, Wyo #12 and Wyo #10, were those that were retained in the breeding program for the reason that they exhibited unique morphological traits (Wyo #12: no tendrils and green cotyledons; Wyo #10: clear seed coat). These were retained and tested in the interest of maintaining phenotypic diversity throughout the breeding program.

Regarding Wyo #11, Wyo #13, and the blend, Wyo #11+Wyo#13, we tested the blend because, in the course of breeding these lines, it became obvious early on that Wyo #11 and Wyo #13 were potentially superior lines (unpublished breeders' notes). Wyo #11 and Wyo #13 have different parentage and are morphologically distinct (wild-type and afilia leaf types, respectively). It has long been known that a mix of genetically superior lines may "overyield," (Harper, 1977), and as recently reviewed by authors who advocate "in-field" diversity in agroecosystems (Vandermeer, 2011; Connor et al., 2011; Denison, 2012). We did not observe over-yielding of the blend of Wyo #11 and Wyo #13. Rather, the blend yielded approximately midway between Wyo #11 and Wyo #13. (Table 1; Figures 1 and 2). Perhaps the merit of the blend may simply be considered further evidence for the superiority of elite breeding lines Wyo #11 and Wyo #13.

Correlations among the four traits, FD, FI, SD, and SI, based upon the both "mean-adjusted index of merit" and "standardized index of merit" were identical and presented in Table 4. Correlations ranged from r = .5132 to r = .6782, with all six pair-wise correlations significant at p = .05 or nearly so.

Table 4. Correlations among traits: Mean–adjusted index of merit, Standardized index of merit

	Mean-adjusted index of merit			Standardized index of merit		
	Irrigated Forage (FI)	Dryland Seed (SD)	Irrigated Seed (SI)	Irrigated Forage (FI)	Dryland Seed (SD)	Irrigated Seed (SI)
Dryland Forage (FD)	.5360 p = .0551	.6782 p = .0156	.6372 p = .0238	.5360 p = .0551	.6782 p = .0156	.6372 p = .0238
Irrigated Forage (FI)		.6607 p = .0188	.5799 p = .0394		.6607 p = .0188	.5799 p = .0394
Dryland Seed (SD)			.5132 p = .0646			.5132 p = .0646

Note. FD = Forage dryland, FI = Forage irrigated, SD = Seed dryland, SI = Seed irrigated.

These may be considered genetic correlations (Crow, 1986; Falconer & Mackay, 1996) because what plants within lines/cultivars have in common is their genes and difference among the diverse lines/cultivars is genetic. The lines/cultivars tested here are highly homozygous (F_9 generation for the Wyoming-bred lines) and are mostly homogeneous for the same alleles at most loci (via identity-by-descent), tracing back to single plant F_3 parentage in the case of the Wyoming-bred lines. Wyo #11 + Wyo #13 is a uniform blend of two lines where plants are highly homozygous, but there is heterogeneity due to the mix. The Palouse-bred cultivars ('Specter' and 'Windham') are highly homozygous and homogeneous based on their breeding history (McPhee & Muehlbauer, 2007; McPhee et al., 2007). 'Common' would also be highly homozygous but no F-generation can be specified because it does not trace back to a hybridization. Rather, 'Common' is an old land-race of Austrian winter pea, and can be considered a mix of inbred purelines.

Importantly, seed of every line/cultivar tested at Lingle and Laramie in 2010-2010 and 2011-2012 was from seed increased at Lingle in 2009-2010, except for seed of 'Specter' and 'Windham' which was fresh commercial seed supplied by Washington State Agricultural Experiment Station in Fall 2010. Use of uniformly produced fresh seed minimizes any differences among lines/cultivars due to age or quality of seed. Thus, performance differences among lines should be mostly due to genetics.

It was concluded that because the four measures of merit (forage dryland, FD; forage irrigated, FI; seed dryland, SD; and seed irrigated, SI) in the both "mean-adjusted index of merit" and "standardized index of merit" are positively correlated, and that no serious compromises or "trade-offs" were manifested among these four traits. It can be also noted that because selection over segregating generations after hybridizations was all conducted in dryland and mostly for seed production (SD), that correlations of SD with FD, FI, SI may be considered to be results of correlated response to selection for adaptation and overall yield potential (Simmonds & Smartt, 1999; Acquaah, 2012). In addition, production of forage and seed under irrigation may provide a measure production on dryland in very good years when moisture conditions are good (significant and timely precipitation) and water is not limiting.

3.3 Heritability of Merit

In quantitative genetics and breeding, variances are additive, such that

$$V_P = V_G + V_E + V_{G \times E} \tag{5}$$

Where, V_P is the phenotypic variance; V_G is the genotypic variance; V_E is the environmental variance; and $V_{G \times E}$ is the genotype by environment interaction variance. From the ANOVAs for merit, Table 3, we must conclude that approximately 70% of the variance among lines/cultivars for merit observed in this study is genetically based. The percent of total sums of squares for lines/cultivars, or V_{LINES}, is 69.58% in the ANOVA for the "means-adjusted index of merit" and 70.08% in the ANOVA for the "standardized index of merit" (Table 3). Either way, 70% of phenotypic (merit) variance, V_P, is genetic variance, V_G. For the remainder of this discussion we shall consider $V_G/V_P = 70\%$.

As discussed above, the advanced elite lines from the Wyoming feed pea breeding program, as well as the check cultivars, can be expected to be highly homozygous at all genetic loci, and for the same alleles at most loci (with the exception of 'Common' which is a "landrace" or "mixed line" consisting of different homozygous genotypes).

Plant breeders (e.g., Fehr, 1987; Sleper & Poehlman, 2006; Simmonds & Smartt, 1999; Brown & Caligari, 2008; Acquaah, 2012) and quantitative geneticists (e.g., Crow, 1986; Falconer & MacKay, 1996) define Broad Sense Heritability, H_B as

$$H_B = V_G/V_P \tag{6}$$

Where, V_G is the genotypic variance, and V_P is the phenotypic variance. In words, Broad Sense Heritability, H_B, is that proportion of phenotypic variance that is genetic, and where H_B ranges from 0 to 1 (or 0% to 100%). In this study, plants within lines/cultivars are highly "correlated" (genetically) due to the inbreeding process that gave rise to lines/cultivars (And, of course, we suppose Mendelian inheritance!).

We conclude that the Broad Sense Heritability, H_B, for merit among the 10 lines/cultivars in this study is 70%. This is actually a rather high heritability for a complex trait, such as plant yield (Falconer & MacKay, 1996). Plant yield is generally considered a continuously varying, polygenic trait that is sensitive to environment, and where genotype by environment interactions are likely (Fehr, 1987; Sleper & Poehlman, 2006; Simmonds & Smartt, 1999; Brown & Caligari, 2008; Acquaah, 2012).

The 30% of variance in ANOVA for merit, as measured by FD, FI, SD, and SI, and using both indices (Table 3) that is not included in the lines/cultivars variance (V_{LINE}) is all error variance (V_{error}). This could include randomly accumulated "noise" contributed by "microenvironmental" factors (e.g., soil heterogeneity among plots) and "macroenvironmental" factors (due to years, locations, dryland vs. irrigation, and forage vs. seed), and genotype by environment interactions and various higher order interactions of genotypes with environment (Brown & Caligari, 2009). However, overall macroenvironmental components of merit (yield), due to production system (dryland vs. irrigated) or product (forage vs. seed), are not main environmental effects here because FD, FI, SD, and SI were all mean-adjusted or standardized and, respectively, make no contribution to phenotypic variance in ANOVA for merit (Table 3).

All of the various V_E and $V_{G \times E}$ components that can contribute to the 30% "error variance", V_{error} in Tables 3 cannot be separated out, one from another. Nevertheless, 30% is the maximal possible contribution of genotype by environment interaction, $V_{G \times E}$, to phenotypic variation, V_P, that is possible in our indices of merit. $V_{G \times E}$, due to interaction of genotypes (lines/cultivars) with components of indices of merit (FD, FI, SD, and SI) can be, at most, 30% of phenotypic variance for merit, and is probably only a fraction of that.

We conclude that overall merit of lines/cultivars is essentially the same across the FD, FI, SD, and SI environments. Even if a substantial proportion of the 30% V_{error} in ANOVAs for merit is due to $V_{G \times E}$, V_{LINES} predominates at 70% of phenotypic variance, V_P. Thus, V_G for merit is most important here, and merit is highly heritable.

Plant breeders and geneticists often try to break down the genetic variance into various components where,

$$V_G = V_A + V_D + V_I \tag{7}$$

Where, V_G is the genetic variance; V_A is the additive genetic variance; V_D is the dominance genetic variance; and V_I is the interaction variance (epistasis).

It is not possible to break out these components here. Nevertheless, we speculate that the superiority of top Wyoming-bred lines in this study is primarily due to accumulation of favorable alleles at multiple loci with favorable epistatic interactions among loci during the breeding process that involved hybridization among diverse parents with subsequent natural and artificial selection for performance both among and within segregating populations on the CGP.

3.4 Which Index of Merit Is Most Useful?

We calculated indices of merit in two ways: first, a mean-adjusted index and, second, a standardized index. ANOVA (Tables 3) separated means of lines/cultivars as illustrated graphically in Figures 2 and 3. The four measures of merit (FD, FI, SD and SI) were highly genetically correlated for both indices (Table 4), and in fact, correlations were identical for the two indices. The standardized index was slightly more efficient in separating means with five mean groups (Figure 2) versus four mean groups for the mean-adjusted index (Figure 1). ANOVA of both indices of merit indicated that a broad-sense heritability of merit, H_B, is 70%. The standardized index might be considered superior because it is theoretically more robust (Acquaah, 2012). However, the mean-adjusted index may be more useful in communicating results to producers because results are presented as percent deviations from a mean of 0. Growers might reasonably expect elite Wyoming-bred lines to produce 20% more than existing cultivars (Figure 2). Most producers would be unfamiliar with dimensionless standardized variables, and might misinterpret Figure 2 to indicate that our top lines are twice as productive as average lines. We conclude that both indices are useful and that the choice of index to present may depend upon the audience.

4. Conclusions

We conclude that breeding in the CGP produced locally-adapted winter pea lines with good yield for both forage and seed, and under both dryland and irrigated conditions. Our elite lines, especially Wyo#11 and Wyo#13, proposed for release were well-adapted to Wyoming and yield well under different conditions. "Indices of Merit" proved to be useful for comparing lines/cultivars for multiple use in sustainable agroecosystems.

Acknowledgements

We are thankful to Dr. James M. Krall and Jerry J. Nachtman for their constructive ideas and technical help in conducting the field research at the Wyoming Agricultural Experiment Station in Lingle.

References

Acquaah, G. (2012). *Principles of plant genetics and breeding* (2nd ed.). Wiley, Chichester, UK. http://dx.doi.org/10.1002/9781118313718

Brown, J., & Caligari, P. D. S. (2008). *An introduction to plant breeding*. Blackwell, Oxford. http://dx.doi.org/10.1002/9781118685228

Connor, D. J., Loomis, R. S., & Cassman, K. G. (2011). *Crop ecology: Productivity and management in agricultural systems* (2nd ed.). Cambridge Univ. Press, Cambridge. http://dx.doi.org/10.1017/cbo9780511974199

Crow, J. F. (1986). *Basic concepts in population, quantitative and evolutionary genetics*. Freeman, New York.

Denison, R. F. (2012). *Darwinian agriculture: How understanding evolution can improve agriculture*. Princeton Univ. Press, Princeton.

Falconer, D. S., & MacKay, T. F. C. (1996). *Introduction to quantitative genetics* (4th ed.). Longman, London.

Fehr, W. R. (1987). Principles of cultivar development. *Theory and technique* (Vol. 1). McGraw-Hill, Inc., New York.

Harper, J. L. (1977). *Population biology of plants*. Academic, London.

Hazel, L. N., & Lush, J. L. (1943). The efficiency of three methods of selection. *J. Hered., 33*, 393-399.

Jost, E., Ribeiro, N. D., Maziero, S. M., Possobom, M. T. D. F., Rosa, D. P., & Domingues, L. S. (2013). Comparison among direct, indirect and index selections on agronomic traits and nutritional quality traits in common bean. *J. Sci. Food Agr., 93*, 1097-1104. http://dx.doi.org/10.1002/jsfa.5856

Luby, J. J., & Shaw, D. V. (2009). Plant breeders' perspectives on improving yield and quality traits in horticultural food crops. *HortScience, 44*, 20-22.

McPhee, K. E., & Muehlbauer, F. J. (2007). Registration of 'Specter' winter feed pea. *Crop Sci., 1*, 118-119. http://dx.doi.org/10.3198/jpr2006.12.0826crc

McPhee, K. E., Chen, C., Wichman, D., & Muehlbauer, F. J. (2007). Registration of 'Windham' winter feed pea. *Crop Sci., 1*, 117-118. http://dx.doi.org/10.3198/jpr2006.12.0828crc

Przystalski, M., Osman, A., Thiemt, E. M., Rolland, B., Ericson, L., Østergård, H., ... Krajewski, P. (2008). Comparing the performance of cereal varieties in organic and non-organic cropping systems in different European countries. *Euphytica, 163*, 417-433. http://dx.doi.org/10.1007/s10681-008-9715-4

Simmonds, N. W., & Smart. J. (1999). *Principles of Crop Improvement* (2nd ed.). Blackwell Science, Oxford.

Sleper, D. A., & Poehlman, J. M. (2006). *Breeding field crops* (5th ed.). Blackwell, Ames, IA. http://dx.doi.org/10.1007/978-94-015-7271-2

Smith, H. F. (1936). A discriminant function for plant selection. *Ann Eugen, 7*, 240-250. http://dx.doi.org/10.1111/j.1469-1809.1936.tb02143.x

Snedecor, G. W., & Cochran, W. G. (1967). *Statistical methods* (6th ed.). Iowa State Univ. Press, Ames.

Vandermeer, J. H. (2011). *The ecology of agroecosystems*. Jones and Bartlett, Sudbury MA.

12

Effect of Cassava Mosaic Disease (CMD) on Yield and Profitability of Cassava and Gari Production Enterprises in Sierra Leone

Alusaine E. Samura[1], Kepifri A. Lakoh[2], Osman Nabay[1], Sahr N. Fomba[2] & James P. C. Koroma[2]

[1] Sierra Leone Agricultural Research Institute (SLARI), Njala Agricultural Research Centre (NARC), Freetown, Sierra Leone

[2] Crop Protection Department, School, of Agriculture, Njala University, Freetown, Sierra Leone

Correspondence: Alusaine E. Samura, Sierra Leone Agricultural Research Institute (SLARI), Njala Agricultural Research Centre (NARC), Freetown, P.O. Box 540, Sierra Leone. E-mail: aesamura@yahoo.com

The research is financed by the Government of Sierra Leone through the West Africa Agricultural Productivity Programme (WAAPP).

Abstract

Cassava Mosaic Disease (CMD) seriously affects cassava (Samura et al., 2013; Fargette et al., 1988). There is limited information on tuberous root yield loss and the profitability of growing improved and local varieties infected with the cassava mosaic virus for tuber and gari production in Sierra Leone. The objectives of the study were to determine yield loss associated with cassava mosaic disease and conduct cost benefit analysis (budgets and break-even analysis) on cassava production using two improved varieties (SLICASS 4 and 6) compared against the local susceptible variety Cocoa, for gari production and Cocoa as a poundable enterprises for the boil and eat market. Activity 1 involved the establishment of a yield loss trial using the paired plot technique. Activity 2 assessed productivity of cassava and gari production. Data collected were analysed using the analytical frame work that incorporates the concept of gross margin analysis as describe by Brown (1979). This included cost benefit analysis (CBA), the net social benefit (NSB) or the excess of total benefit over total cost represents the net present value (NPV) and The Internal Rate of Returns (IRR).

The yield loss associated with the local variety Cocoa under this system was 4.27 t/ha which is equivalent to 38.92% yield loss resulting from the cassava mosaic disease infection. SLICASS 4 and SLICASS 6 recorded positive returns to production of tubers and gari. The local variety Cocoa enterprises for gari under the same condition recorded a negative returns even in the 6 year. However Cocoa for the boil and eat market had the highest profit level. The implication of this study is that breeding effort should be geared towards high yielding mosaic resistant varieties that are poundable more profitable and suits the cultural and domestic demand of the producers, processors and consumers.

Keywords: cassava mosaic disease, cassava genotypes, yields loss, profitability

1. Introduction

Cassava remains to be a major food security crops in Sierra Leone and is regarded as the most important root and tuber crop in the country (Fomba et al., 2012). Despite positive attempts by government to revive the agricultural sector, post-war levels of productivity are still very low, estimated at 7 t/ha (FAO, 2014). The tuber, which is a vital part of several local cuisines in West Africa, does not only serve as a major input in the production of food items, but also serves as a highly viscous and dense starchy input for several industries (in-edibles). The leaves are mainly used as vegetables (when cooked) and are sometimes used as cattle/sheep fodder. Over the last three decades (pre and post-war Sierra Leone), various types of research activities have been ongoing in the country. The Njala Agricultural Research Center (NARC), formally, Institute of Agricultural Research, has been conducting various types of research activities on cassava in the country. All of these strides were geared towards coming up with high yielding, disease free, drought tolerant and consumer friendly varieties for onward dissemination to smallholder farmers in the country. However, in spite of all these efforts, there continues to be severe challenges affecting the crop's productivity and hence, the rate of varietal adoption on farmer fields. The most important disease and leading constraint of cassava production in Africa is the African Cassava Mosaic Disease (ACMD) (Fregene et al., 2,000; Berry & Rey, 2001). Farmers in Sierra Leone continue to cultivate local varieties infected with ACMD. A necessary first step to improved crop varietal selection by potential adopters (farmers) is to provide empirical answers to some fundamental socioeconomic issues that would encourage farmers to adopt new technologies. Some include identification of genotypes with traits that meets farmers demand. Also, the potential profits farmers stand to gain should they adopt the improved varieties against following their traditional varieties. Furthermore, and in line with the value chain approach promoted by the West African Agricultural Productivity Program (WAAPP), these improved crop varieties should be promoted beyond the tuber stage, to more value added products or enterprises like chips, gari, fufu, toa etc. Understanding the potential gains (in financial terms) that entrepreneurs stand to gain should they utilize the improved varieties to process gari, fufu or other up-stream products in vital for adoption.

It is against this background this research is carried out. The objectives of this research are therefore listed below:

1) To determine the effects of ACMD on the yield and yield components of cassava in Sierra Leone.

2) Conduct cost benefit analysis (budgets and break-even analysis) on cassava production in Sierra Leone for SLICASS 4 and SLICASS 6 compared against the local variety (Cocoa) used for gari and Cocoa used as a poundable variety.

3) Conduct cost benefit analysis of gari production as an upstream enterprise.

2. Method

Two activities were undertaken to determine the loss associated with the cultivation of cassava mosaic infected varieties. The first activity was geared towards determining yield loss in terms of tuberous root yield as a result of the African cassava mosaic disease. The second activity compared income derived from processing ACMD resistant and ACMD susceptible varieties per unit area of cassava.

2.1 Activity 1: The Effects of Cassava Mosaic Disease (ACMD) on Yield and Yield Components of Infected and Symptomless Cuttings Derived from the Local Cassava Variety (Cocoa) in Sierra Leone

2.1.1 Preparation of Cuttings

The local variety (Cocoa) known to be infected by the Cassava Mosaic Virus was used for this trial. Symptomless plantlets derived from the apical meristem of neem treated cassava cuttings were raised from three generations and multiplied for use in the cassava mosaic free plots. Before planting, cuttings were cut into 30 cm length and treated for 10 min in solution of benomyl (2.5 g a.i/L of H_2O and planted in polythene bags in the screen house for symptom observation before establishment field trials.

2.1.2 Experimental Design

The paired plot technique was used to compare cassava yield of mosaic infected and the mosaic free in three replications.

2.1.3 Data Collection

Data were collected from all plants. Data included plant height measured in centimeters from the surface of the pot to the top most expanded leaf using a well calibrated meter rule. Leaf number was calculated by counting the number of leaves at each sampling period. Cassava Mosaic Disease assessment was done using the 1-5 scale (IITA, 1990) where 1 represents no symptom expression and 5, severe symptom expression. Tuberous root yield was determined by harvesting forty plants to determine the weight per plot, and ten plants to determine the number of tubers, weight of tubers per plant. Leaf fresh weight was determine by weighing ten plants to determine the weight per plant, while the whole plot was used to estimate weight of leaf per plot and number of bundles per plot.

2.1.4 Data Analysis

Cassava Mosaic virus infected plots and mosaic symptomless plot were compared using analysis of variance and mean separated using the least significant difference test (LSD). Genstat statistical software version 8 was used to analyse the data.

2.2 Activity 2: Profitability of Cassava Mosaic Infected Variety and Cassava Mosaic Resistant variety in Gari Production Enterprise

2.2.1 Data Collection

Primary and secondary data were collected. Primary data collected include a detail investigation on the production systems, utilization and marketing of three cassava varieties *i.e.* Cocoa (Local variety), SLICASS 6 and SLICASS 4 (improved varieties). Secondary information includes the relevant literature from other studies on cassava cultivation, processing and marketing. Primary data were collected in 2013 cropping season in collaboration with Village Hope, an American Non-Governmental organisation working with the Njala Agricultural Research Center (NARC) on cassava processing. This study involved thirty farmers, ten farmers each growing the mosaic resistant cassava variety Slicass 4 and 6 and another ten set of farmers growing the local cassava mosaic infected variety, Cocoa. Cassava tubers were bought from farmers' fields and harvested and transported to the processing centre using a Ford F250 truck. Harvested area from each group of farmers was calculated, tubers weighed before processing and after peeling, to determine the peeling loss. After grating and pressing mashed tubers, decanted weight of mash and total weight of gari produced were also recorded. Detailed records of the running costs inputs, and operational cost on the cultivation, processing and marketing of those varieties were taken on each batch processed. This was done to compare their inputs cost and gross farm gate benefits between the two cassava varieties.

2.2.2 Method of Data Analysis

Data collected were analysed using profitability, cost benefit ratio and simple descriptive statistics analysis. Profitability analysis uses the analytical frame work that incorporates the concept of gross margin analysis as describe by Brown (1979) and profit was employed to estimate the costs and returns of cassava and gari production in the study area. The Gross Margin formula is represented as:

$$GM = GI - TVC \tag{1}$$

Where,

GM = Gross margin; GI = Gross income; TVC = Total variable cost.

The profitability was also represented symbolically by,

$$\pi = TR - TC \tag{2}$$

Where,

π = profit; TR = Total revenue/gross income; TC = Total cost [total fixed cost (TFC) + Total variable cost (TVC)].

To calculate the Gross Margin (GM), the Total Variable Cost (TVC) was computed by aggregating the cost of cultivation to have the roots, processing and marketing components. Processing costs included the cost of carrying out the activities in the process flow of producing the products. For instance, gari production involved cost of roots and its transportation/handling (loading and off-loading) charges, peeling, washing, grating, pulverizing, and toasting (frying). Similarly, marketing costs involved bagging, cost of packaging materials (bags, polyethylene). The enterprise Total Revenue (TR) was computed by multiplying the quantity (Q) of processed product from 1 hectare of cassava farm by the price (P) *i.e.* Q×P. To assess the profitability of the two different cassava varieties in the study area, the means of the two varieties were compared using single factor analysis of variance (ANOVA).

The following assumptions were true for the study:

Table 1. Economic assumptions used for profitability analysis of gari enterprise using improved and local varieties from 1 ha of cassava farms

Assumption	Cocoa	Slicass 4	Slicass 6
Cost/Bag	15,000	10,000	10,000
Number of bags processed	20	20	20
Average weight per bag (Kg)	48	55	51.8
Average weight of tubers processed	950	1,100	1,037.7
Average yield per ha	7.1	24.82	23.18
Gari/tuber ration	0.18	0.20	0.19
Truck mpg	12	12	12
Cost of gasoline per gallon	25,000	25,000	25,000
Cost per mile for gas	2,083	2,083	2,083
Multiplier	1.5	1.5	1.5
Total cost per mile	3,125	3,125	3,125
Truck mile	30	30	30
Number of managers	3	3	3
Number of staff	19	19	19
Selling price	2,500	2,500	2,500

From the institute's breeding activities over the last ten years, the Sierra Leone Agricultural Research Institute has formally released fourteen improved cassava varieties. These varieties are hybrid forms with several improved traits ranging from pest and disease resistance, improved yield, high dry matter content poundability and more. Due to some data limitations, we would focus on the two most popular varieties SLICASS 4 and SLICASS 6 out of the fourteen SLICASS cassava varieties already released for the research. The main data requirements for this analysis were:

1) Input and output data for the different varieties of cassava;

2) Price data of the different inputs;

3) Market prices;

4) Input and output data for gari production and marketing.

For this research, input and output figures were obtained from following a group of 30 farmers through two growing seasons and keeping track of input quantities and harvests. These farmers grew all varieties tested independently. These farmers were based in the Njala Community, Mano dase and the surrounding villages and hence, received prompt technical support from researchers when needed. Price data were obtained from ongoing market prices of the respective inputs and the crops.

2.2.3 Net Present Value

In cost benefit analysis (CBA), the net social benefit (NSB) or the excess of total benefit over total cost represents the net present value (NPV). Therefore, in determining the NPV of a technology of improved cassava varieties, the cost (C) and benefit (B) is needed. The NSB is calculated by subtracting the cost stream from the benefit stream and is represented as follows: NSN = B – C (Gittinger, 1995).

In determining the NPV, applied a discount rate to the identified costs and benefits. It is necessary to discount costs and benefit occurring later relative to those occurring sooner. This is because, money received now can be invested and converted into a larger future amount and because people generally prefer to receive income now rather in the future (Hanley & Spash, 1995). A future value into present values is known as discounting. Discounting requires knowledge of a rate known as a discount rate. The discount rate does not necessarily equate to the rate of interest offered at banks. The formula for discounting is as follows:

$$PV = \frac{X_t}{(1 + r)^t} \qquad (3)$$

Where,

PV is the present value; X_t is the amount of money in year t; r is the rate gof discount (expressed as a proportion, i.e. 10% = 0.1), and t is the number of years from the present date.

When all benefits and costs are converted to present values, comparison is possible. Comparison is commonly made using three decision-making criteria: net present value, internal rate of return, and the benefit-cost ratio.

Internal Rate of Return

An alternative approach to assessing an investment is the calculation of the internal rate of return (IRR) of the investment. The IRR is the discount rate that will make the net present value equal to zero. In simpler terms, the IRR is the interest rate that will make the investment just break even Ingersoll Jr and Ross (1992). If the IRR exceeds the minimum acceptable discount rate or the opportunity cost of money, the project is worth further consideration.

Investors would have to invest NPV more (a total of NPV + already invested amount) to get the cash flows shown at the said interest rate. Therefore the project has a value of NPV for investors. The interest rate is called the cost of capital, because it is the opportunity cost of funds—The rate investors can earn on alternative investments. A project is chosen if it costs less than the PV of its cash flows. More generally: take a project if its Net Present Value is positive.

3. Results

3.1 Activity 1: The Effects of Cassava Mosaic Disease (ACMD) on Yield and Yield Components of Infected and Symptomless Cuttings Derived from the Local Cassava Variety (Cocoa) in Sierra Leone

3.1.1 Disease Severity Expression of Cassava Mosaic Treated and Non-Treated Cassava Cuttings over Time

Cassava Mosaic Disease (CMD) treated cuttings in the protected plot showed highly significant ($P_{0.05} < 0.001$) reduction in the disease symptom expressed compared to the unprotected plot. Mean severity score for the protected plot was 1.1 while the unprotected plot was 2.9. Manifestation of symptoms appeared between 6 to 9 months after planting for the protected plot with cuttings raised from symptomless plants. In the ACMV infected plot disease symptom expression was as early as 1 month after planting. Disease severity significantly increased with time with 12 months after planting (MAP) accounting for the highest severity levels of 3.7 in the unprotected plot. In all the period of assessment the unprotected plot had higher severity scores compared to the protected plot (Table 2).

Table 2. ACMD severity of cassava mosaic treated and non-treated cassava cuttings as affected by months after planting in 2014

| Treatment | Months after planting (MAP) | | | | | Mean |
	1	3	6	9	12	
ACMD protected plot	1.0	1.0	1.0	1.2	1.3	1.1
ACMD infected plot	2.0	2.0	3.3	3.3	3.7	2.9
Mean	1.5	1.5	2.2	2.3	2.5	
$LSD_{0.05}$ (Treatment)**	0.16					
$LSD_{0.05}$ (MAP)**	0.25					
$LSD_{0.05}$ (Treatment × MAP)**	0.35					
CV (%)	10.4					

Note. ** = Significantly different at $P_{0.05} < 0.001$.

3.1.2 Yield and Yield Component of Cassava Mosaic Disease Treated and Non-Treated Cassava Cuttings

(1) Mean Number of Tubers per Plant

The protected plot with symptomless cuttings produced an average of 3.67 tubers per plant significantly ($P_{0.05} < 0.002$) higher than the unprotected plot which produced an average of 2.0 tubers (Table 3).

(2) Number of Tubers per Plot

A similar trend was observed in terms of total number of tubers produced per plot as in the number of tubers produced per plant. The protected plot produced an average of 107 tubers per plot when harvested at 12 months after planting and was significantly ($P_{0.05} < 0.011$) higher than the unprotected plot which produced an average of 71.7 tubers (Table 3).

(3) Average Weight of Tubers per Plant

The protected plot with symptomless cuttings produced an average weight of 0.41 kg per plant not significantly ($P_{0.05} < 0.093$) higher than the unprotected plot which produced an average 0.37 kg per plant (Table 3).

(4) Weight of Tubers per Plot

The protected plot produced an average weight of 43.9 kg per plot when harvested at 12 months after planting and was significantly($P_{0.05} < 0.007$) higher than the unprotected plot which produced an average of 27.1 kg (Table 3).

(5) Tuberous Root Yield (t/ha).

Tuberous root yield for the protected plot, from which symptomless plant were raised, produced 10.97 t/ha significantly ($P_{0.05} < 0.001$) higher than the ACMD infected plot which produced 6.70 t/ha.

(6) Yield Loss Associated with Cassava Mosaic Disease from the Local Variety Cocoa

Result from the tuberous yield of infected and symptomless plots revealed that symptomless plot had 10.97 t/ha while the infected plot had 6.7 t/ha. The yield loss associated with the local variety Cocoa under this system was 4.27 t/ha which is equivalent to 38.92% yield loss resulting from the cassava mosaic disease infection.

Table 3. Yield and yield component of cassava mosaic treated and non-treated cassava cuttings

Treatment	Average Number of tubers per plant	Number of tubers per plot	Average Weight of tubers per plant	Weight of tubers per plot	Yield (t/ha)
ACMD protected plot	3.67	107	0.41	43.9	10.97
ACMD infected plot	2.0	71.7	0.37	27.1	6.7
Mean	2.8	89.3	0.39	35.5	8.8
$LSD_{0.05}$ Treatment	0.286**	16.29*	6.06 NS	1.95**	0.66**
CV (%)	2.9	5.2	3.8	4.9	2.1

Note. * = Significantly different at P < 0.005; ** = significantly different at $P_{0.005} < 0.001$; NS =Not significantly different.

3.2 Activity 2: Profitability of Cassava Mosaic Infected Variety and Cassava Mosaic Resistant Varieties in Gari Production

In this section, the key results obtained from our analysis include cassava production and profitability, followed by gari profitability. This includes net annual cash flow for the different varieties assessed and the net present value of the enterprises. Four scenarios were examined. These were Slicass 4; Slicass 6, Cocoa used for the gari market and Cocoa used for the poundability market (boil and eat market). The main thrust of this economic analysis is to identify the profitability of the different enterprises by comparing profitability of the mosaic resistant varieties (SLAICASS 4 and 6) to the local mosaic infected variety (Cocoa) as stated above.

3.2.1 Net Annual Cash Flow of SLICASS 6, 4 and Cocoa for Gari Production and Cocoa for the Boil and Eat Market in Cassava Production

Table 4 shows the net annual cash flow that farmers get when they grow cassava for any of the four enterprises (SLICASS 6, 4 and Cocoa for gari production and cocoa for the boil and eat market). It also shows the break even points. That is, the year the different enterprises start recording positive returns. The result indicated that farmers growing SLICASS 4 and SLICASS 6 at a buying price of Le 8,000 per 50 kg bag (Le160/kg) for the gari industry started recording positive returns from the first year of operation, while the local variety Cocoa enterprises for gari under the same condition recorded a negative returns even in the 6 year. However Cocoa for the boil and eat market at a price of Le 40,000 commanded had the highest profit level compared to SLICASS 4 and SLICASS 6.

Table 4. Net annual cash flow of improved cassava genotype SLICAS 6, 4 and Cocoa used for gari and Cocoa used for it poundability

	Cash Flow (Le) Showing Break-Even Point					
	Year 1	Year 2	Year 3	Year 4	Year 5	Year 6
Slicass 6	241,905	230,385	219,415	208,966	199,016	189,539
Slicass 4	409,524	390,023	371,450	353,762	336,916	320,873
Cocoa-Garri	-777,143	-740,136	-704,891	-671,325	-639,357	-608,912
Cocoa-Poundability	3,763,810	3,584,580	3,413,886	3,251,320	3,096,495	2,949,043

Note. For the, Cocoa variety, market prices vary depending on the type of demand. Huge volumes are sold to the gari industry at Le 8,000 per 50 kg (Le160/kg) bag while small stocks are sold at the local market at Le 40,000 per bag (Le 800/ kg).

Exchange rate: 1$ = Le 5,582.23.

3.2.2 Net Present Value (NPV) at Different Interest Rates for SLICASS 4, 6 and the Local Cassava Variety Cocoa for Gari and Cocoa for Poundability

Market dynamics change regularly. This makes profit estimates not an accurate measure that can be used to advise adoption of an enterprise or technology. A more proper measure is the Net Present Value. That is the value of an enterprise in today's conditions. The table 5 below presents NPV estimates for the four enterprises considered in this study using the tubers, leaves and stems as marketable upstream products.

Table 5. Net present value at different interest rates for SLICAS 4, 6 and the local cassava variety Cocoa for gari and Cocoa for poundability

Cassava Varieties	Annual discount (%) (Le)					
	10%	12%	14%	16%	18%	20%
Slicass 4	1,312,074	1,219,898	1,136,509	1,060,877	992,117	727,656
Slicass 6	756,766	702,639	653,691	609,315	568,987	414,082
Cocoa for gari	-2,619,249	-2,442,061	-2,281,623	-2,135,983	-2,003,451	-2,003,451
Cocoa for Poundability	13,370,041	12,424,528	11,570,957	10,798,355	10,097,267	9,459,524

Note. Exchange rate: 1$ = Le 5,582.23.

At the going market interest rate (between 10 and 20%) SLICASS 4 SLICASS 6 and the Cocoa used as an enterprise for poundable variety were profitable with the poundable Cocoa enterprise having the highest value. The Cocoa enterprise when used for gari was not profitable at any given interest rate having the lowest value. This means that this enterprise is clearly unfavorable. The SLICASS 4 enterprise had a higher value compared to SLICASS 6. This is further projected in the Figure 1 below:

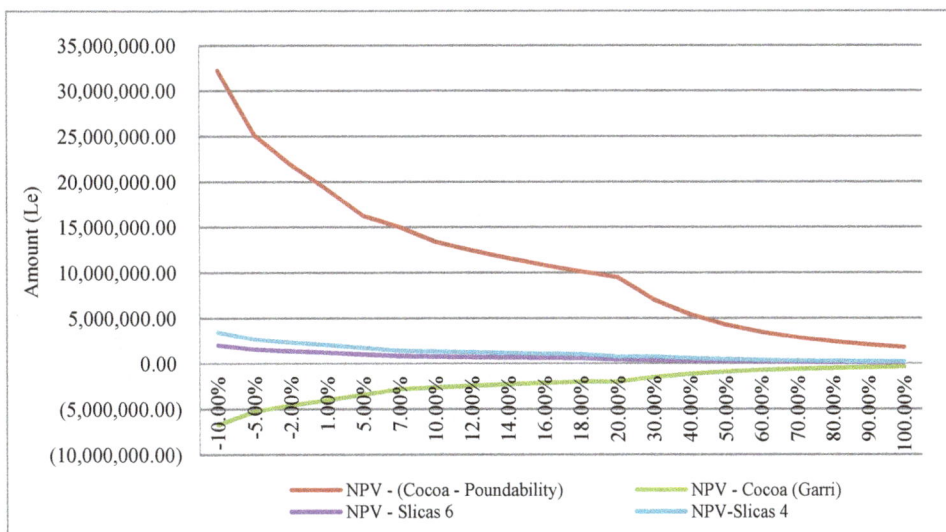

Figure 1. Graph showing NPV for SLICAS 4, 6, Cocoa used for gari and Cocoa used for poundability

Note. Exchange rate at the time of the study 1$ = Le 5,582.234.

3.2.3 Net Annual Cash Flow for Gari Processing and Sales in Freetown, Njala and with World Food Programme (WFP)

This section presents data on gari profitability (Market Effect) irrespective of the variety used. The assumption here is that there was little difference in the fresh tuber and gari conversion ratio for Cocoa, Slicass 4 and 6. Therefore for equal amount of fresh tuber processed, estimate of gari production was similar for each variety.

(1) Gari Profitability (Market Effect) Cash Flow for Gari Enterprise at World Food Programme (WFP), Freetown and Njala Prices

Based on cost of gari per kg at Njala, Freetown and the World Food programme (WFP) price at the time of the study which was Le 2,500, le 3,500 and Le 2,500 respectively for each location, it was not profitable to sell at Le 2,500 at Njala due to the high cost of packaging, and transportation incurred at the processing centre. Under such condition profit can only be realized at the seventh month of operation unlike the situation in Freetown where cash flow was Le13, 363,045.7 at the first year at the price of Le 3,500 per kg despite the same level of expenditure as Njala. The WFP scenario presented a better option since the product is sold in bulk with no packaging cost, marketing cost and transportation cost. Under this circumstance the cash flow was calculated at Le 18,149,125.7 in year 1 (Table 6).

Table 6. Gari profitability (Market Effect) cash flow for gari enterprise at world food programme, Freetown and Njala prices (by year)

Year(s)	Cash Flows (Le)		
	WFP	**Freetown**	**Njala**
1	**18,149,125.7**	**13,363,045.7**	-20,139,162.3
2	60298251.39	50726091.39	-16278324.61
3	102447377.1	88089137.09	-12417486.91
4	144596502.8	125452182.8	-8556649.216
5	186745628.5	162815228.5	-4695811.52
6	228894754.2	200178274.2	-834973.824
7	271043879.9	237541319.9	**3025863.872**
8	313193005.6	274904365.6	6886701.568
9	355342131.3	312267411.3	10747539.26
10	397491257	349630457	14608376.96
11	439640382.7	386993502.7	18469214.66
12	481789508.4	424356548.4	22330052.35

Note. Exchange rate: 1$ = Le 5,582.23.

This is further projected on a monthly basis where profit of Le 586,990 was observed in the 7th month of operations for the WFP scenario and a profit of Le 908,697 observed in the 8th month of operation for the Freetown market (Table 7).

Table 7. Cash flows for gari enterprise for WFP and Freetown markets based on locations (by month)

Month	Cash Flows (Le)	
	WFP	**Freetown**
1	-20,487,573	-20,886,413
2	-16,975,146	-17,772,826
3	-13,462,719	-14,659,239
4	-9,950,291	-11,545,651
5	-6,437,864	-8,432,064
6	-2,925,437	-5,318,477
7	**586,990**	-2,204,890
8	4,099,417	**908,697**
9	7,611,844	4,022,284
10	11,124,271	7,135,871
11	14,636,699	10,249,459
12	18,149,126	13,363,046

Note. Exchange rate: 1$ = Le 5,582.23.

(2) Break-Even Point from Total Revenue (TR) and Total Cost (TC) for Gari Production

Break- even point for processing of cassava to gari as an enterprise can only be attained at the 8th year of operation primarily due to high capital investment. At this point total revenue derived from gari was Le1, 200,000,000 while total cost was Le 1,193,113,298 when 1 ton of fresh tubers were processed daily for the 8 year period (Table 8).

Table 8. Break-even point from Total Revenue (TR) and Total Cost (TC) for gari production using the local variety Cocoa

Years	Total Revenue	Total Cost
0		24,000,000
1	150,000,000	170,139,162.3
2	300,000,000	316,278,324.6
3	450,000,000	462,417,486.9
4	600,000,000	608,556,649.2
5	750,000,000	754,695,811.5
6	900,000,000	900,834,973.8
7	1,050,000,000	1,046,974,136
8	1,200,000,000	1,193,113,298
9	1,350,000,000	1,339,252,461
10	1,500,000,000	1,485,391,623
11	1,650,000,000	1,631,530,785
12	1,800,000,000	1,777,669,948

(3) Net Present Value (NPV) for Gari Production and Marketing in Different Locations (WFP, Freetown and Local)

Net Present Values (NPV) shows that it is profitable to market gari in Freetown and or supply to the World Food Programme at an interest rate of 10% to 60% while local sale at Njala would not be profitable even if the enterprise is subsidized (Figure 2).

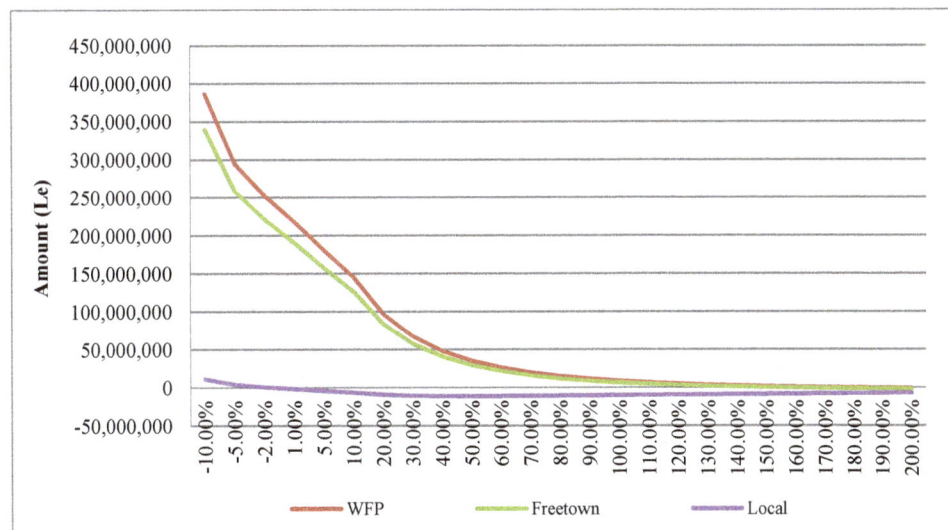

Figure 2. NPV for garri production and marketing in different locations (WFP, Freetown and local)

4. Discussion

Net Cash Flow which is the profits of the business plus non-cash expenses was used to determine profitability of gari enterprise from mosaic infected and mosaic resistant cassava varieties. In this study an attempts was made to show the effect of cassava mosaic disease on root yield production and production of gari from a local infected Cocoa variety and resistant improved cassava varieties (SLICASS 4 and 6) per hectare. The steps carried out for gari production are consistent with modern gari processing technologies (Sanni et al., 2009; Ijigbade et al., 2014).

In Sierra Leone, gari enterprise gave positive returns for improved cassava varieties compared to the susceptible local variety Cocoa. This result compliment report on profitability of gari in Ghana which confirmed that gari production business is a profitable and viable enterprise which can drive high income generation based on a benefit cost ratio of 1.33 and profitability ratio of 0.33 (Ijigbade et al., 2014).

Nweke et al. (2002) further stated that farmers who planted improved varieties (TMS varieties) in Nigeria produced yield which were 40% higher than that of local varieties even when grown without fertilizer. Report from IITA also indicates that improved varieties have contributed an extra 1.4 million tons of gari per year than would have been available from local varieties. In the same study profitability of TME varieties depended on the available grating technology.

Financial analysis shows that farmers who plant local varieties and grate manually earned the least revenue per ton of gari while farmers who plant local varieties and use mechanical grating earned higher. Improved varieties using mechanical grated earned the highest per ton of gari (Nweke et al., 2002).

Further analysis using the Net Present Value (NPV) which is the difference between the present value of the future cash flows from an investment and the amount of investment was done to check the viability of the enterprise. In this study NPV at different interest rates for SLICASS 4, 6 and Cocoa (the local cassava variety) using an interest rate of 10%, it shows that SLICASS 4 enterprise has more value than SLICASS 6 and the traditional enterprises. The local variety as an enterprise had negative NPV at 10% interest rate. This means that this enterprise is clearly unfavorable. A zero net present value means the project repays original investment plus the required rate of return. A positive net present value means a better return, and a negative net present value means a worse return, than the return from zero net present value. An intrest rate ranging from 10% to 20% was used to assess the profitability of the enterprises based on the current intrest rate on loan levied by commercial banks; however a wider intrest rate was expressed to determine different senarios at which the various enterises could be profitable. Based on cost of gari per kg at Njala, Freetown and the World Food program (WFP) price at the time of the study which was Le 2,500, Le 3,500 and Le 2,500 respectively, it was not profitable to sell at Le 2,500 at Njala due to the high cost of packaging, transportation incurred at the processing center. Under such condition profit can only be realized at the seventh month of operation unlike the situation in Freetown where cash flow was Le13, 363,045.7 at the first year at Le the price per kg was Le 3,500 despite the same level of expenditure as Njala. The WFP scenario presented a better option since the product is sold in bulk with no packaging cost, marketing cost and transportation cost.

Attaining such economic benefit from improved cassava varieties come with huge challenges which creates bottle necks in the processing of cassava to gari. Constraints such as harvesting of tubers, transportation of tubers within farm and to processing centers, peeling of tubers whiles ensuring minimal tuber rot makes the who enterprise a challenge.

5. Conclusion

Yield loss as a result of cassava mosaic disease was estimated in this study and revealed that number of tubers increased from 2.0 to 3.67 tubers per plant with an increase in weight from 0.3 kg to 0.41 kg per plant when symptomless cutting derived from the local infected variety Cocoa was used. The yield loss associated with the local variety Cocoa under this system was 4.27 t/ha which is equivalent to 38.92% yield loss resulting from the cassava mosaic disease infection.

The assumption used for the gari enterprise was based on commercial production of cassava using tractors and modern input and took into account yield potential for each variety assessed. Calculations based on Net Cash Flow analysis shows that the.SLICASS 4 and SLICASS 6 enterprises start recording positive returns earlier than the local cassava enterprises which recorded negative returns even in the 6 year. The same variety Cocoa used as poundable variety enterprise for the boil and eat market had the highest returns.

Further analysis using the Net Present Value (NPV) for SLICAS 4, 6 and Cocoa (the local cassava variety) using an interest rate of 10%, it shows that growing and processing improve varieties such as SLICASS 4 and SLICASS 6 was more profitable for a gari enterprise compared to the local mosaic infected variety Cocoa. However growing cocoa for the boil and eat market was the most profitable enterprise. Price influenced the overall profitability of gari enterprise. Based on cost of gari per kg at Njala, Freetown and the World Food program (WFP) price at the time of the study which was Le 2,500, le 3,500 and Le 2,500 respectively, it was not profitable to sell at Le 2,500 at Njala due to the high cost of packaging, transportation incurred at the processing center.

Acknowledgements

The authors wish to acknowledge the Government of Sierra Leone through the Ministry of Agriculture Forestry and Food security (MAFFS) The World Bank and The West African Agricultural Productivity Programme

(WAAPP) for the financial and technical support provided for this project. I wish to acknowledge staff of Njala University, Dr Jim A. B. Whyte of International Institutes of Tropical Agriculture and Dr Abdul Raman Conteh, Director, Njala Agricultural Research Center for their technical input and supervision throughout this study. I wish to thank all collegues who participated in the preparation of manuscript.

References

Berry, S., & Rey, M. E. C. (2001). Molecular evidence for diverse populations of cassava-infecting begomoviruses in southern Africa. *Archive of Virology, 146*, 1795-1802. http://dx.doi.org/10.1007/s0070 50170065

Brown, R. S., Caves, D. W., & Christensen, L. R. (1979). Modelling the structure of cost and production for multiproduct firms. *Southern Economic Journal*, 256-273. http://dx.doi.org/10.2307/1057018

FAO of the United Nations. (2014). *FAOSTAT*. Rome, Italy. Retrieved September 11, 2016, from http://faostat.fao.org

Fomba, S. N., Massaquoi, F. B., Samura, A. E., Fornah, D. S., Benya, M. T., Dixon, A. G. O., ... Sanni, L. O. (2012). Innovative systems to improve small to medium scale cassava processing in Sierra Leone. In Y. U. Okechukwu & P. Ntawuruhunga (Eds.), *Proceedings of the 11th Triennial Symposium of the ISTRC* (pp. 603-616). AB Held at Memling Hotel, Kinshasa, D. R. Congo, October 4-8, 2010.

Fregene, M., Bernal, A., Duque, M., Dixon, A., & Tohme, J. (2000). AFLP analysis of African cassava (*Manihot esculenta* Crantz) germplasm resistant to the cassava mosaic disease (CMD). *Theor. Appl. Genet., 100*, 678-685. http://dx.doi.org/10.1007/s001220051339

Gittinger, J. (1995). Economic analysis of agricultural projects. *Economic Development Institute Series in Economic Development*. John Hopkins University Press. http://dx.doi.org/10.1017/s0014479700022894

Hanley, N., Spash, C., & Walker, L. (1995). Problems in valuing the benefits of biodiversity protection. *Environmental and Resource Economics, 5*(3), 249-272. http://dx.doi.org/10.1007/bf00691519

Ijigbade, J. O., Fatuase, A. I., & Omisope, E. T. (2014). Conduct and Profitability of Gari production for increased food security in Ondo State, Nigeria. *IOSR Journal of Humanities and Social Science, 19*, 89-95. http://dx.doi.org/10.9790/0837-19758995

Ingersoll Jr, J. E., & Ross, S. A. (1992). Waiting to invest: Investment and uncertainty. *Journal of Business*, 1-29. http://dx.doi.org/10.1086/296555

International Institiue of Tropical Agriculture. (1990). *Cassava in Tropical Africa, A Reference Manual*. Ibadan, Nigeria: International Institute of Tropical Agriculture (IITA).

Nweke, F. I., Spencer, D. S. C., & Lynam, J. K. (2002). *The cassava transformation. Africa's best kept secret*. East Lansing: Michigan State University. http://dx.doi.org/10.5860/choice.39-6428

Sanni, L. O., Oluwatobi, O., Onadipe, P. I., Mussagy, M. D., Abass, A., &Dixon, A. G. O. (2009). *Successes and challenges of cassava enterprises in West Africa: A case study of Nigeria, Bénin, and Sierra Leone*. Ibadan, Nigeria: International Institute of Tropical Agriculture.

13

Effects of Rainfall and Temperature Oscillations on Maize Yields in Buea Sub-Division, Cameroon

N. Balgah Sounders[1], Tata Emmanuel Sunjo[1] & Mojoko Fiona Mbella[1]

[1] Department of Geography, University of Buea, Cameroon

Correspondence: Tata Emmanuel Sunjo, Department of Geography, University of Buea, Cameroon. E-mail: tataemmanuel@ymail.com

Abstract

There has been increasing concerns about the continuous variability in the climatic parameters of rainfall and temperature due to their manifold impacts. Some of these effects are observed through changes in crop yields such as maize in most parts of Sub-Saharan Africa which lacks the capital and technological viabilities to deal with the situation. This paper therefore examines the effects of the growing climatic oscillations on maize production. The study used primary and secondary data in order to provide insights on the quantitative effects of rainfall and temperature on maize yields in Buea Sub-Division. The climatic and crop trend analyses were done using simple regressions, means, and standard deviations. These were done using 2010 Excel Software. The impacts of varying rainfall and temperature on maize yields were determined using the logistic regression analysis in Stata 10 statistical software. Based on the analysis, results show that there has been growing rainfall and temperature fluctuations over Buea. This has been x-rayed through the increasing temperatures, slight declines in rainfall amounts, and the unpredictability of the sensation and the departure of the rains. Other effects have been observed through short dry spells especially in the months of April as well as increasing flooding of some farmlands in the months of August and September. Results further show that the unpredictability of the commencement of the rains has shifted the sowing season of maize by an average of four weeks and because of this situation, maize yields have increased during the minor season more than yields in the main season. Other climatic impacts were observed through increasing maize attacks from pests and diseases. As the way forward, there is the need for the development of maize germplasms that are heat-tolerant and need for the concentration of maize in the second season when soil moisture to ensure maize seed germination, growth and maturity is assured.

Keywords: adaptations, Buea, Cameroon, pest-diseases, maize yields, rainfall and temperature oscillations

1. Introduction

Over one and the half centuries, the world has continued to witness growing concentration of greenhouse gases (GHG) largely from human activities such as industrialisation, deforestation and unsustainable agricultural activities. The results have been increasing climate change and variability and the associated impacts on human livelihoods. These climate change and variability scenarios observed through increasing temperatures, changing precipitation patterns, and increase in the frequency of climatic hazards (droughts and floods) have had untold impacts especially on the economies of developing countries that lack the required capital and technical packages to deal with the problem. Climatic models indicate that the future effects of these growing climatic changes will be more complicated if current changes continue unabated (IPCC, 2007). While the global distribution of the impacts ensuing from this growing unpredictable climatic conditions are more regionalised, the localisation of these change and variability effects present a herculean task in ensuring sustainable food security (Tata & Lambi, 2015). One of the hardest hit sectors of the economies of the developing communities is agriculture. This is more so because agriculture especially in much of Sub-Saharan Africa is dominantly rain fed (Rao, Verchot, & Laarman, 2007). With agriculture significantly contributing to the GDPs of these countries, the growing impacts of climate change and climate variability on agriculture could thus be a significant rupture to efforts of raising a significant number of populations of developing countries above the poverty line of less than one dollar per day. The changes in precipitation patterns alone will significantly increase the likelihood of crop failures such as maize and production declines (Nelson et al., 2009).

After Cameroon gained independence in 1961, the government's involvement in the agricultural sector significantly increased the cultivation of crops especially cash crops which led to an increase in farmers' incomes. However, following the introduction of the Structural Adjustment Programmes (SAP) which reduced government intervention in a number of sectors of the Cameroonian economy as was the case with other developing countries in Africa by Breton Woods Institutions against the wishes of farmers and some politicians, farmers were left on their own and at the mercy of demand and supply market forces. This resulted in significant drops in the outputs and income from cash crops such as coffee and cocoa. The response from the farmers was the abandonment of these cash crops in favour of food crops such as maize, beans, and potatoes. Amongst these food crops, maize has remained one of the most widely cultivated crops. According to the Food and Agricultural Organisation Statistics (FAOSTAT, 2010), maize is produced on nearly 100 million hectares in developing countries, with almost 70 % of the total maize production in the developing world coming from low and lower middle income countries such as Cameroon. Maize is the most important cereal that is cultivated in most parts of Sub-Saharan Africa.

In Buea Sub-Division like in most parts of Cameroon, maize is the principal staple food crop accounting for a significant proportion of calorie intake by the population. It does not only provide a sustained and secured food supply in terms of high yields, it also remains a significant boost to the income of the peasant farmers given that much of the maize yields are often sold in local and sub-regional markets. Thus, maize production provides a safety valve against frequent food insecurity which, most often, arises from climatic shocks such as droughts, floods and tornadoes, and climatically enhanced spread of pests and diseases. This is because, in the event of these climatic and climatic-assisted caprices, maize yields are often better off than yields of crops such as beans, potatoes, and a host of market gardening crops which are widely cultivated especially in the humid and sub-humid tropical Africa. Improving maize yields involves a combination of factors such as research on hybrid species, improving soil fertility through fertilizer application, and combating pests and diseases with the use of pesticides, insecticides and fungicides among others. However, even though climate plays a key underpinning role in maize yields, temporal variations in climatic parameters of rainfall and temperature could usher in direct and indirect adverse effects on maize production. Crop simulation models indicate that by 2050 in Sub-Saharan Africa, average rice, wheat, and maize yields will decline by up to 15, 22, and 10 %, respectively, as a result of climate change (IFPRI, 2009). In this regard, this study investigates the effects of such climatic oscillations on maize by using statistical analysis.

2. Materials and Methods

2.1 The Study Area

Buea Sub-Division is one of the Sub-Divisions that make up the Fako administrative area of Cameroon. The sub-division occupies a surface area of 870 km^2 and is located between latitudes 4°12′ and 4°31′ North of the Equator and longitudes 9°9′ to longitude 9°12′ east of the Greenwich meridian (Mojoko, 2011). Buea Sub-division is bounded to the West by Mount Cameroon, to the east by Tiko Sub-Division, to the north by Muyuka Sub-Division and to the south by Limbe Sub-Division (Figure 1). The humid tropical climatic conditions alongside the rich volcanic soils provide ideal conditions for the cultivation of commercial crops mostly in plantations by the Cameroon Development Corporation as well as food crops by the local population. The area has unique topographic and geologic features due to its location on the eastern slopes of an active Mount Cameroon strato-volcano. Buea Municipality rest at an elevation of 1000 m above sea level on the south-eastern slopes of Mount Cameroon (4095 m), the highest in West Africa hosting a tan estimated population of over 176,000 inhabitants (Regional Institute of Statistics, Buea cited in Tosam, 2012).

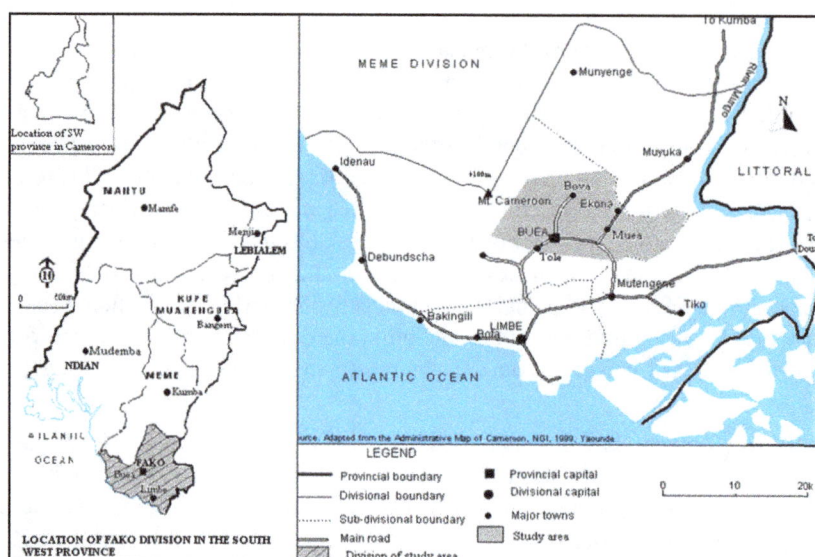

Figure 1. Location of Buea Sub-Division within Fako Division of Cameroon

2.2 Data Collection and Techniques of Analysis

A holistic approach was assumed to evaluate the effects of rainfall and temperature on maize output in the Buea Municipality of Cameroon. This study used the standard social science methodology involving data collection and analysis. The data were collected from primary and secondary sources. Primary data were obtained through focused group discussions with maize farmers from which 150 farmers were randomly drawn in order to acquire data relating to their adaptation strategies to the growing variable rainfall and temperature conditions in Buea Sub-Division. Also, interviews were randomly conducted with stakeholders involve in farming activities in the study area such as the agricultural extension workers and some experts in the Buea Sub-Divisional Delegation of Agriculture and Rural Development. This was accompanied by field visits to some maize farms in the area in order to appreciate some of the climatic-oriented problems of maize production and the various adaptation measures used by maize farmers. The information obtained from these sources centred on the various adaptation measures of maize farmers to the growing annual and seasonal variability in the climatic conditions of rainfall and temperature.

Secondary data sources were equally consulted. In this regard, data on the seasonal and annual yields of maize were obtained from the Divisional Delegation of Agriculture and Rural Development for Buea Sub-division. These data were used to establish the seasonal and average trend and average maize yields within the study area. This government establishment equally provided secondary data relating to the seasonal prevalence of pests and diseases affecting maize production in the area. Furthermore, data on climatic records on rainfall and temperature were obtained from the Cameroon Development Corporation (CDC) Head Office at Bota which records climatic parameters in the different locations within the Mount Cameron Area. The study equally relied significantly on secondary data sources related to the trend study of climatic behaviour of the Mount Cameroon Region in general (Fraser et al., 1998; Orock, 2011; Nkemasung, 2014). These data were equally used in establishing the rate of fluctuations in the climatic parameters using the coefficient of variations, anomalies, and averages. The results were used to establish, through logistic regression analysis, the extent to which the fluctuations in these climatic parameters affect maize yields within this humid and montane tropical environment.

The collected data were analysed quantitatively using statistical software packages such as 2010 Microsoft Excel for linear trend analysis, averages, and the Person Product Moment Correlation Coefficient to establish the relationship between variations in maize yields and changes in hectares of farm land cultivated. Logistic regression analyses on the seasonal and annual hydro-climatic effects on cereal yields were done using the Stata 10 statistical software. These correlation coefficient and regression analyses were statistically tested at 95% level of significance with a threshold probability of 0.05 for the cereals considered in the study. The results of these analyses were presented in tables and figures.

3. Results and Discussion

3.1 An Analysis of the Climatic Conditions over Buea

A number of studies have been conducted on the prevailing climatic conditions over Buea Sub-Division and beyond (Fraser et al., 1998; Orock, 2011; Nkemasong, 2014). Buea Sub-Division is found within the humid tropical region with characteristics of the A-climatic type according to the Koppen Classification System. This gives the study area two distinctive seasons; the long rainy season which ranges from mid-March to mid-November, and the short four months dry season expanding from mid-November to Mid-March. The high annual mean minimum and maximum temperatures of 18 °C and 28 °C respectively give an annual average temperature of 23 °C and a narrow temperature range of 10 °C. Rainfall trend analysis using simple regression show a general decline in rainfall amounts over the years with more negative deviations observed since the beginning of the 21st Century (Figure 2).

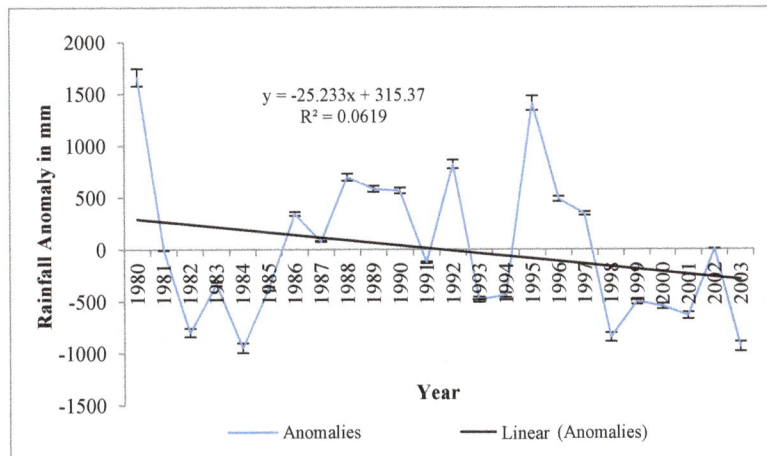

Figure 2. Rainfall anomaly in Buea Sub-Division

Source: CDC Meteorological Station (2014).

The climate of Buea is significantly determined by a combination of global, regional and local factors which are related to the Walker Circulation Models and the seasonal movement of the Inter-tropical Convergence Belt over West and Central Africa respectively. The local conditions which significantly shape the climatic conditions over Buea include the relief factor arising from the sharp rising of the Mount Cameroon from the Atlantic Ocean, the high intensity of the Warm Guinea Currents and dense primary and secondary vegetation cover. These local factors are determined by the proximity of the study area (with an average distance of 36 kilometres) opposite the expansive Atlantic Ocean in the south. These factors combine to make the area a high energy environment characterised by high temperatures of over 24 °C and high mean rainfall amounts of over 2500mm (Fraser et al., 1998; Nkemasong, 2014). These results on temperature behaviour in Buea in particular and the Mount Cameroon Region in general show a general trend of warming (Figure 3) since the dawn of the 20th Century. These results are in accordance with those at the regional level such as across maize mega environments within sub-Saharan Africa conducted at the country level (IPCC, 2007; Burke et al., 2009).

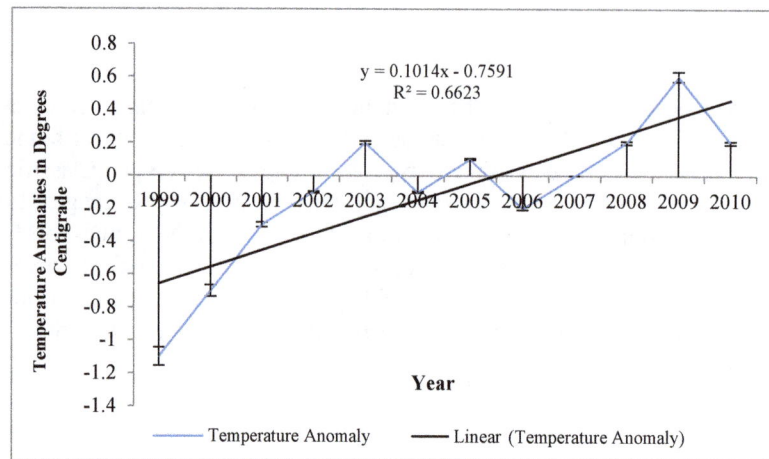

Figure 3. Temperature variability in Buea Sub-Division

Source: CDC Meteorological Station (2014).

3.2 Maize Production in Buea

Maize is one of the most widely produced crops in Buea Sub-Division. This is attributed to the fact that unlike other crops, maize plant does not require much care especially at the maturity stage. Maize cultivation is often done in two main seasons in the course of the year. The first season which is the main season use to run from the 3rd week of March to the last week of June while the minor season use to run from the last week of July to the 3rd Week of October. However, with growing climatic oscillations, the sowing season has been shifted due to the late sensation of the rains to the last week of April for the main maize season, and to the first two to three weeks of August for the second season of maize cultivation. The results have been that, since the second season of 2011, maize yields have been observed to be greater during the second season as opposed to the main season situation (Figure 4). Although total and mean output of yields have remained high in the main season when compared to statistics for the second season (Table 1), the trend analyses between 2006 and 2013 show that with continuous climatic variability, the trend is going to be reversed.

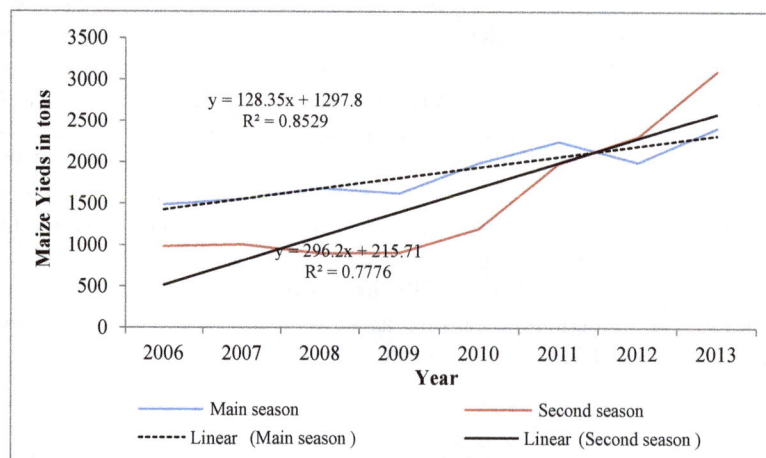

Figure 4. Variations and trends of seasonal maize yields over Buea over the last 8 years

It should be remarked that this reversal situation in maize yields between the main season and second season is not directly linked to the role of climatic parameter. Instead, the uncertainties in the sensation of the rains especially during the main season have made some farmers to engage more hectares of farmlands for the cultivation of crops which require less moisture as it is the case with market gardening crops such as tomatoes and vegetables. This accounts for the gradual increase in the hectares of land cultivated with maize during the second season as opposed to the main season. Although there is a positive relationship between maize yields and hectares cultivated during the main season, this relationship is not statistically significant (Table 1). On the contrary, the relationship between the second season and annual yields and their respective hectares cultivated is strongly positive (0.9 and 0.97 respectively) and statistically significant.

Table 1. Statistical summaries of data on maize yields and hectares cultivated

Parameter	Main season	Second season	Annual
Total: Yields in tons	15003	12389	27392
Hectares	4639	4580	9219
Mean: Yields in tons	1875.4	1548.6	3424
Hectares	579.9	572.5	1152.4
Correlation Coefficient	0.51	0.90	0.97
Degree of Freedom	6	6	6
Calculated T-value	1.5	4.93	9.7
Probability Value	2.45	2.45	2.45
Conclusion	Not significant	Significant	Significant

Note. 0.05 level of significance.

3.3 Climatic Impacts on Maize Yields

The observed oscillations in rainfall and temperature could have direct and indirect implications for maize production. On seasonal basis, the unreliability in the beginning of the rains and the reductions in rainfall amounts are observed to have negative impacts on maize yields putting aside other confounders such as reductions in hectares of maize cultivated lands during this season. Decreasing rainfall amounts combined with increasing temperatures have implications for maize production especially in relation to soil moisture and pests-diseases prevalence. The growing erratic nature in the beginning and progress in rainfall amounts have reduced the number of maize growing days. This has therefore resulted in the qualitative reduction in maize yields. Although the analysis on maize trends show an increasing trend, it is important to remark that this increase is related to increases in the number of hectares cultivated and not the qualitative improvements in maize yields. This is why a negative relationship which is not statistically significant has been observed between oscillations in rainfall and temperature, and maize production (Table 2). This negative relationship must have thus occurred by chance. However, a positive and statistically significant relation is observed between changes in the number of hectares cultivated and the yields of maize (Tables 1 and 2).

Table 2. Analysis of multiple regressions on the effects of variations in rainfall, temperature and hectares on maize yields

Maize Yields Association with:	Coefficient	t	P > t	[95% Conf. Interval]
Rainfall	-0.2897714	-0.97	0.388	-1.12-0.54
Temperature	-11.19685	-0.02	0.986	-1641.75-1619.36
Hectares	3.330709	7.54	0.002	2.11-4.56

The occurrence by chance of the relationship between rainfall amounts and maize yields has been justified by field surveys through observations and interviews which reveal that the uncertainties in the beginning of the rains and the increasing temperatures have gradually shifted the sowing period from the historical third week of March to the last week of April and sometimes the first week of May which receive minimum optimal soil moisture amounts necessary for maize germination and growth. This often reduces the growing maize season by some four to five weeks thus resulting in a decline in maize yields as reported by 75% of the sample farmers. Such shifts especially in the main maize growing season equally affect the minor maize growing season due to over lapping or concomitant shifts in the sowing season from July and August to the latter half of the month of September. Consequently, this gives just about 6 weeks for maize to grow during this minor season prior to the arrival of the dry season in the Month of November. However, the shifts in the sowing season have resulted to the increase in the maize yields in the minor season much more than the yields in the main season since 2011.

In some instances, farmers reported extra losses associated with repeated planting particularly during the main season. These were mostly farmers who continue to stick to the traditional sowing period of the third week of March. This is because, by this time, they often erroneously sow in anticipation of the fact that the rains will come at the conventional time. When these rains fail to come at this time, coupled with the high temperatures, the planted seed become dry and dead in the soil thus necessitating repeated planting of the seeds in the last week of April when the rains must have fallen effective enough to increase soil moisture required for maize seed germination and growth. There is thus the need for sensitisation by the various agricultural institutions on the effective period to plant for every season in the midst of these variable rainfall and temperature conditions.

The observed negative relationship between increasing warming and maize yields over Buea have been supported by a number of studies in different parts of the world. Battisti and Naylor (2009) observed that growing season temperatures will exceed the most extreme seasonal temperatures recorded in the past century. As such, with the use of crop production data and climatic records, a 6 °C increase in temperature during the grain filling period resulted in a 10% yield loss in the US Corn Belt (Thomson et al., 1966 cited in Cairns et al., n.d.). This has equally been supported by later studies of Dale (1983) cited in (Cairns et al., n.d.) when he showed, in the same region, a negative correlation between maize yields and accumulated degrees of daily maximum temperatures above 32 °C during the grain filling period. In their study on the use of statistical models in predicting crop yields to climate change in sub-Saharan Africa, Lobell and Burke (2010) suggested that an increase in temperature of 2 °C would result in a greater reduction in maize yields within sub-Saharan Africa unlike a decrease in precipitation by 20%. In a related study, Lobell et al. (2011) cited in Cairns et al. (n.d.) observed that for every degree day above 30 °C, grain yield was reduced by 1% and 1.7% under optimal rainfed and drought conditions, respectively.

Based on these analysis and discussions, it is clear that climatic effects on maize production arising from continuous warming will be greater that those arising from the slight decreases in rainfall amounts observed over Buea. It is therefore necessary that in the research on seed improvements using biotechnology, heat tolerant to heat stress by maize plants should be incorporated in the maize germplasm. In this context, heat stress is considered to be temperatures above a threshold level that results in irreversible damage to crop growth and development (Cairns et al., n.d.). This is usually a function of the intensity, the duration and the rate of increase in temperature. A number of studies support these results on the impact of slight increase of temperatures on maize outputs in Buea, thereby strongly suggesting that maize growing regions of sub-Saharan Africa will encounter increased growing season temperatures and frequency of droughts (IPCC, 2007).

3.4 Climatic Impacts on Maize Pests and Diseases

Variability in climatic parameters especially temperatures in Buea do not only affect the anatomical and physiological welfare of the maize plant. They are equally observed to have indirect effects on maize especially through pests and diseases. Given that they do not use their metabolic activities to ensure thermo-regulation, these insects such as stem borers are often classified as poikilothermic (cold blooded), relying significantly on the surrounding environmental temperatures for their activeness. Temperature is thus the single most important environmental factor influencing the behaviour, distribution, development and survival, and reproduction of insect (Cairns et al., n.d.). This is why the life cycle predictions of insects and pests affecting crops are often calculated on accumulated degree days, which is a function of both time and temperature. The rate of propagation of the population of a particular pest or insect is often significantly determined by the extent to which temperatures are increasing or decreasing. In this light, it has been estimated that a 2 °C increase in temperature has the potential to increase the number of insect life cycles during the crop season by one to five times (Petzoldt & Seaman, 2005; Bale et al., 2002; Porter et al., 1991 cited in Cairns et al., n.d.). The observed temperature rises and the frequent dry spells in the month of April over Buea have thus been noted to favour the proliferation of insects and related crop damages as well as aflatoxin and fumonisinmy cotoxins in maize. The increasing attacks on maize by these insect-pests and their associated pathogens such as maize streak virus, corn stunt complex that are vectored by different species of leaf hoppers have been attributed to their extensive geographical distribution as a results of global warming.

3.5 Farmers' Resilience and Adaptations to Oscillating Climatic Conditions in Buea

Despite the identified as well as the prospective effects of oscillations in rainfall and temperature parameters, food security within the Buea Municipality is guaranteed. This is significantly related to the high diversification of crops cultivated by farmers. Located within the humid and montane tropical environment, the Buea Municipality is ideal for the cultivation of crops of many varieties ranging from cereals, market gardening crops, tubers/root crops, to tree crops. Apart from the above discussed climatic parameters, there are other factors which contribute to high crop yields and diversity in the area. Some of these include the rich fertile volcanic soils on the slopes of the Mount Cameroon which is an active strato-volcano and the increase in the use of chemicals in order to minimise pest-disease attacks on crops. Other adaptations are observed through the use of genetically modified hybrid seedlings by some farmers which often mature faster than the traditional maize seeds. However, these genetically modified maize seeds are not used by all the farmers due to cultural inertia. Some farmers noted that they prefer the use of their tradition maize seeds on grounds that the harvested maize is tasteful. No wonder, 90% of the sampled farmers who plant these genetically modified maize seeds noted that they cultivate maize mostly for commercial purposes and not for home consumption.

Another important adaptation to the erratic rainfall commencement and increasing temperatures is observed through the shifts in the planting dates of maize by farmers (Table 3). A total of 63% of the surveyed farmers accepted to have shifted their sowing dates of maize from the conventional third week of March to the last week of April during the main season, and from the last week of July to the second week of September during the minor maize growing season. Other adaptation measures include changing the crop types cultivated, and the shift to off farm jobs. The later has been growing given that the increasing urbanisation of the Town of Buea which has been accelerated by the creation of the University of Buea (Ntansi, 2004; Mojoko, 2011) has provided jobs in the formal and informal sectors. A total of 19.2% of the respondents noted that with the off farmers opportunities associated with the urbanisation of Buea, they have been largely engaged in these job opportunities. Some of them only receive incomes from farming activities through the renting of their farms to those who can continue cultivating crops on them in the midst of the erratic rainfall beginnings and increasing temperatures. However, the rate at which these adaptation measures will continue to be sustainable in the next 50 years in the face of changing climatic patterns will significantly depend on extension services to and farmers' access to credit and secured land tenures.

Table 3. Sampled farmers' responses to the various adaptation strategies to erratic rainfall commencement and increasing temperatures in Buea Sub-Division

Adaptation Measure	Erratic Beginning of Rains (%)	Increasing temperature (%)
Diversification of Crop	25.5	14.3
Change of planting date	45	17
Find off-farm jobs	10.2	9
Farm size reduction	1	1.1
Change of crops	4	8.7
Planting of short-season variety	12.1	13.9
No adaptation	2.2	48
Total	100	100

4. Conclusion

The humid equatorial and montane climatic conditions that characterised the Buea Municipality provide ideal conditions for the growth of maize which is widely consumed in varied forms by the population. However, the spatio-temporal inconsistencies in climatic conditions continue to dictate, amongst other factors, the pace of crop production in general and maize cultivation in particular. This is more so because the variations in the different climatic elements (especially soil moisture from rainfall and temperature) usher in different effects on maize yields. In this light, the observed slight reductions in rainfall and the increasing temperatures (Nkemasong, 2011) in the Mount Cameroon Region as a whole have brought about significant reductions in maize yields particularly in during the first season of cultivation. On the contrary, these effects have increased maize yields during the second season which is considered as a subsidiary season of cultivation. In the face of these effects, farmers on their part have developed a series of adaptation strategies even though some of them are simply makeshift adaptations. If there are concerns about ensuring food security in the region as well as the Central African Region as whole, then there is the need for expert intervention in order to empower the farmers on robust adaptation measures. Such efforts should focus among other things on the need for the development of new maize varieties alongside improved management options and other agricultural innovations. In a retrospective survey, the use of new varieties alongside improved management options showed to have counteracted maize yield losses by up to 40% (Thornton et al., 2009).

References

Battisti, D. S., & Naylor, R. L. (2009). Historical warnings of future food Insecurity with unprecedented seasonal heat. *Science, 323*, 240-244. https://doi.org/10.1126/science.1164363

Burke, M. B., Lobell, D. B., & Guarino, L. (2009) Shifts in African crop climates by 2050, and the implications for crop improvements and genetic resources conservation. *Global Environ. Change, 19*, 317-325. https://doi.org/10.1016/j.gloenvcha.2009.04.003

Cairns, J. E., Sonder, K., Zaidi, P. H., Verhulst, N., Mahuku, G., Babu, R., ... Prasanna, B. M. (n.d.). *Maize production in a changing climate: Impacts, adaptation and mitigation strategies.*

Charity, M., Shakespear, H., & Lawrence, M. (2013). The Effects of Climate Change and Variability on Food Security in Zimbabwe: A Socio-Economic and Political Analysis. *International Journal of Humanities and Social Science, 3*(6), 270-286.

FAO. (2007). *Adaptation to Climate Change in Agriculture, Forestry and Fisheries: Perspective, Framework and Priorities*. Food and Agriculture Organization of the United Nations Rome. Retrieved from ftp://ftp.fao.org/docrep/fao/009/j9271e/j9271e.pdf

FAOSTAT. (2010). *FAO Statistical Database*. Food and Agricultural Organization of the United Nations (FAO). Retrieved from http://faostat.fao.org

Fosu-Mensah, B. Y. (2012). Modelling Maize (*Zea mays* L.) Productivity and Impact of Climate Change on Yield and Nutrient Utilization in Sub-Humid Ghana. In P. L.G. Vlek (Ed.), *Ecology and Development Series* (No. 87).

Fraser, P. J., Hall, J. B., & Healey, J. R. (1998). *Climate of the Mount Cameroon Region*. Long and Medium Term Rainfall, Temperature and Sunshine Data, University of Wales, Bangor, Mount Cameroon Project and Cameroon Development Corporation, School of Agriculture and Forest Sciences Publication Number 16.

Inter-Governmental Panel on Climate Change (IPCC). (2007). Summary for Policymakers. *Working Group II Climate Change 2007: Climate Change Impacts, Adaptation and Vulnerability*. IPCC, Geneva.

Inter-Governmental Panel on Climate Change (IPCC). (2009). *Fourth Assessment Report: Synthesis*. Retrieved November 17, 2007, from http://www.ipcc.ch/pdf/assessment-report/ar4/syr/ar4_syr.pdf

International Food Policy Research Institute (IFPRI). (2006). *How Will Agriculture Adapt to a Shifting Climate?* IFPRI Forum, December, 2006.

Lambi, C. M., & Molua, E. L. (2007). The Economic Impact of Climate Change on Agriculture in Cameroon. *World Bank Policy Research Working Paper No. 4364*. Retrieved from http://papers.ssrn.com/sol3/papers.cfm?abstract_id=1016260

Lobell, D. B., & Burke, M. B. (2010). On the Use of Statistical Models to Predict Crop Yield Responses to Climate Change. *Agric. Forest Metero., 150*, 1443-1452. https://doi.org/10.1016/j.agrformet.2010.07.008

Mojoko, F. M. (2011). *Urban Development in Fako Division* (Unpublished M.Sc. Thesis). Department of Geography, University of Buea.

Nelson, G. C., Rosegrant, M. W., Koo, J., Robertson, R., Sulser, T., Zhu, T., ... Lee, D. (2009). *Climate Change Impact on Agriculture and Costs of Adaptation*. Food Policy Report. Washington, D.C: International Food Policy Research Institute.

Nkemasong, N. A. (2014). *Climate Variability and Implications on Hydrological Systems within the Southern Volcanic Province of Cameroon* (Unpublished Ph.D. Thesis). Department of Geography, University of Buea, Cameroon.

Orock, F. T. (2011). *Wetland Utilisation, Problems and Management on the North Eastern and Southern Slopes of Mount Cameroon* (Unpublished Ph.D. Thesis). Department of Geography, University of Buea, Cameroon.

Rosenzweig, C., & Hillel, D. (1995). Potential Impacts of Climate Change on Agriculture and Food Supply. *Consequences, 1*(2). Retrieved May 12, 2011, from http://www.gcrio.org/CONSEQUENCES/summer95/agriculture.html

Tata, E. S., & Lambi, M. C. (2015). Hydro-Climatic Implications on Market Gardening Production in the Bui-Noketunjia Wetlands. *African Journal of Social Sciences, 6*(2), 4-17.

Thornton, P. K., Jones, P. G., Alagarswamy, G., & Andersen, J. (2009). Spatial variation of crop yield response to climate change in East Africa. *Global Environ. Change, 19*, 54-65. https://doi.org/10.1016/j.gloenvcha.2008.08.005

Tosam, H. N. (2012). An Analysis of Spatial Population Pattern and the Provision of Socio-Economic Infrastructure in Buea Sub-Division. *African Journal of Social Sciences, 3*(1), 116-128.

14

Cultivar, Planting Date, and Row Spacing Effects on Mungbean Seed Composition

Harbans L. Bhardwaj[1] & Anwar A. Hamama[1]

[1] Agricultural Research Station, Virginia State University, Petersburg, USA

Correspondence: Harbans L. Bhardwaj, Agricultural Research Station, Virginia State University, Petersburg, Virginia 23806, USA. E-mail: hbhardwaj@vsu.edu

This paper is a contribution of Virginia State University, Agricultural Research Station. The use of any trade names or vendors does not imply approval to the exclusion of other products or vendors that may also be suitable.

Abstract

Mungbean [(*Vigna radiata* (L.) R. Wilczek, Fabaceae] is becoming an important food crop in the United States of America. This crop has previously been produced in the US states of Texas and Oklahoma but this production is currently not significant. Recent efforts have established that mungbean can be easily produced in Virginia, located in the mid-Atlantic region of the United States of America. However, there is a complete lack of information related to nutritional quality of mungbean produced in this region. We grew mungbean during 2012 and 2013 using two cultivars (Berken and TexSprout), two planting dates (early and late July), and two row spacings (0.375 and 0.75 m) to characterize composition of mungbean seed produced in Virginia. Mungbean seeds produced in this study averaged 1.59, 24.3, and 4.91% oil, protein, and sugars, respectively. These mungbean seeds also contained 38.8, 61.2, 5.79, and 55.1% in saturated, unsaturated, mono-unsaturated, and poly-unsaturated fatty acids, respectively. Predominant fatty acids in the mungbean seed were C16:0 (26.1%), C18:0 (6.11%), C18:2 (36.8%), and C18:3 (18.3%). Iron and zinc contents of the mungbean seed were 8.42 and 3.88 mg·100 g^{-1}. Concentrations of fructose, glucose, sucrose, raffinose, stachyose, and verbascose sugars in mungbean seed were 0.45, 0.30, 0.70, 0.24, 0.84, and 2.37%, respectively. Effects of cultivars, planting dates, and row spacings on mungbean seed composition were, generally, not significant. Overall, mungbean seed compared well with nutritional quality of kidney bean, pinto bean, navy bean, and tepary bean.

Keywords: *Vigna radiata*, food legumes, fatty acids, minerals, sugars

1. Introduction

Mungbean [(*Vigna radiata* (L.) R. Wilczek], a member of the Fabaceae family (also known as the Leguminosae family) along with the common pea, chickpea, soybean, alfalfa, and other crops, is a native of India-Burma (Myanmar) region of Asia and is grown principally for its protein-rich edible seeds for use as food or livestock feed. This plant and its production strategy are very similar to soybean. The seeds have potential as human food (cooked beans or sprouts) and livestock feed. The plant, being a legume, can fix atmospheric N via Symbiotic N Fixation and also can also be used as forage or hay (Oplinger et al., 1990).

Mungbean is one of the most important food legume crops in the Asia. It is also gaining importance in other parts of the world, such as Australia and Canada. The United States imported 12,731, 13,474, and 13,672 Mg of mungbean, respectively during 2012, 2013, and 2014, respectively (ERS, 2015). Although previous research indicated that mungbean has potential as a short-duration summer crop in the mid-Atlantic region of the United States (Bhardwaj & Hamama, 2015; Bhardwaj et al., 1999), there is a lack of information about nutritional quality of mungbean produced in this region. In a previous study (Bhardwaj & Hamama, 2015), we observed that cultivar and planting date effects were not significant for seed yield, however, narrow row spacing resulted in significantly higher seed yield and protein concentration over wide row spacing (1.76 vs. 0.86 Mg ha^{-1} yield and 24.9 vs. 23.7% protein, respectively for 0.375 and 0.75 m row spacings). We also observed that average values for mungbean seed yield, seed size, and concentrations of protein, sugars, and oil were 1.31 Mg ha^{-1}, 7.08 g seed^{-100}, 24.3%, 4.91%, and 1.59%, respectively.

The objective of the current study was to characterize seed composition of mungbean seed produced in experiments conducted during 2012 and 2013 in Virginia (Bhardwaj & Hamama, 2015).

2. Materials and Methods

2.1. Plant Material and Production

Seeds of two mungbean cultivars (Berken and TexSprout) were used in this study. Berken (PI-662958) is a small seeded cultivar whereas TexSprout (PI-536545) is a large seeded cultivar. These cultivars were grown during 2012 and 2013 (Bhardwaj & Hamama, 2015) by planting on July 7 and July 27 during 2012 and July 8 and July 23 during 2013 and by using two inter-row spacings (0.375 m and 0.75 m) at Randolph Farm at Virginia State University (37°15′N, 77°31′W). The experiment during each year was designed as a split-plot with planting dates as main plots, cultivars as subplots, and row spacings as sub-subplots. Each plot consisted of four rows spaced either 0.375 or 0.75 m apart with 1.5 m distance between plots. Each experiment consisted of four replications per planting date. The rows were 3.6 m long. About 100 seeds were planted in each row with a tractor-driven research planter. The seed depth was 0.02 to 0.03 m. These plots received 30 kg·h^{-1} of NPK. The soil type was Abel sandy loam (fine-loamy, mixed, thermic Aquatic Hapludult). Experimental area received a preplant incorporated treatment of Trifluralin herbicide at the rate of 1 L·ha^{-1}. The plots were manually weeded once. All plots were harvested on 24 Oct. during 2012 and on 31 Oct. during 2013.

2.2 Analysis of Seed Composition

Mineral contents, including nitrogen (N), in mungbean seed were determined according to AOAC methods (AOAC, 1995) by A&L Eastern Agricultural Laboratory, Richmond, Virginia. Total protein was calculated by multiplying N content with protein factor 6.25. The concentration of minerals was expressed as mg·100^{-1} g.

The oil was extracted from ground seeds (2 g) three times at room temperature by homogenization with hexane/isopropanol (3:2, v/v) as described by Hamama et al. (2005). The oil percentage (g·100^{-1} g dry basis) was determined gravimetrically (Bhardwaj et al., 2004) after drying in a vacuum oven at 40 °C and stored under N$_2$ at -10 °C until analyzed.

The fatty acid methyl esters (FAME) of the oil were prepared by acid-catalyzed transesterification technique as previously described (AOAC, 1995). Analyses of FAME were carried out as described by Hamama et al. (2005) in a SupelcoWax 10 capillary column (25 m × 0.25 mm i.d. and 0.25 μm film thickness (Supelco, Inc., Bellefonte, PA) in a Varian Model Vista 6000 Gas Chromatograph equipped with a flame ionization detector and a Spectra Physics Model 4290 integrator. Helium was used as a carrier gas at 25 cm·s^{-1}, with a split ratio of 1:100. The column temperature was isothermal at 210 °C. The injector and detector temperatures were 250 and 260 °C, respectively. Fatty acids were identified by reference to the retention of FAME standards and quantitated by the aid of heptadecanoic acid (17:0) as an internal standard. The concentration of each FA was calculated as the percentage (w/w) of the total fatty acids.

Sugars were extracted from ground sample (1 g) and analyzed by HPLC following the methods optimized by Johansen et al. (1996). Sugars in the extracts were identified by comparing their retention times with standard sugars. For quantification, trehalose was used as internal standard and the sugar concentration was expressed as g· 100 g^{-1} oil-free meal.

Seed composition traits of mungbean produced in this study were compared to those in the literature for kidney bean, pinto bean, navy bean, and tepary bean for evaluating the nutritional quality of mungbean.

All data were analyzed using version 9.1 of SAS (SAS, 2014) using ANOVA with 5 percent level of significance.

Table 1. Cultivar effects on concentrations of fatty acids, minerals, and sugars in Mungbean seed produced in Virginia during 2012 and 2013

	Berken	TexSprout	LSD (0.05)
Oil (%)	1.59*	1.60*	ns
Protein (%)	24.7	23.9	ns
Total Sugars (%)	4.92	4.90	ns
Fatty acids (%)			
C14:0	0.256	0.266	ns
C16:0	26.015	26.247	ns
C18:0	6.010	6.113	ns
C18:1	5.713	4.785	ns
C18:2	36.680	36.999	ns
C18:3	18.214	18.336	ns
C20:0	1.855	1.921	ns
C20:1	0.374	0.333	0.039
C22:0	2.935	2.950	ns
C24:0	1.417	1.486	ns
Minerals (mg/100g)			
P	479	484	ns
K	1349	1356	ns
S	206	210	ns
Ca	144	146	ns
Mg	182	186	ns
Na	12.3	12.3	ns
Fe	8.43	8.42	ns
Al	1.56	1.69	ns
Mn	1.78	2.10	ns
Cu	1.20	1.20	ns
Zn	3.81	3.94	ns
B	1.32	1.32	ns
Sugars (%)			
Fructose	0.450	0.456	ns
Glucose	0.294	0.306	ns
Sucrose	0.632	0.765	ns
Raffinose	0.273	0.217	ns
Stachyose	0.844	0.835	ns
Verbascose	2.423	2.324	ns

Note. * Means over two years, two planting dates, two inter-row spacings, and two replications.

3. Results and Discussion

Concentrations of various constituents of mungbean seed were, generally, not significantly affected by cultivars, planting dates, and inter-row spacings (Tables 1, 2, and 3). However, the small-seeded cultivar Berken had significantly higher concentration of C20:1 fatty acid (0.37% of oil) as compared to the large-seeded cultivar TexSprout (0.33% of oil) (Table 1). We believe that this lack of significance resulted from inclusion of only two cultivars in this study, due to limited availability of mungbean cultivars, and expect that significant differences would be observed among cultivars if a large number of cultivars were evaluated. Similar results were observed from a previous study with a limited number of white lupin cultivars (Bhardwaj & Hamama, 2013). Effects of planting dates were significant for concentrations of C16:0, C22:0, and C24:0 fatty acids (Table 2). Earlier planting resulted in significantly lower concentrations of these three fatty acids (26.0 vs. 26.3% C16:0, 2.87 vs. 3.02% C22:0, and 1.32 vs. 1.59% C24:0) in mungbean seeds. Narrow inter-row spacing of 0.375 m resulted in significantly lower concentration of sulfur (204 mg·100 g^{-1}) as compared to the 0.75 m row spacing of (212 mg·100 g^{-1}) (Table 3). As previously reported (Bhardwaj & Hamama, 2015), narrow inter-row spacing resulted in significantly higher concentration of protein (24.9 vs. 23.7%), while early planting resulted in significantly lower concentrations of sugar (4.38 vs. 5.51%) and oil (1.24 vs. 1.99%) in mungbean seeds produced in Virginia.

These results demonstrate that, in general, mungbean seed composition was not affected by agronomic factors in our limited studies in mid-Atlantic region of USA. However, our results establish a baseline for mungbean nutritional quality for comparison with future studies.

Table 2. Planting date effects on concentrations of fatty acids, minerals, and sugars in mungbean seed produced in Virginia during 2012 and 2013

	Early July	Late July	LSD$_{0.05}$
Oil (%)	1.24*	1.99*	0.001
Protein (%)	24.2	24.4	ns
Total Sugars (%)	4.38	5.51	0.001
Fatty acids (%)			
C14:0	0.252	0.272	ns
C16:0	26.0	26.3	0.015
C18:0	6.40	5.78	ns
C18:1	5.58	4.87	ns
C18:2	36.3	37.5	ns
C18:3	18.3	18.2	ns
C20:0	1.97	1.79	ns
C20:1	0.352	0.356	ns
C22:0	2.87	3.02	0.019
C24:0	1.32	1.59	0.002
Minerals (mg/100g)			
P	468	497	ns
K	1321	1388	ns
S	209	207	ns
Ca	149	140	ns
Mg	184	183	ns
Na	12.4	12.2	ns
Fe	8.35	8.50	ns
Al	1.78	1.45	ns
Mn	1.94	1.93	ns
Cu	1.24	1.16	ns
Zn	3.92	3.82	ns
B	1.32	1.31	ns
Sugars (%)			
Fructose	0.450	0.456	ns
Glucose	0.294	0.306	ns
Sucrose	0.632	0.765	ns
Raffinose	0.273	0.217	ns
Stachyose	0.844	0.835	ns
Verbascose	2.423	2.324	ns

* Means over two years, two cultivars, two inter-row spacings, and two replications.

Table 3. Inter-row spacing effects on concentrations of fatty acids, minerals, and sugars in mungbean seed produced in Virginia during 2012 and 2013

	0.375 m	0.75 m	$LSD_{0.05}$
Oil (%)	1.55*	1.64*	ns
Protein (%)	24.9	23.7	0.001
Total Sugars (%)	4.90	4.92	ns
Fatty acids (%)			
C14:0	0.261	0.262	ns
C16:0	26.207	26.056	ns
C18:0	6.109	6.104	ns
C18:1	4.767	5.731	ns
C18:2	37.152	36.526	ns
C18:3	18.557	17.993	ns
C20:0	1.847	1.929	ns
C20:1	0.354	0.353	ns
C22:0	2.982	2.902	ns
C24:0	1.484	1.419	ns
Minerals (mg/100g)			
P	474	489	ns
K	1360	1346	ns
S	204	212	0.007
Ca	149	140	ns
Mg	183	185	ns
Na	11.8	12.8	ns
Fe	8.23	8.62	ns
Al	1.62	1.63	ns
Mn	1.84	2.04	ns
Cu	1.16	1.24	ns
Zn	3.89	3.86	ns
B	1.31	1.33	ns
Sugars (%)			
Fructose	0.385	0.521	ns
Glucose	0.265	0.334	ns
Sucrose	0.752	0.646	ns
Raffinose	0.214	0.276	ns
Stachyose	0.852	0.828	ns
Verbascose	2.434	2.313	ns

Note. * Means over two years, two cultivars, two planting dates, and two replications.

3.1 Seed Composition of Mungbean and Other Food Legumes

We compared the seed composition of mungbean with that of literature values for kidney bean, pinto bean, navy bean, and Tepary bean (Table 4). Comparisons of mungbean seed composition values to those available for other food legumes (USDA, 2015; Bhardwaj & Hamama, 2004, 2005) indicated that concentrations of C16:0, C18:0, C18:2, C20:1, saturated, and poly-unsaturated fatty acids were, generally, higher in mungbean seeds whereas concentrations of C18:1, unsaturated, and mono-unsaturated fatty acids were lower in mungbean seeds. Concentration of C18:3 (The Omega-3 fatty acid) in mungbean seeds was lower than that in seeds of kidney bean, pinto bean, and navy bean but higher than that in the seeds of Tepary bean. Overall, these results indicate that nutritional quality of mungbean produced in mid-Atlantic region of USA is quite similar to that of other food legumes and is acceptable.

Table 4. Seed composition traits of mungbean compared to literature values for kidney bean, pinto bean, navy bean, and tepary bean

	Mungbean		Kidney bean[3]	Pinto bean[3]	Navy bean[3]	Tepary bean[3]
	Virginia grown[1]	Literature[2]				
Oil (%)	1.59	1.15	1.1	1.1	1.3	1.8
Protein (%)	24.3	23.9	22.5	20.9	22.3	23.9
Total Sugars (%)	4.91	6.60	2.23	2.11	3.88	n/a
Fatty acids (%)						
C14:0	0.26	n/a	n/a	n/a	n/a	n/a
C16:0	26.1	n/a	12.8	20.3	24.2	23.3
C18:0	6.11	n/a	1.7	0.4	1.6	6.1
C18:1	5.25	n/a	7.7	20.3	8.7	22.2
C18:2	36.8	n/a	21.5	15.0	23.5	31.5
C18:3	18.3	n/a	33.8	21.0	19.7	10.9
C20:0	1.89	n/a	n/a	n/a	n/a	1.5
C20:1	0.35	n/a	0.00	0.00	n/a	0.8
C22:0	2.94	n/a	n/a	n/a	n/a	1.0
C24:0	1.45	n/a	n/a	n/a	n/a	1.4
Saturated	38.8	34.8	14.5	20.8	25.9	33.4
Unsaturated	60.9	65.2	85.5	79.2	74.1	66.6
MUFA	5.79	1.61	7.7	20.3	8.7	24.2
PUFA	55.1	38.4	55.3	36.0	43.1	42.3
Minerals (mg/100g)						
P	481	367	406	418	443	451
K	1353	1246	1359	1328	1140	1531
S	208	n/a	n/a	n/a	n/a	311
Ca	145	132	83	121	155	184
Mg	184	189	138	159	173	192
Na	12.3	15	n/a	n/a	n/a	n/a
Fe	8.42	6.74	6.7	5.9	6.4	11
Al	1.62	n/a	n/a	n/a	n/a	n/a
Mn	1.94	n/a	1.1	1.1	1.3	2.8
Cu	1.20	n/a	0.7	0.8	0.9	1.2
Zn	3.88	2.68	2.8	2.5	2.5	4.3
B	1.32	n/a	n/a	n/a	n/a	1.2
Sugars (%)						
Fructose	0.45	n/a	n/a	n/a	n/a	n/a
Glucose	0.30	n/a	n/a	n/a	n/a	n/a
Sucrose	0.70	n/a	n/a	n/a	n/a	n/a
Raffinose	0.24	n/a	n/a	n/a	n/a	n/a
Stachyose	0.84	n/a	n/a	n/a	n/a	n/a
Verbascose	2.37	n/a	n/a	n/a	n/a	n/a

Note. [1]: Means over two years, two planting dates, two spacings, and two cultivars grown during 2012 and 2013 at Petersburg, Virginia.

[2]: Adapted from: US Department of Agriculture, Agricultural Research Service, Nutrient Data Laboratory. USDA National Nutrient Database for Standard Reference, Release 28. Version Current: September 2015. Internet: http://www.ars.usda.gov/ba/bhnrc/ndl

[3]: Adapted from: Bhardwaj and Hamama (2004, 2005).

4. Conclusions

Results of experiments conducted over two years indicate that mungbean can be easily produced in the United States of America especially the mid-Atlantic region. The nutritional quality of mungbean produced in this region was acceptable in comparison with that of other food legumes such as kidney bean, pinto bean, navy bean, and Tepary bean.

References

AOAC. (1995). *Official Methods of Analysis* (16th ed.). Association of Official Analytical Chemists. Arlington, VA. Retrieved June 19, 2016, from http://www.abebooks.com/servlet/BookDetailsPL?bi=12646263720&se archurl=tn%3Dofficial%2Bmethods%2Banalysis%2Bassociation%2Bofficial%2Banalytical%2Bchemists

Bhardwaj, H. L., & Hamama, A. A. (2004). Protein and Mineral Composition of Tepary Bean Seed. *HortScience, 39,* 1363-1365. Retrieved from http://www.ashs.org

Bhardwaj, H. L., & Hamama, A. A. (2005). Oil and Fatty Acid Composition of Tepary Bean Seed. *HortScience, 40,* 1436-1438. Retrieved from http://www.ashs.org

Bhardwaj, H. L., & Hamama, A. A. (2013). Cultivar and Growing Location Effects on Fatty Acids, Minerals, and Sugars in Green Seeds of White Lupin (*Lupinus albus* L.). *The Open Horticulture Journal, 6,* 1-8. Retrieved from http://benthamopen.com/TOHORTJ/home

Bhardwaj, H. L., & Hamama, A. A. (2015). Cultivar, Planting Date, and Row Spacing Effects on Mungbean Performance in Virginia. *HortScience, 50,* 1309-1311. Retrieved from http://www.ashs.org

Bhardwaj, H. L., Hamama, A. A., & van Santen, E. (2004). Fatty acids and oil content in white lupin seed as affected by production practices. *J. Amer. Oil Chemists Soc., 81,* 1035-1038. http://dx.doi.org/10.1007/s11746-004-1018-0

Bhardwaj, H. L., Rangappa, M., & Hamama, A. A. (1999). Chickpea, Faba Bean, Lupin, Mungbean, and Pigeonpea: Potential New Crops for the Mid-Atlantic Region of the United States. In J. Janick (Ed.), *Perspectives on new crops and new uses* (pp. 202-205). ASHS Press, Alexandria, VA.

ERS. (2015). *Vegetable and Pulse Data.* US Department of Agriculture, Economic Research Service, Washington, D.C. Retrieved March 9, 2016, from http://www.ers.usda.gov/data-products/vegetables-and-pulses-data/data-by-commodity-imports-and-Exports.aspx?reportPath=/TradeR3/TradeTables&programAre a=veg&stat_year=2007&top=5&HardCopy=True&RowsPerPage=25&groupName=DryBeans&commodity Name=Mung and urd beans

Hamama, A. A., Bhardwaj, H. L., & Starner, D. E. (2005). Genotype and Growing Location effects on Phytosterols in Canola Oil. *J. Am. Oil Chem. Soc., 80,* 1121-1126. http://dx.doi.org/10.1007/s11746-003-0829-3

Johansen, H. N., Gilts, V., & Knudsen, K. E. N. (1996). Influence of Extraction Solvent and Temperature on the Quantitative Determination of Oligosaccharides from Plant Materials by High-Performance Liquid Chromatography. *J. Agric. Food Chem., 44,* 1470-1474. http://dx.doi.org/10.1021/jf950482b

Oplinger, E. S., Hardman, L. L., Kaminski, A. R., & Combs, S. M. (1990). J.D. Doll. Mungbean. *Alternative Field Crops Manual.* University of Wisconsin-Extension, Cooperative Extension, Madison, WI. Retrieved from https://hort.purdue.edu/newcrop/afcm/mungbean.html

SAS. (2014). *SAS for Windows version 9.4.* SAS Institute, Cary, N.C. Retrieved from http://www.sas.com/en_us/software/sas9.html

USDA. (2015). *USDA National Nutrient Database for Standard Reference, Release 28.* Version Current: September 2015. US Department of Agriculture, Agricultural Research Service, Nutrient Data Laboratory. Retrieved from http://www.ars.usda.gov/ba/bhnrc/ndl

15

Mixed Cropping System on Diversity and Density of Plant Parasitic Nematodes

Kingsley Osei[1], Haruna Braimah[1], Umar Sanda Issa[1] & Yaw Danso[1]

[1] Crops Research Institute, Kumasi, Ghana

Correspondence: Kingsley Osei, Crops Research Institute, P.O. Box 3785, Kumasi, Ghana. E-mail: oseikingsley4@gmail.com

Abstract

The potential of mixed cropping system on the diversity and suppression of nematodes was investigated at two locations in Ghana. The treatments in the study were; sole plantain, sole cassava and plantain+cassava systems replicated five and four times in a randomized complete block design (RCBD) at Kwadaso in the Ashanti and Assin Foso in the Central region of Ghana respectively. Growth parameters (height and girth) and components of yield (No. of suckers/plant, bunch weight/plant, No. of hands/plant, No. of fingers/plant) were studied on plantain in addition to No. of weevils per plant. On cassava, total biomass, tuber number and tuber weight (yield) were analyzed using GenStat software and means were separated with Fisher's least significance test at $\alpha = 0.05$. There were no differences in height and girth of plantain at Assin Foso. However, plant height was 25% and girth 13% more under sole plantain system over the mixed cropping system at Kwadaso. The sole plantain system recorded 60% and 75% more suckers than the Plantain-Cassava system at both locations. Mixed and sole cropping systems did not influence the diversity of nematode community but significant differences were observed in the density of the nematode taxa encountered under the two systems. Throughout the investigation at both locations, it was observed that the mixed cropping system recorded significantly ($P < 0.05$) lower nematode population densities in comparison to sole cropping system. It is therefore true that an agro-ecological strategy for pests and diseases control is the growing of a mixture of crops differing in their susceptibility to pests and pathogens

Keywords: mixed cropping system, monoculture system, nematode species

1. Introduction

Genetic uniformity of monocultures has been reported to predispose crops to pests and diseases outbreaks (Meung et al., 2003). Since the introduction of plant diversity increases the number of individual functional traits and potential ecosystem services (Hajjar et al., 2008; Malezieux et al., 2009) an agro-ecological strategy for pests and diseases control is the growing of a mixture of crops differing in their susceptibility to pests and pathogens (Smithson & Lenne, 1996; Wolfe, 1985). Mixed culture is recommended for reduction of risk of total crop failure, production of a variety of produce, and improvement of soil fertility where legumes are included which ultimately improve yield of associated crops.

Plant parasitic nematodes (PPN) are among the most important pests of crops worldwide (Yadav & Sehgal, 2010). Mono and mixed cropping systems are affected by PPN parasitism however; the type of mixed cropping system adopted might influence the diversity and density of PPN community. In Ghana, crops may be grown together in mixed cultures often following a system that has been long established and generally successful. It has been observed that crop losses due to pests and diseases are on the ascendency as improved crop cultivars are cultivated in monoculture systems. The greatest incentive to the Ghanaian farmer in practicing mixed culture therefore is food security.

Plantain (*Musa* spp.) is an important food source for many people in the tropics and sub-tropics of the world (Kainga & Seiyabo, 2012). In Ghana, plantain is a starchy staple crop of considerable importance, which contributes about 13% of the Agricultural Gross Domestic Product (GNA, 2007). About 90% of production is consumed locally because plantain is ranked high in food preference (Schill et al., 1996). Consequently it serves as an important source of family income as a result of its high price compared with other starchy staples (Dadzie & Wainwright, 1995). In addition, its production provides job opportunities (Robinson, 2000).

Cassava, *Manihot esculenta* on the other hand, is the most important vegetatively propagated food crop and the second most important food staple in terms of calories per capita in Africa (Nweke et al., 2002). The major nutritional component of cassava is carbohydrate. In Ghana, cassava accounts for a daily calorie intake of 30% and is grown by almost every farming family (FAO, 2006). The importance of cassava to many Africans is epitomized in Ewe (a language spoken in Ghana, Togo and Benin) for the plant, *agble*, meaning "there is life" (Manu-Aduening, 2005). The crop plays an important role in Ghana's economy; it contributes 22% of the Agriculture Gross Domestic Product (Al-hassan, 1989).

Nematode species which are detrimental to plantain are those which destroy the primary roots, disrupting the anchorage system and resulting in toppling of the plants. *Radopholus similis*, *Pratylenchus* species and *Helicotylenchus multicintus* are the most widespread and important (Gowen et al., 2005). Plant-parasitic nematodes most frequently found associated with cassava are *Meloidogyne* spp., *Pratylenchus brachyurus*, *Rotylenchulus reniformis* and *Helicotylenchus dihysteria* (Coyne et al., 2003). These nematode species appear however, of limited importance, with little evidence of significant effect on the crop. However, the significance of nematodes in the cultivation of cassava cannot be overemphasized as some nematodes may interact with other pathogenic organisms in the development of disease complexes (Bridge et al., 2005).

The banana weevil, *Cosmopolites sordidus* is another major pest of plantain. The weevil can confuse the diagnosis of a nematode problem because symptoms of damage are similar (Gowen et al., 2005). With fungi (*Cylindrocarpon* spp., *Fusarium* spp., *Rhizoctonia* spp. and *Cylindrocladium* spp.) the problem becomes even more complex as nematodes and fungi occur within the same cells and infestations result in the same types of discoloration and necrosis (Jones, 2000; Riséde & Simoneau, 2004). In preliminary studies on cassava in Nigeria, the presence of *M. incognita* substantially increased the incidence and severity of damage to storage roots by *Botrydiplodia theobromae* a causal agent of root rot (Dixon et al., 2003).

We must understand and manage these complex organisms so that we may continue to develop and sustain our food production systems (Barker et al., 1994). The potential of farming systems to manage nematode populations below the economic threshold level (ETL) must be investigated. Such a strategy might reduce the over reliance on synthetic chemicals which are detrimental to man and the environment. Therefore, the objective of this study was to evaluate the effect of plantain-cassava mixed culture on the diversity and density of plant-parasitic nematodes in southern Ghana.

2. Materials and methods

2.1 Treatments and Experimental Procedure

Three treatments; sole plantain, sole cassava and mixed plantain-cassava were replicated five and four times in a randomized complete block design (RCBD) at Kwadaso in the Ashanti and Assin Foso in the Central region of Ghana respectively. Both locations are in the forest belt and experience a bi-modal rainfall pattern. The variety of cassava used was "Doku" while the plantain was a local variety. Completely decomposed poultry manure was used to fertilize plantain (in both the sole and mixed culture plots) at a rate of 900 g/plant at planting time. A plot measured 12×12 m. At Kwadaso, plantain was planted on June 15, 2010 and cassava on June 23, 2010. Both cassava and plantain were planted at Assin Fosu on July 15, 2010. Plantain was planted at a spacing of 3×3 m while cassava was at 1×1 m at both locations.

2.2 Soil and Root Sampling for Nematode Assay

Soil samples were collected at two time periods; at the start of the trial before the planting of plantain and cassava and during 14 months after planting when cassava was harvested and about 70% of the first plantain crop had been harvested. Soil samples were randomly collected with a 5 cm soil auger to a depth of 20 cm. Roots of plantain were also sampled at the time of soil sampling. The soil samples, 200 cm^3 per treatment and 5 g of root samples were extracted using the modified Baermann funnel method. After 24 h of extraction, samples were fixed with TAF (Formalin-37% formaldehyde 7.6 ml, Tri-ethylamine 2 ml and distilled water 90.4 ml) and second, third and fourth stage nematodes were mounted on aluminium double-coverglass slides and specimens were identified (CIH, 1978) under a stereo microscope at magnification 100x using morphological characteristics such as the spear (stylet), head skeleton, lumen of the oesophagus, excretory pore and spicules.

2.3 Data Analysis

Statistical analysis was performed using Genstat 8.1 software. Yield, being continuous data was not transformed but nematode count data was log (ln (x + 1)) transformed to improve homogeneity of variance before analysis. Significant mean separation was determined with Fisher's least significance test at $\alpha = 0.05$.

3. Results and Discussion

Growth parameters (height and girth) and components of yield (No. of suckers/plant, bunch weight/plant, No. of hands/plant, No. of fingers/plant) were studied on plantain in addition to No. of weevils per plant. The different farming systems; Sole plantain (mono-cropping) and Plantain + Cassava (mixed cropping) did not show differences in plant growth parameters (height and girth) of plantain at Assin Foso. However, differences were recorded at Kwadaso. Plant height was 25% and girth 13% more under sole plantain system over the mixed cropping system at Kwadaso.

Table 1a. Plantain growth parameters, components of yield and weevil infestation at Assin Foso

Farming system	Height (m)	Girth (cm)	No. suckers	Bunch weight (kg)	No. hands	No. fingers	No. weevils
Sole Plantain	2.05	43.60	5.00	21,792	21,371	95,433	696 (2.8)
Plantain +Cassava	2.02	37.80	2.00	19,592	20,942	82,225	346 (2.5)
Mean	**2.035**	**40.7**	**3.5**	**20,692**	**21,156.5**	**88,829**	**521 (2.65)**
Lsd (5%)	0.97	7.03	1.93	8,292.8	6,386.7	28,844	245 (0.2)

Table 1b. Plantain growth parameters, components of yield and weevil infestation at Kwadaso

Farming System	Height (m)	Girth (cm)	No. suckers	Bunch weight (kg)	No. hands	No. fingers	No. weevils
Sole plantain	2.47	42.37	4.00	32,173	24,417	98,611	312 (1.9)
Plantain +Cassava	1.85	35.99	1.00	26,567	23,067	82,629	227 (1.6)
Mean	**2.16**	**39.18**	**2.5**	**29,370**	**23,742**	**90,620**	**269.5 (1.75)**
Lsd (5%)	0.14	3.29	1.34	7,953.9	3,265.6	3,865.7	94 (1.1)

Similarly, no differences were recorded in bunch weight (yield) at both locations. However, significant differences were observed regarding the No. of suckers/plant and No. of weevils/plant at both locations. The Sole plantain system recorded 60% and 75% more suckers than the Plantain-Cassava system at both locations (Tables 1a and 1b). The mixed cropping system negatively affected sucker production.

The significantly low weevil population recorded under the Plantain-Cassava system was 101% less than the population recorded under the Sole plantain system at Assin Foso (Table 1a). Similarly, the Plantain-Cassava system recorded 34% less weevils than in the Sole plantain system at Kwadaso (Table 1b). The results of the current study corroborate the finding that "the growing of a mixture of crops is an agro-ecological way of controlling pests and diseases" (Smithson & Lenne, 1996).

Three parameters; total biomass, tuber number and tuber weight (yield) were studied on cassava under the two farming systems (Sole cropping and mixed cropping). Differences observed were not significant in any of the parameters studied (Figures 1a and 1b).

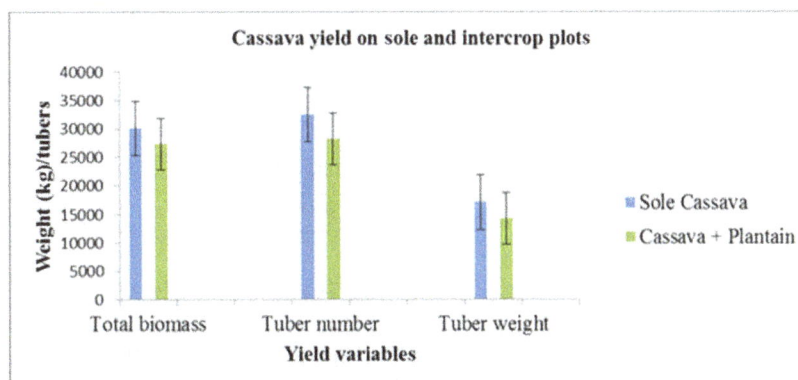

Figure 1a. Cassava yield in sole and intercrop systems at Assin Foso

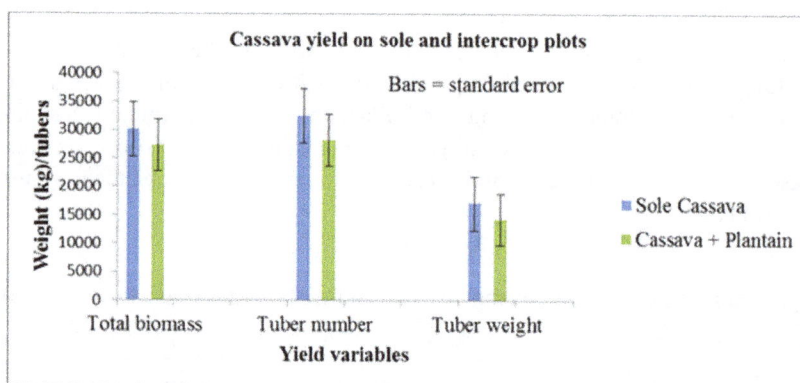

Figure 1b. Cassava yield in sole and intercrop systems at Kwadaso

Four plant parasitic nematodes belonging to the Order: Tylenchida were encountered at both locations at the beginning and at the end of experiment. The nematodes encountered were: *Meloidogyne* spp., *Pratylenchus coffeae, Rotylenchulus reniformis* and *Helicotylenchus multicintus*. From the initial soil samples, nematode population/200 cm^3 soil was comparatively higher at Assin Fosu with *R. reniformis* being the most abundant while *H. multicintus* predominated at Kwadaso (Figure 2).

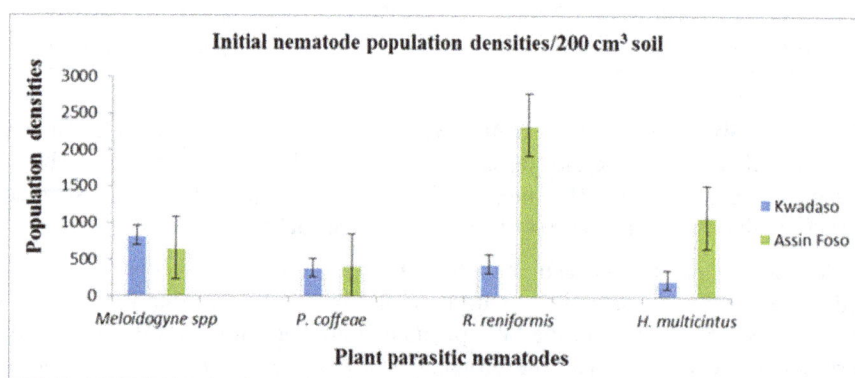

Figure 2. Initial nematode population densities/200 cm^3 soil

Throughout the investigation at both locations, it was observed that the mixed cropping system recorded significantly (P < 0.05) lower population densities regarding; *Meloidogyne* spp., *P. coffeae* and *R. reniformis* in comparison to sole cropping treatment. However, there were no differences in population of *H. multicintus* in both treatments at both locations. The sole cropping system recorded (82 and 637) % more *Meloidogyne* spp. and *R. reniformis* than the mixed cropping system at Assin Foso while (351 and 280) % more *Meloidogyne* spp. and *R. reniformis* were recorded at Kwadaso respectively (Tables 2a and 2b).

Table 2a. Plant parasitic nematodes/200 cm^3 in soil samples at harvest at Assin Foso

Farming system	*H. multicintus*	*Meloidogyne* spp.	*P. coffeae*	*R. reniformis*
Sole plantain	9.5 (0.58)[1] b	51.0 (1.56) c	42.0 (1.55) c	33.2 (1.47) b
Plantain + Cassava	9.2 (0.42) b	28.0 (1.42) b	14.2 (0.31) b	4.5 (0.29) a
Sole cassava	1.0 (0.00) a	8.8 (0.91) a	4.0 (0.42) a	9.0 (0.56) a
Mean	**6.6 (0.33)**	**29.3 (1.29)**	**20.07(0.76)**	**15.57 (0.77)**
Lsd (5%)	0.37 (0.29)	0.18 (0.01)	0.10 (0.07)	0.03 (0.06)

Note. Values are means of 4 replications. Means followed by the same letters do not differ significantly. [1]Log transformed [ln (x + 1)] data used in analysis in parenthesis.

Table 2b. Plant parasitic nematodes/200 cm^3 in soil samples at harvest at Kwadaso

Farming system	H. multicintus	Meloidogyne spp.	P. coffeae	R. reniformis
Sole plantain	71.4 (1.81)[1] b	307.0 (2.42) b	261.0 (2.12) b	293.0 (2.37) b
Plantain + Cassava	29.8 (1.45) b	68.0 (1.59) a	12.0 (0.88) a	77.0 (1.85) ab
Sole Cassava	1.0 (0.00) a	55.0 (1.65) a	7.0 (0.62) a	28.0 (1.11) a
Mean	**34.1 (1.09)**	**143.3 (1.89)**	**93.3 (1.21)**	**132.7 (1.78)**
Lsd (5%)	0.002 (0.001)	0.012 (0.014)	0.052 (0.002)	0.032 (0.015)

Note. Values are means of 5 replications. Means followed by the same letters do not differ significantly. [1]Log transformed [ln (x + 1)] data used in analysis in parenthesis.

The sole cassava system recorded significantly least population densities in all the four nematodes encountered at both locations which did not result in any differences in the parameters studied on cassava (Tables 2a and 2b). All the four nematode species have been reported to be associated with plantain and cassava cultivation (Coyne et al., 2003; Gowen et al., 2005). The insignificantly low nematode numbers recovered from the rhizosphere of cassava has confirmed the fact that nematodes are of limited importance, with little evidence of significant effect on the crop (Bridge et al., 2005).

Table 3a. Plant parasitic nematode population densities/5 g plantain root at Assin Foso

Treatment	H. multicintus	Meloidogyne spp.	P. coffeae	R. reniformis
Sole plantain	210 (1.26)[1]b	166 (2.15) b	104 (1.98) b	24 (1.18) b
Plantain+ cassava	4.2 (0.57) a	24 (1.18) a	18 (1.10) a	4.0 (0.56) a
Mean	**126 (0.92)**	**95 (1.67)**	**61 (1.54)**	**14 (0.87)**
Lsd (5%)	0.088 (0.071)	0.096 (0.115)	0.050 (0.034)	0.074 (0.02)

Note. Values are means of 4 replications. Means followed by the same letters are not significantly different. [1]Log transformed [ln (x + 1)] data used in analysis in parenthesis.

Table 3b. Plant parasitic nematode population densities/5 g plantain root at Kwadaso

Treatment	H. multicintus	Meloidogyne spp.	P. coffeae	R. reniformis
Sole plantain	148 (2.05)[1] b	147 (2.10) b	297 (2.19) b	66 (1.59) b
Plantain+ cassava	22 (0.79) a	34 (1.14) a	76 (1.25) a	8 (0.64) a
Mean	**85 (1.42)**	**90 (1.62)**	**186.5 (1.72)**	**37 (1.12)**
Lsd (5%)	0.040 (0.017)	0.027 (0.033)	0.216 (0.019)	0.047 (0.005)

Note. Values are means of 5 replications. Means followed by the same letters do not differ significantly. [1]Log transformed [ln (x + 1)] data used in analysis in parenthesis.

Nematode population densities recovered from plantain roots under the sole cropping system were significantly (P < 0.05) higher than under the mixed cropping system. For instance, *Meloidogyne* spp. and *Pratylenchus coffeae* populations under the sole plantain system were (592 and 332%) and (478 and 291%) higher than under the mixed cropping system at Assin Foso and Kwadaso respectively (Tables 3a and 3b). Severe infestation of plantain root system by plant parasitic nematodes often result in toppling over of plants particularly at fruiting stage (Gowen et al., 2005) leading to significant yield losses.

Acknowledgements

Authors are grateful to the West Africa Agricultural Productivity Programme (WAAPP) for funding the study.

References

Al-Hassan, R. M. (1989). Cassava in the economy of Ghana. In: Status of data on cassava in major producing countries of Africa. *COSCA working paper No. 3*. Collaborative study on Cassava in Africa, IITA, Ibadan, Nigeria.

Barker, K. R., Hussey, R. S., & Krusberg, L. R. (1994). Plant and soil nematodes: societal impact and focus for the future. *J. Nematol., 26*, 127-137.

Bridge, J., Coyne, D. L., & Kwoseh, C. K. (2005). Nematode parasites of tropical root and tuber crops. In M. Luc, R. A. Sikora, & J. Bridge (Eds.), *Plant Parasitic Nematodes in Subtropical and Tropical Agriculture* (2nd ed., pp. 221-258). CAB International Publishing, Wallingford, UK. http://dx.doi.org/10.1079/9780851997278.0221

CIH. (1978). *Commonwealth Institute of Helminthology Description of plant parasitic nematodes*. CAB International, Wallingford, UK.

Coyne, D. L., Talwana, H. A. L., & Maslen, N. R. (2003). Plant-parasitic nematodes associated with root and tuber crops in Uganda. *Afr. Plt. Protect., 9*, 87-98.

Dadzie, B. K., & Wainwright, H. (1995). Plantain utilization in Ghana. *Trop Sci., 35*, 405-410.

Dixon, A. G. O., Bandyopadhyay, R., Coyne, D., Ferguson, M., Ferris, R. S. B., Hanna, R.,…Ortiz, R. (2003). Cassava: From poor farmer's crop to pacesetter of Africa rural development. *Chronica Hortica, 43*, 8-15.

FAO. (2006). Food Outlook. *No. 1 June Outlook.*

GNA. (2007). *Ghana News Agency Report 2007*. Plantain production in Ghana increases

Gowen, S. R., Quénéhervé, P., & Fogain, R. (2005). Nematode parasites of bananas and plantains. In M. Luc, R. A. Sikora, & J. Bridge (Eds.), *Plant Parasitic Nematodes in Subtropical and Tropical Agriculture* (2nd ed., pp. 611-643). CAB International Publishing, Wallingford, UK. http://dx.doi.org/10.1079/9780851997278.0611

Hajjar, R., Jarvis, D. I., & Gemmill-Herren, B. (2008). The utility of crop genetic diversity in maintaining ecosystem services. *Agric. Ecosystems & Ent., 123*, 261-270. http://dx.doi.org/10.1016/j.agee.2007.08.003

Jones, D. R. (2000). *Diseases of Banana, Abaca and Enset*. CAB International, Wallingford, UK.

Kainga, P. E., & Seiyabo, I. T. (2012). Economics of plantain production in Yenagoa local government area of Bayelsa State. *J. Agric.and Soc Research, 12*(1), 114-123.

Malezieux, E., Crozat, Y., Dupraz, C., Laurans, M., Makwoski, D., Ozier-Lafontaine, H., … Valantin-Morrison, M. (2009). Mixing plant species in cropping systems: Concepts, tools and models. A review. *Agron. for Sustainable Devt., 29*, 43-62. http://dx.doi.org/10.1051/agro:2007057

Manu-Aduening, J. (2005). *Participatory Breeding for superior mosaic-resistant cassava in Ghana* (pp. 33-38, PhD. Thesis). University of Greenwich, UK.

Meung, H., Zhu, Y. Y., Revilla-Molina, I., Fan, J. X., Chen, H. R., Pangga, I., … Mew, T. W. (2003). Using genetic diversity to achieve sustainable rice disease management. *Plt. Disease, 87*, 1156-1169. http://dx.doi.org/10.1094/PDIS.2003.87.10.1156

Nweke, F. I., Spencer, D. S. C., & Lynam, J. K. (2002). *The cassava transformation, Africa's best-kept secret*. East Lansing, MI, USA: Michigan University Press.

Riséde, J. M., & Simoneau, P. (2004). Pathogenic and genetic diversity of soil borne isolates of Cylindrocladium from banana cropping systems. *Eur. J. Plt. Pathol., 110*, 139-154. http://dx.doi.org/10.1023/B:EJPP.0000015337.54178.c0

Robinson, J. C. (2000). Banana productivity: The impact of agronomic practices. *Proceedings of the International Symposium on Banana and Plantain for Africa, Acta Hort., 540 ISHS 2000* (pp. 247-258).

Schill, P., Gold, C. S., & Afreh-Nuamah, K. (1996). Assessment and characterization of constraints in plantain production in Ghana as an example for West Africa. Plantain and banana production and research in West and Central Africa. *Proceedings of a regional workshop, held at Onne, River State, Nigeria, IITA, Ibadan, Nigeria* (pp. 45-51).

Smithson, J. B., & Lenne, J. M. (1996). Varietal mixtures: A viable strategy for sustainable productivity in subsistence agriculture. *Ann. Applied Biol., 128*, 127-158. http://dx.doi.org/10.1111/j.1744-7348.1996.tb07096.x

Wolfe, M. S. (1985). The current status and prospects of multiline cultivars and variety mixtures for disease resistance. *Ann. Rev. Phytopathol., 23*, 251-273. http://dx.doi.org/10.1146/annurev.py.23.090185.001343

Yadev, S. M., & Sehgal, M. (2010). Management of plant parasitic nematodes through chickpea-groundnut cropping system. *Pakis. J. Nema., 28*(2), 361-362.

16

Adoption and Use of Precision Agriculture in Brazil: Perception of Growers and Service Dealership

Emerson Borghi[1], Junior C. Avanzi[2], Leandro Bortolon[3], Ariovaldo Luchiari Junior[4] & Elisandra S. O. Bortolon[3]

[1] Embrapa Milho e Sorgo, Sete Lagoas, Brazil

[2] University of São Paulo, Pirassununga, Brazil

[3] Embrapa Pesca e Aquicultura, Palmas, Brazil

[4] Embrapa Informática Agropecuária, Campinas, Brazil

Correspondence: Leandro Bortolon, Embrapa Pesca e Aquicultura, Palmas, TO, Brazil.
E-mail: leandro.bortolon@embrapa.br

The research is financed by Fundação Agrisus, Embrapa.

Abstract

Precision agriculture (PA) is growing considerably in Brazil. However, there is a lack of information regarding to PA adoption and use in the country. This study sought to: (i) investigate the perception of growers and service dealership about PA technologies; (ii) identify constraints to PA adoption; (iii) obtain information that might be useful to motivate producers and agronomists to use PA technologies in the crop production systems. A web-based survey approach method was used to collect data from farmers and services dealership involved with PA in several crop production regions of Brazil. We found that the growth of PA was linked to the agronomic and economic gains observed in the field; however, in some situations, the producers still can not measure the real PA impact in producer system. Economic aspects coupled with the difficulty to use of software and equipment proportioned by the lack of technical training of field teams, may be the main factors limiting the PA expansion in many producing regions of Brazil. Precision agriculture work carried out by dealership in Brazil is quite recent. The most services offered is gridding soil sampling, field mapping for lime and fertilizer application at variable rate. Many producers already have PA equipment loaded on their machines, but little explored, also restricting to fertilizers and lime application. Looking at the currently existing technologies and services offered by dealership, the PA use in Brazil could be better exploited, and therefore, a more rational use of non-renewable resources.

Keywords: soybean, precision agriculture, maize, Brazilian agriculture

1. Introduction

Precision agriculture (PA) involves the development and adoption of some techniques to improve the management of agricultural systems, aiming to optimize inputs applications such as fertilizers, pesticides, seeds and irrigation resources to reduce inputs costs and maximize the crop production (Bora et al., 2012), besides to reduce environmental impacts (Bramley et al., 2008). In several crop production Brazilian regions, PA has been played an important role in crop production systems, mainly due to the technical and economic benefits that PA provides over the years. Costa and Guilhoto (2011) stressed that the benefits of PA adoption, impacts directly on social and economic benefits of Brazilian agricultural economy. However, the benefic effects of PA are more restricted to large cropped areas, usually operated by major companies linked to crop production. Pierpaoli et al. (2013) found that the size of cropped area is the most important parameter to farmers when they have to decide to adopt PA, due to the higher possibility to increase income. Then, properties with large cropped area has more potential to be capable to invest large amount of resources, time and learning in order to use PA technologies compared to properties with small cropped area (Adrian et al., 2005).

Fertilizer optimization has been the major target to use PA in Brazil (Costa & Guilhoto, 2011). However, Bora et al. (2012) showed that in North Dakota, farmers that adopted GPS systems or automatic steering observed reduction of fuel consumption and machine operation time. Besides the economical aspects, negative environmental impacts arising should be reduced with PA adoption due to a more rational use of inputs in crop production systems (Bramley et al., 2008). Australian farmer's point of view about the low adoption of PA in Australia relays on technology frustration and the lack of technical support in the field. They pointed out that technology costs was not the overriding factor to PA adoption (Mandel et al., 2010). On the other hand, Batte and Arnholt (2003) analyzed six farms in Ohio (US) that recently adopted PA technologies and the profitability was the major factor that motivated farmers to adopt PA, although even not all farmers surveyed have observed the global profitability linked to PA adoption. The farmers surveyed also pointed out that on-farm research, quality information generated by PA to support decision and risk reduction in the environmental contamination were the major concerns to adopt PA. More than a decade ago Swinton and Lowenberg-Deboer (2001) concluded that PA adoption/expansion would increase slowly in area with high population and, in area with less cropland available unless the environmental benefits would be very well reported.

Reichardt and Jürgens (2009) found in German conditions that some major issues related to PA adoption were lack of technical support to PA tools and lack of knowledge to manage correctly the data to apply them correctly in the crop production system. For German farmers, the systems incompatibility among several companies' suppliers stills the major constraint to PA adoption. Batte and Arnhorld (2003) concluded that to increase PA adoption, the development of more simple technologies is the most important contribution to support farmers in the decision making process. In Alabama, farmers PA adoption is related to well establish farmers with large cropped areas and more educated level (Adrian et al., 2005).

Precision agriculture adoption survey has been conducted over the years in United States. In 2011, 85% of the respondents reported that they have been used at least one PA technology in the crop production system (Whipker & Erickson, 2011, 2013). Similar results were found in the previous survey in the same region (Whipker & Akridge, 2009). Although similarity among results was found regarding to PA technology use, Whipker and Erickson (2013) found that the use of GPS guided systems with autocontrol/autosteer was the major used over years. This fact emphasizes that implementation of new PA technologies is more suitable to be embraced for farmers that already use any PA technology available. For instance, PA technology adoption by new users has been increasing annually with 76 and 83% in 2007 and 2008 respectively (Whipker and Akridge 2007; 2008).

There is little information available regarding to PA adoption in Brazil. Silva et al. (2011) demonstrated that PA adoption in sugarcane production increased sugarcane yield and quality, and also increased profitability to farmers. On the point of view of sugarcane industry, the reduction in the environmental impact was the major issue to take attention. The use of PA technologies will be essential to sustainability of Brazilian agribusiness and mainly to achieve higher crop yield while reduce environmental impact (Silva et al., 2011). Precision technology adoption use can affect directly the economy at regional and large scale (Costa & Guilhoto, 2013). A study that evaluated scenarios such as i) increase in crop yield; ii) input reduction; iii) increase in crop yield and reduction in inputs, and; iv) increase in crop yield and increase in inputs, concluded that the major impact was on increase of crop yield, that impact directly in social benefits (employment raised) and economics benefits (increase of income) to Brazilian economy (Costa & Guilhoto, 2013). The benefits of inputs reductions is solely to increase farmers income and it is not reflect in economic benefits to the society; then, the benefits to the society must be analyzed on the point of view of reduction to environmental impact (Costa & Guilhoto, 2013). Silva et al. (2007) attempting to clarifying the costs of PA technology in order to increase the adoption by farmers, carried out a comparative analysis of the costs and economic profitability involved in implementing PA and conventional farming practices in the state of Mato Grosso do Sul. Even though PA presented higher effective operational cost comparing to traditional agriculture due to technical assistance, maintenance of sophisticated equipment, yield and soil mapping, for example, the unitary cost (*i.e.*, the cost per kilogram) in the precision agriculture system was lower than the cost in the traditional system.

The benefits of PA adoption are widely known and transferred to agricultural systems. Although crop yield is one the major factor that impact PA adoption in some cases, is important to scientist to understand how the perception of farmers and agronomists about PA are in the crop production systems. Research with farmers, agronomists and users about their perceptions of PA technologies are limited in Brazil and there is a gap of information and knowledge that must be filled. Then, our study sought to investigate the perception of growers and service dealership about PA technologies and to identify constraints to PA adoption and also to obtain information that might be useful to motivate producers and agronomists to use PA technologies in the crop production systems.

2. Method

A web-based survey was used to collect data from farmers and services dealership involved with precision agriculture in several crop production regions of Brazil. Survey used was similar found in Silva et al. (2011) to evaluate PA adoption in Sao Paulo State (Brazil) to sugarcane production system.

We previously identified users of PA technologies to obtain their perceptions and support needed and followed the same approach of Diekmann and Batte (2010) adjusted for Brazilian conditions. After this process, we develop a web-based survey to obtain information from two groups: 1) farmers that use AP technology, and; 2) professionals and companies that are technical support providers or farmer consultants. The division in two groups was the same approach used by Reichardt and Jürgens (2009). Web-based survey was developed based on the same approach in Whipker and Akridge (2009) adapted to Brazilian conditions. Both web-based surveyes were accessed by respondents with the follow links: https://sites.google.com/site/agriculturadeprecisaotocantins/ which was developed to farmers to answer it; https://docs.google.com/spreadsheet/viewform?formkey=dG9UUG 51WEhtTVZQSXJCR0ZNTVRna3c6MQ to be answered by professionals and companies that are technical support providers or farmer consultants. The sampling frame used to select the respondents was lists of individuals from Precision Agriculture Network coordinated by Embrapa (Brazilian Agricultural Research Corporation) and stakeholders, machinery companies' technical representatives and crop production sales personnel. Similar approach to sampling respondents was used to Larson et al. (2008) and Dieckman and Batte (2010). For each group separately an e-mail was sent them explaining the survey goals and the respective link to the web-survey. Based on prior published data that involved survey (mail or web), we expected about one third of the invited to respond the web-survey (Larson et al., 2008; Dieckman & Batte, 2010).

2.1 Web-Survey to Farmers

The survey included questions about respondent's as follow: year started PA in the farm; total area used with PA considering lime and fertilizers application with variable rate; use PA in the decision making process for crop and soil management; soil sampling using grid or GPS; soil sampling grid size; soil sampling in soil layers; how PA is adopted in the farm (own equipment or contract the service); list of PA equipment's used in the farm; farming operation used as variable rate application (VRA); observations related to PA use (*e.g.* crop production cost reduction); problems found regarding to equipment's maintenance and software support; PA technical support; investments; observations in increase crop yield; constraints to adopt PA at farm and regional scale.

2.2 Web-Survey to Professionals and Companies that Are Technical Support Providers or Farmer Consultants

The survey included the following questions: major company activity; how long the company is on the PA market; average size of the farms assisted; PA market grow since started to work with PA; total area in hectares the service dealership assist; PA area in the farms assisted; service most required by farmers; average soil sampling grid size; percent of PA service in the company income; PA impact on crop production reduction costs to assisted farmers; PA equipment's available to be used in the farmers that are assisted; company grow expectation in the next years; expectations to increase PA adoption at regional scale and the constraints observed and found to consolidate PA as crop and soil management practice to increase nutrient use efficiency.

2.3 Data Analysis

Primary data obtained from the web-survey were analyzed considering percentages of questions answered (Silva et al., 2011). A total of 250 e-mails were sent to farmers and 10% were answered from several states such as Goiás (GO), Rio Grande do Sul (RS), Paraná (PR), Maranhão (MA) and Tocantins (TO). Although returned answers was relatively low, it was similar that was found in similar studies with survey research (Whipker & Akridge, 2009; Holland et al., 2013; Watcharaanantapong et al., 2014; Whipker & Erickson, 2013). The answer from professionals and companies that are technical support providers or farmer consultants reached a large number of crop production region. The respondent companies were from states of São Paulo (SP), Mato Grosso (MT), Bahia (BA), Paraná (PR) and Rio Grande do Sul (RS). Answers from companies were from several regions of Brazil despite the company headquarter was located in one of the listed states above, giving more accurate results.

3. Results and Discussion

3.1 Vision of Farmers that Are PA Users

Analyzing farmer respondents we identified that soybean and maize are the most common crops that PA is used (Table 1). Both crops represent the major Brazilian commodities. Silva et al. (2007) already demonstrated that PA technology may guarantee higher production, decreasing the unitary cost and, consequently, making the system more rewarding on a long-term basis for soybean and maize. On the other hand, the adoption of PA technologies by farmers in area smaller than 200 hectares were found in crops such as cotton, pasture, beans, sugarcane, wheat and coffee (Table 1). It is important to highlight that large producers of cotton or sugarcane, for example, were not contemplated in this study. Silva et al. (2011) investigating the PA adoption from sugar-ethanol companies figured out an area cultivated with sugarcane, using PA technology, much larger than those given for this crop.

Precision agriculture in Brazil is relatively recent were 67% of the respondents stated that are PA users between two and five years (Figure 1a) and about 20% of the farmers respondents are PA users for more than eight years. Maybe for PA is recently widely adopted, farmers do not realize the technology cost-benefit, with more than half of respondents use less than 2,000 ha with PA (Figure 1b). According to Brazilian official statistics, areas up to 2,000 ha represent over 99% of the number of Brazilian farms, which represent about 57.2% of farmland (Incra, 2012). Forty-four percent of the respondents use more than 2,000 ha with precision agriculture technology. These data suggest that technology adoption is mainly with larger producers, similar with reported by Adrian et al. (2005). Among the farmers who already are PA users, about 67 percent of respondents answered they have designed the whole farm area with some PA technology (data not showed).

Table 1. Crops and area used for precision agriculture in both surveyed farms and web-based survey

Crops	Area (ha)				
	Less than 200	200 to 500	500 to 800	800 to 1000	More than 1,000
Soybean	0%	22%	11%	0%	67%
Maize	22%	22%	0%	0%	56%
Cotton	78%	0%	0%	0%	22%
Pasture	56%	0%	11%	0%	33%
Beans	67%	11%	0%	0%	22%
Sugarcane	78%	0%	0%	0%	22%
Wheat	67%	11%	0%	0%	22%
Coffee	78%	0%	0%	0%	22%

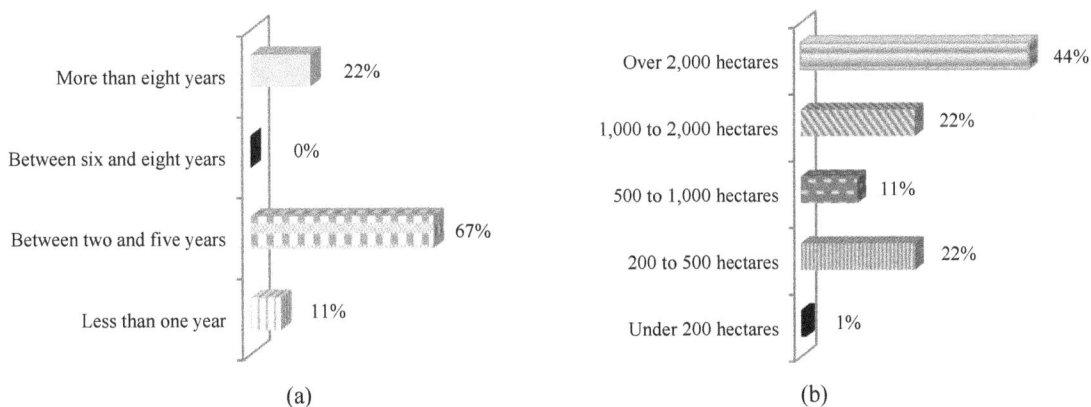

(a)   (b)

Figure 1. Frequency distribution of (a) years of PA adoption in the farm, and (b) PA area adopted in the farm (b). Data from questionnaires in the crop season 2011/12

All the survey respondents collect soil samples using grid (Figure 2a), representing the major tool used among the PA technologies available, although not all dealership offer this service. The cost related to soil sampling grid and analysis remains relatively high, resulting in soil sampling grid over five hectares and mostly to soil fertility evaluation at 0-20 topsoil layer.

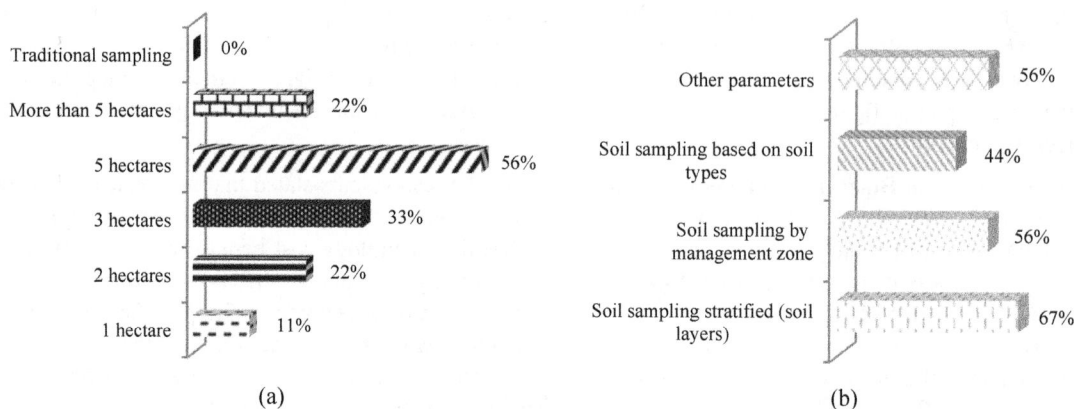

(a)   (b)

Figure 2. Soil sampling grid size used by farmers (a), and parameters for georeferenced soil sampling in addition to the grid size (b). Data from questionnaires in the crop season 2011/12

Some farmers also carried soil sampling in soil layers (67%); however, management zones and soil type are less used (Figure 2b). Brazilian soils are highly weathering, for which a soil sample in layers below 20 cm are performed mainly by farmers with higher input, with the objective to apply lime or gypsum to achieve better yields and to prevent yield loss due to drought.

Basically PA service has been provided by technical support providers or companies; however, it was observed that farmers are increasing PA. Total answers about this topic, 33% stated that all PA service (soil sampling, field mapping, results interpretation and recommendation) was carried out by dealership; although 45% informed that technical support companies are responsible to soil sampling and field mapping, and the results interpretation and recommendation are organized by farmer's technical team (Figure 3).

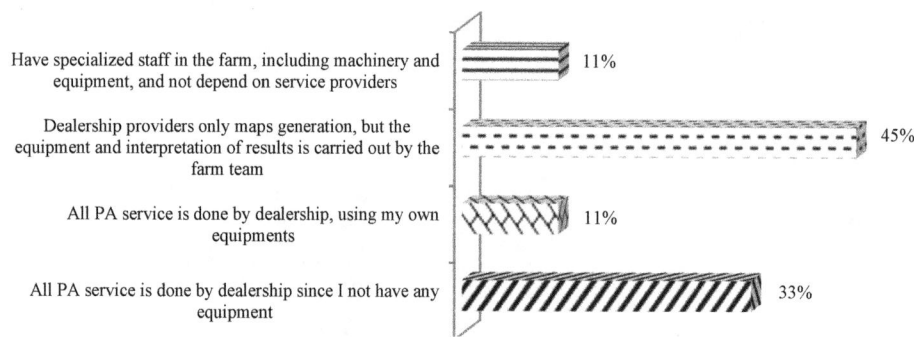

Figure 3. Precision agriculture technical support. Data from questionnaires in the crop season 2011/12

Regarding the use of variable rate application (VRA), without specifying which one (fertilizer, lime, pesticides or seeding), 44% of farmers surveyed used in more than 2,000 hectares for this service (Figure 4a). About 22% of respondents had planned between 1,000 and 2,000 ha and between 200 and 500 ha. These data are similar with those found for the areas designed for PA on farms (Figure 1b). Such information suggests that the incorporation of new areas by the PA users is through the VRA service. Therefore, PA technology of VRA along with soil sampling represents the major services adopted in Brazil.

All PA users adopt controller in soil amendment practices, followed by application of fertilizers (Figure 4b). Variable seeding rates is the least significant service in its category. In order to performing VRA, farmers had been using several equipment's to ancillary them. Manual control systems (light bar) were the most popular type of guidance system (Figure 5). In the USA survey, GPS guidance systems for custom application showed considerable advancements in PA technology. In past surveys, the use of manual control system increased, until the 2009 survey when their popularity reached its peak (Whipker & Akridge, 2009). Automatic control systems (autosteer) showed trending upwards in recent surveys, representing nowadays 47.5% of the guidance system technology (Holland et al., 2013). In our results, we noticed many farmers employing both systems (likely each machine has different systems), although manual control systems still are more common use (Figure 5).

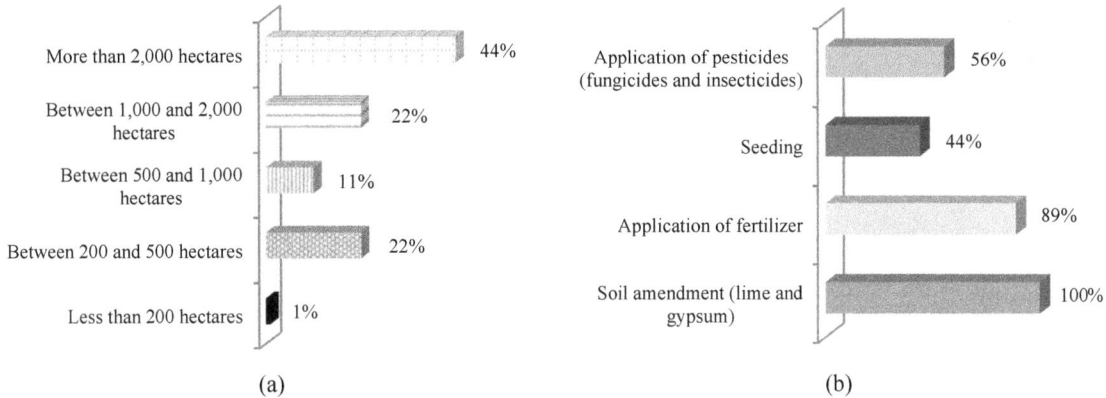

(a)   (b)

Figure 4. Area designed to variable rate application (a), and type of operation that uses variable rate application in the farm (b). Data from questionnaires answered by farmers in the crop season 2011/12

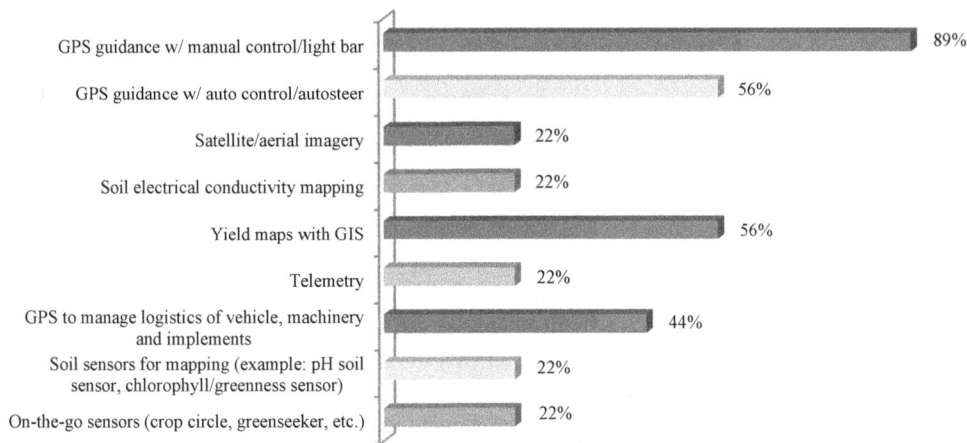

Figure 5. Which precision technology have you been used in the field. Data from questionnaires in the crop season 2011/12

After the control system, yield maps were the most used precision technologies (Figure 5). GPS for logistics also had been used for almost 50% of the PA adopters. Telemetry for field-to-home office communications was one of the biggest increases observed for the USA 2013 survey reaching 15.2% (Holland et al., 2013), which was much less than observed in Brazilian survey. Although farmers observed a cost reduction in the crop production costs due to PA adoption, most of farmers did not know in which situations PA had been impacted in the crop production costs (Table 2). Most of respondents would not be able to determine or measure the real impact in the cost production reduction considering lime, fertilizer and herbicide application or maintenance and consulting with software and equipment (Table 2). This information is corroborated with data showed in Figure 6a, where only one third of respondents informed that observed increase in crop yield was lower than 5% after PA adoption. More than 22% of respondents confirmed that the increase in crop yield ranged from 6 to 10% and in equal proportion the crop yield observed was over 40%. Farmers that did not observed increase on crop yield and do not know the real economic benefits of PA adoption, usually opted to invest in machinery and equipment (Figure 6a).

Table 2. Crop production reduction costs due to precision agriculture adoption. Data from questionnaires in the crop season 2011/12

	Lime application	Fertilizer application	Herbicides application	Maintenance and consulting with software and equipment
Less than 10%	22	34	44	33
10 to 20%	22	22	0	0
20 to 30%	12	0	0	0
More than 30%	0	0	0	0
Do not know	44	44	56	67

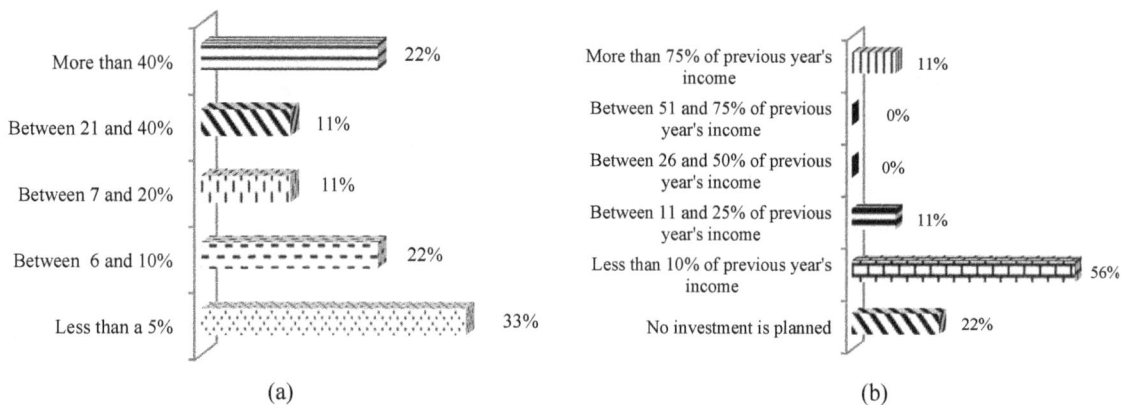

Figure 6. Crop yield observed due to PA adoption (a), and investments in precision technology for the next years (b). Data from questionnaires answered by farmers in the crop season 2011/12

According to investments expectations for the further years, 56% of the farmers answered that they will invest less than 10% of the gross income past year-based in PA. On the other hand, 22% answered that no investments in PA were planned for the next growing season. Results showed PA can increase crop yield from respondent's perspective; however, economic balance needs to be clarified to farmers because it was not clear to respondents if PA adoption in fact is more profitable. Silva et al. (2007) showed that in maize and soybean crops in Brazil the unitary effective cost was lower with use of PA technologies; however, we have to point out that many farmers do not make a detailed description and control in their production costs, making dificult to realize the financial benefits of PA. Based on this scenario, adding cost of production of each growing season into account makes PA adoption very susceptible to commodities market. Then, farmers' decision to adopt PA relayed on costs and economic balance at short term. For most respondent's, to continue or to invest in PA is highly dependent of commodities prices for the next growing season.

There were limitations to increase PA adoption and respondent's disagreed that PA costs are higher than the benefits observed (Table 3), even they could not measure a reduction in the production cost (Table 2) it was possible to see benefits with PA adoption. Farmer's respondent also disagreed that there are constraints to increase PA use in agriculture such as topography, soil type; they also stressed out that timing from gather information and maps generation is relatively short and acceptable to decision making process in a growing season. Another very important point raised from the results was that the amount of information regarding to new products, technologies and training for field team to use PA technology still need to be improved.

The companies providing PA service have showed good relationships with producers, supplying information such as costs management and benefits with the PA adoption. However, analyzing the limitations (Table 3), the farmers agree that this service generates a security recommendations, although the producers did not know quantify (Figure 6a). Farmers also reported that PA cost is still high, especially in the acquisition of equipment and software, but the amounts charged are not excessive and are consistent with the technology.

For producer's perspective, nowadays equipment and software used in the PA are barriers to growth and investment in technology. The incompatibility between different equipment and hardware device (*e.g.* data formats, information sharing) has limited the use; Reichardt and Jürgens (2009) reported similar obstacles. In addition, the manuals are quite complex and lack of training of field teams, together with the lack of skilled labor trained to operate machines and equipment with embedded PA systems, contribute for low investment in PA or new users over the years.

Regarding to service provides, farmer's respondents agree that it is too difficult to keep updated with constant new technologies in PA, since dealership provides the PA service or the option to purchase of machinery, but did provide free technology upgrades. Furthermore, respondents also agree that data collection can suffer climatic or operational interference that were not fully corrected by software, thus compromising data accuracy and therefore the recommendation. The lack of PA monitoring for dealerships related in updating and maintenance of software and equipment, training and technical accompaniments, make the field teams remain limited on proper handling of equipment. Pieces for timely replacement are also important obstacles.

Table 3. Limitations that constraint PA adoption to increase responded from farmers from several Brazilian regions in the crop season 2011/2012. Values in percent (%) calculated from answers signed in each specific item

Limitations	1	2	3	4	5
PA costs to producers are higher than the benefits observed	22	45	0	11	22
Soil type in the field reduce PA profitability	56	11	0	11	22
Field topography limit PA use	67	11	0	11	11
Long time between gathering information and map processing for decision making process	45	11	11	22	11
Confidence on the recommendations based on field division grid-based	12	0	33	33	22
Benefits of PA adoption in the own business	11	0	33	45	11
Values applied in PA are not excessive and are coherent with technology applied	0	22	22	45	11
Constraint to find qualified personnel to handling equipment	11	11	33	33	11
High costs software and equipment acquisition	11	0	33	45	11
Constraint to convince the profitability increase with PA adoption	33	22	0	33	11
High costs with personnel	11	11	11	56	11
PA equipments are very changeable and the costs are high	11	11	11	56	11
Incompatibility between different software's in the market	0	33	11	45	11
Incompatibility between different software's and research-based recommendations	0	33	22	33	11
Equipment and software with very complex manuals to understand	11	11	22	45	11
Constraint to personnel training and support to handling software and equipments	0	22	33	33	11
There are software and equipment's that are not accurate to use in PA	0	11	22	56	11
Data gathering with interferences (climatic, operational, etc.) that makes difficult the results accuracy	11	22	22	33	11
Incompatibility between equipment and technologies constraint the ability to offer new products and services to clients	0	22	22	45	11
Companies do not offer software maintenance	0	33	22	33	11
Companies do not make available software update and/or new free updates via internet	0	33	11	45	11
Little information regarding to new products/technologies in PA	0	45	11	33	11
My team do not receive training about the correct use of software/equipment	11	45	11	22	11
No costumer service response regarding to complaint about use and maintenance product	22	33	11	22	11
Constraint to find replacement parts for my equipment's	11	22	22	33	11
Companies do not give cost management and benefits to market access	11	22	11	22	33
There is none tool available in the market at the same PA level to decision making process and planning for inputs acquisition and use	0	11	22	45	22

Note. 1: Completely disagree; 2: Partially disagree; 3: Fully agree; 4: Partially agree; 5: Not agree/not disagree.

Although more than half want to invest less than 10% of the previous year's income (Figure 6b) in PA tools, hindering great adhesion to PA technologies and still be many limitations to overcome (Table 3), farmers expect improvements in some precision agriculture technologies, or new PA technology options (Figure 7). Farmers have great expectation arising new tools for recommendation and application of fertilizer and lime. In the same scale of expectations (78%), an integrated interpretation considering data analysis in different database is expected. With a lower frequency of responses, but expected still with 67 percent, the variable rate seeding, increase in technologies of automatic applications, and sensors for application of variable rates, where this technology can perform controlled applications including formulated fertilizers.

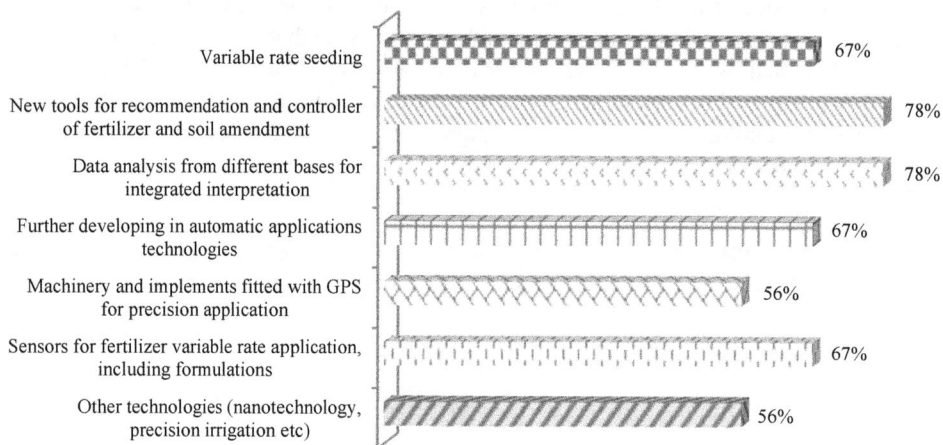

Figure 7. Expectation of new useful precision technologies for your agribusiness. Data from questionnaires answered by farmers in the crop season 2011/12

3.2 Vision of Professionals and Companies that Provides Technical Support

Before of twenty-first century, the difficulty in access to knowledge about PA technology and especially, the cost of purchase new equipment or tools has contributed to prevent the growth of PA. The advent of dealerships, particularly in mechanization and agricultural automation segments, offering the technology already available embedded in machinery and outsourcing of PA sector by technicality companies provided great leap in the use and dissemination of the benefits of PA in Brazilian agriculture. This statement is corroborated with Figure 8a, which shows the PA use in the regions served by the dealerships. 50 percent of respondents reported that the total area of properties that employ PA is between 1,000 and 2,000 hectares. On the other hand, 25% of companies assist farmers with area smaller than 200 hectares.

All dealership answered that the time operation in the PA market only from two to five year. This information suggest the PA work carried out by service providers companies is relatively recent in Brazil. Regarding to services offered, all companies mentioned performing georeferenced soil sampling and prepare maps for lime and fertilizer variable rate applications (Figure 8b). Holland et al. (2013) reported soil sampling with GPS was the most popular use in the USA 2013 survey. Besides the above-mentioned service, half of responding dealership also offering agronomic consulting services, and only a quarter offers product for buying or technical assistance for PA equipment (Figure 8b).

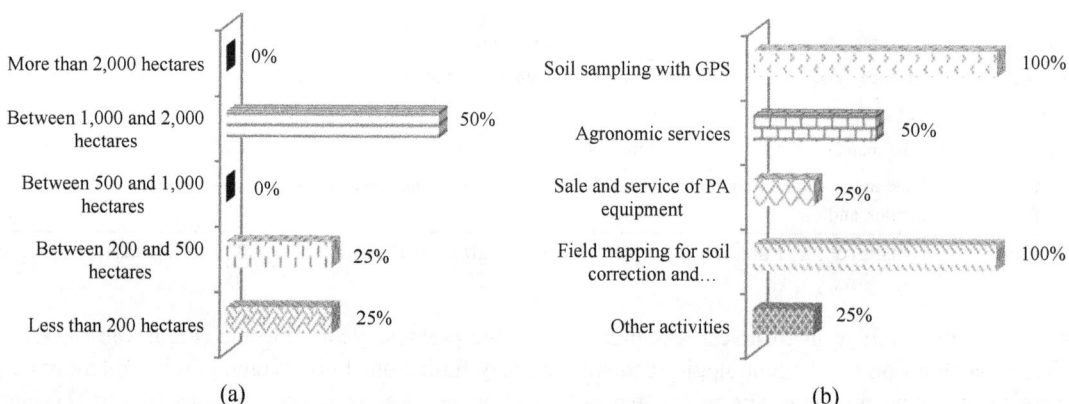

(a)   (b)

Figure 8. Property size assisted by dealership (a) and; company main activities, more than one answer (b). Data from questionnaires answered by dealership in the crop season 2011/12

Costumers have dedicated between 20-40% of the total cropland to use with PA (Figure 9). The other half of the costumers is divided among those who use less than 20% of the area and between 40-60% of the area. Even among the producers who adopt the PA, the dealerships reported that none answered that in the farm is used more than 60% of the area with this technology. As reported by producers (Table 2), this fact may be related to the difficulty of skilled labor, employee training, cost to purchase equipment and software, incompatibility among others. Thus, it can be noted that there is still great scope for PA expansion, considering that the area increase with technology occurs mainly among the producers who already employ some PA technology (Whipker and Erickson 2013).

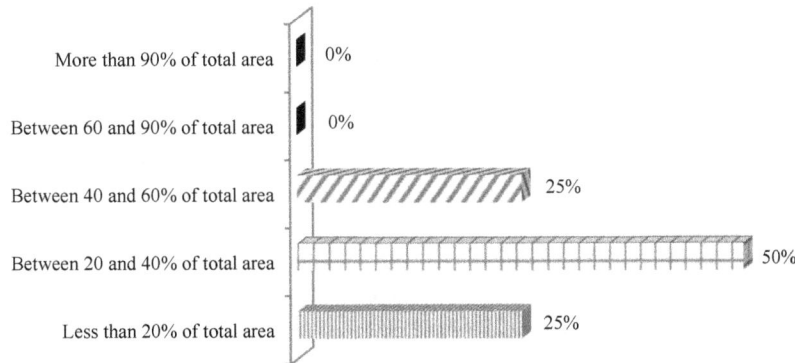

Figure 9. Percentage of area that customers intended to PA. Data from questionnaires answered by dealership in the crop season 2011/12

Both for producers and for dealerships, the use of PA is restricted to the use of some technologies, far below the potential that the PA can offer. The most service sought by customers is the controller to lime and fertilizer application. Dealership reported that all its customers have sought the PA service for lime VRA and 75% of farmers still sought companies to fertilizer VRA. Such percentage is higher that reported by Holland et al. (2013) for American farmers.

Grid size of soil sampling was quite varied; however, 50% of the dealership responded that their customers choose to grids of 5 hectares (Figure 10a), corroborating with producers response (Figure 2a). For the recent America PA survey, the most common was the grid sample between 1 and 2 hectares in size (Holland et al., 2013). The smallest grid requested by producers to the companies was to 2 hectares (25% of respondents), although the dealership also offer grid size of 1 hectare. Due to high cost to denser sampling, map generation and variable rate application, producers opt for larger grid sizes in order to minimize costs, even though there is a direct relationship between the grid size and the cropland variability, which may affect the amount and accuracy of the fertilizers application.

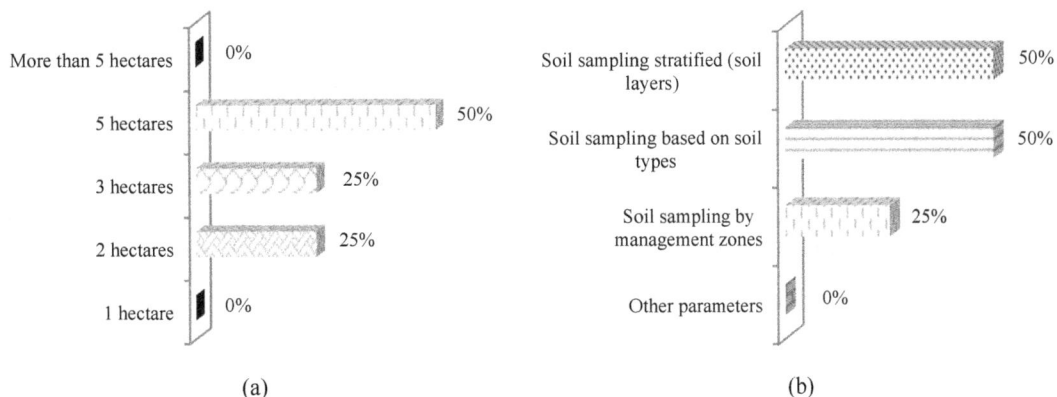

(a)   (b)

Figure 10. Soil sample grid size (a) and; other parameters for perform georeferenced soil sampling (b). Data from questionnaires answered by dealership in the crop season 2011/12

In addition to the sampling grid, dealerships often offer soil sampling georeferenced taking into account other parameters (Figure 10b). In this analysis, 50 percent of survey respondents reported that perform soil samples by layer (0-20 and 20-40 cm) considering the different types of soils on properties. Only 25% of dealership reported that they perform georeferenced soil sampling taking into consideration of management zones. Holland et al. (2013) reported 54% of dealership offered sampling following a grid pattern and 35% offered sampling by management zone. By the currently existing technologies in Brazilian producing regions, the use of management zones for decision making process is still little explored by both dealership and producers, showing great potential for improvement of management more aware of the soil and inputs.

Survey showed farmers already have in their PA machines coupled equipment such as GPS, light bar, harvest monitors etc.; however, they perform some PA work with dealership. Companies reported that the vast majority of producers also have GPS with manual control/light bar guidance system for fertilizer and lime application (75%). To a lesser extent (25%), survey respondents reported that, in addition to GPS, they also found in the producer's machines equipment like harvest monitor with GIS and autopilot in order to guide the operations of mechanized land preparation and cultivation in the fields. Even using some PA tool many producers do not yet have any PA equipment (25%), although already perform soil sampling and lime and fertilizers at variable rates application.

Farmers who already have equipment in their machines perform basically the lime and fertilizers VRA. Only a small proportion of dealership customers perform application of pesticides using the equipment available in machinery (25%). For these answers, it appears that the PA service offered by the dealership on these customers is restricted to soil sampling and field mapping for lime and fertilizers at variable rates application.

Although the producers did not know how to inform the actual reduction in production costs (Table 1), the service providers reported that three-quarters of its customers achieved a reduction in production costs between 11 and 20% (Figure 11a) and for the rest of this reduction was less than 10%. This view by dealership may be related to its marketing strategy to convince new users to acquire some service offered by companies (Table 3). Concerning to growth of the PA market (Figure 11b) the wide majority of companies reported an increase between 6 and 10%.

Most companies already have, since before 2010, in their service list the fertilizers and lime application (Table 4), having GPS navigation to perform this operation. However, many producers make this application with its own machinery and employees (Figure 3). Within this topic application technology, the pesticide application service at a variable rate is still very incipient and is a place where businesses can grow, bringing enormous environmental benefits.

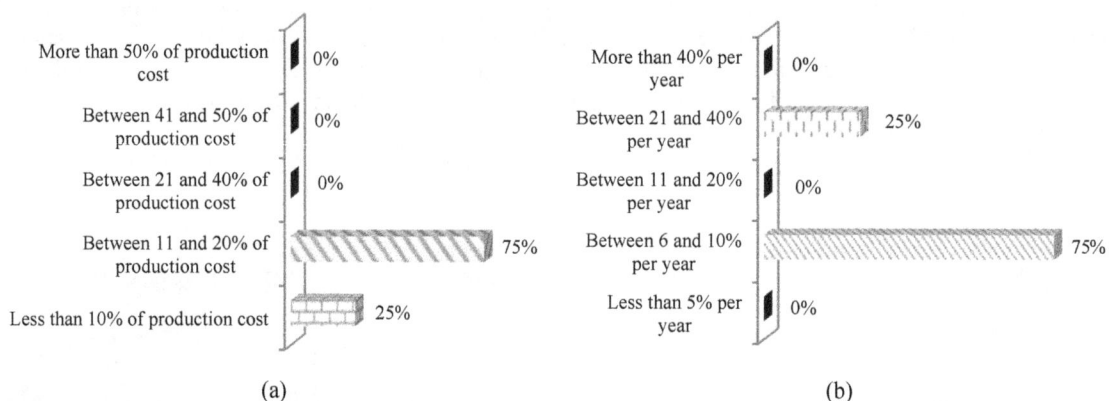

Figure 11. Average reduction of precision technology that dealership observed in their customers (a) and, growing of PA market (b). Data from questionnaires answered by dealership in the crop season 2011/12

Table 4. Products/services offered by companies that providing services and trends. Values in percentages calculated from the responses marked on each item. Data from questionnaires answered by dealership in the crop season 2011/12

Products/Services	Since 2010	Offer by 2011	Intend to offer	Don't know
Soil sampling with GPS	78	0	0	22
Soil sampling for layer with GPS	33	11	11	45
Variable rate application				
Fertilizers	56	0	11	33
Lime	56	0	11	33
Pesticides	12	0	44	44
Controller-driven (GPS) for application				
Fertilizers	45	0	22	33
Lime	45	0	22	33
Pesticides	12	0	44	44
Yield maps	45	11	11	33
Sales / technical support / rental				
Yield monitor without GPS	11	11	33	45
Yield monitor with GPS	0	0	33	67
Softwares and equipments	11	11	22	56
Field maps: fertility, yield, monitoring of pests, diseases, weeds etc.	22	0	33	45
Recommendation of fertilizers, lime and pesticides through field maps	78	0	0	22
Sale / technical support for aerial images/satellite	22	0	33	45
Controller for variable rate application, single nutrient	45	0	22	33
Controller for variable rate application, multiple nutrients	12	0	44	44

We can also notice that the PA branch companies focus their efforts on providing services rather than the sale or rental of PA products. This may be related to the fact that large companies of machinery and equipment now offer this technology at the moment of the machinery sale. This information can be a big opportunity for companies offer a rental service to small farmers who plan to take PA technologies on their property, since they do not need to purchase new machines with load technology. This is another option for service providers expand its operations.

Dealership also responded one series of questions about the limitations for increasing the adhesion to PA technologies. Although a small portion of farmers reported that they agreed that the producer cost was higher than the benefits (Table 3), the vast majority of companies and service providers reported that totally disagree that the costs are higher than the benefits (Table 5). This lower perception of farmers can be due to many of them do not have a detailed control of the cost with and without application of PA tools.

Comparing the results from Table 3 and Table 5, we obserced similarity between the answers given by producers and service providers. However, some points should be highlighted. For example, around 45% of farmers respondent figured out benefits of PA adoption in the own business (Table 3). This percentage could be higher according to the service providers' vision, or even among those who realized this benefit it could be more significant. Once the companies responded that 75% completely disagree that their customers who could have benefits from PA technologies are already users. While 25 percent fully agree (Table 5).

Table 5. Limitations that constraint PA adoption to increase responded from dealerships from several Brazilian regions in the crop season 2011/2012. Values in percent (%) calculated from answers signed in each specific item.

Limitations	1	2	3	4	5
PA costs to producers are higher than the benefits observed	75	25	0	0	0
Soil type in the field reduce PA profitability	50	50	0	0	0
Field topography limit PA use	75	25	0	0	0
Long time between gathering information and map processing for decision making process	0	50	0	25	25
Confidence on the recommendations based on field division grid-based	0	0	25	75	0
All customers who could have benefits with PA already are PA users	75	0	25	0	0
Values applied in PA are not excessive and are coherent with technology applied	0	50	50	0	0
Constraint to find qualified personnel to handling equipment	0	0	75	25	0
High costs software and equipment acquisition	0	25	50	25	0
Constraint to convince the profitability increase with PA adoption	25	25	25	25	0
High costs with personnel	0	25	50	25	0
PA equipments are very changeable and the costs are high	0	0	25	75	0
Incompatibility between different software's in the market	0	25	50	25	0
Incompatibility between different software's and research-based recommendations	0	50	25	25	0
Equipment and software with very complex manuals to understand	0	50	25	25	0
Constraint to personnel training and support to handling software and equipments	0	0	75	25	0
There are software and equipment's that are not accurate to use in PA	0	50	25	25	0
Data gathering with interferences (climatic, operational, etc.) that makes difficult the results accuracy	25	25	0	50	0
Incompatibility between equipment and technologies constraint the ability to offer new products and services to clients	0	50	25	0	25
Companies do not offer software maintenance	25	25	25	25	0
Companies do not make available software update and/or new free updates via internet	25	50	25	0	0
New companies have difficult to introduce new PA products	0	25	50	25	0
Software companies offer training about the correct use of software/equipment	0	50	25	25	0
Software companies offer service response regarding to complaint about use and maintenance product	25	25	0	50	0
Constraint to keep replacement parts for customers equipment's	50	50	0	0	0
With PA tools the companies can plan inputs acquisition and use	0	25	75	0	0
There is negotiation on the costs of services provided to farmers	0	0	75	25	0

Note. 1: Completely disagree; 2: Partially disagree; 3: Fully agree; 4: Partially agree; 5: Not agree/not disagree.

A point that is worth emphasized is that according to the producers, one limitation is finding replacement spare parts (Table 3). However, the service providers as a whole disagree with this statement, reporting not having trouble keeping replacement parts in stock (Table 5).

Services providers companies were also surveyed about expectation of new PA technologies (Figure 12). In general they showed lower expectations when compared with producers. However, it is noteworthy that a higher expectation regarding to technologies for implementing the variable rate seeding. Such service also have been reported in the last USA survey, being among the services that more is expected to grow by 2018 (Widmar & Erickson, 2015).

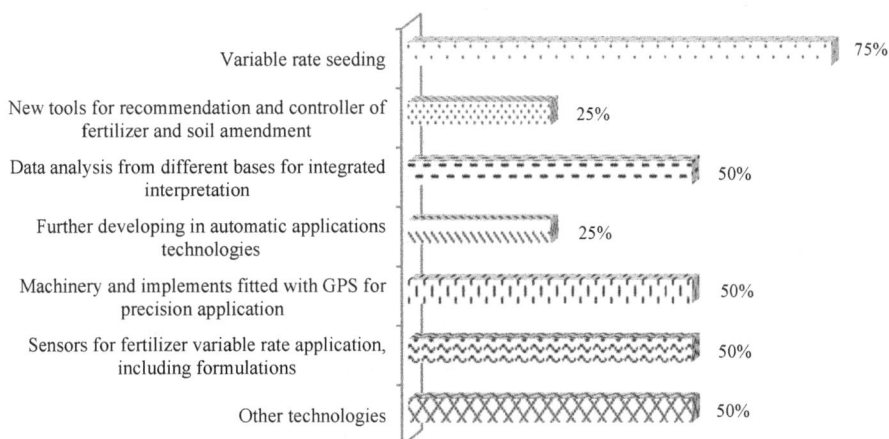

Figure 12. Expectation of new useful precision technologies for future. Data from questionnaires answered by farmers in the crop season 2011/12

While most farmers expect to invest less than 10% of previous year's income (Figure 6b), 50% of service provides respondents expect to reinvest between 11 and 25% of its profit. The other half are divided equally between those wishing to reinvest less than 10%, and the more daring, those how expected to reinvest their profit in the range of 26-50%. This information suggests that service providers are more hopeful about the growth of PA activity in Brazil, likely managing to reach new users.

Results showed the complexity of how PA has been used in Brazil. Since farmers and consultants answered questionnaire from a wide range Brazil region, it was noted that PA is still recent in Brazil, which means that are many opportunities to grow. Based on our results we observed that PA adoption is a quite restrictive and PA needs some adjustments to increase its adoption in Brazil, even with knowledge and benefits that PA brings to crop production system. Precision agriculture adoption could be increased if the environmental benefits due to PA adoption would be better explained to user, besides the economic benefits (Swinton and Lowenberg-Deboer 2001).

4. Conclusion

The growth of precision agriculture is due to the agronomic and economic gains already known in the field; however, in some situations, the producers continue not able to measure the real PA impact in its agribusiness. The information obtained, the economic aspect, coupled with the difficulty in the use of software and equipment proportioned by the lack of technical training of field teams, may be the main factors limiting the PA expansion in many producing regions of Brazil.

Precision agriculture work carried out by dealership in Brazil is quite recent. The most services offered is soil sampling with grids between 2-5 hectares in size, field mapping for lime and fertilizer application at variable rate. Many producers already have PA equipment loaded on their machines, but little explored, also restricting to fertilizers and lime application. Looking at the currently existing technologies and services offered by dealership, the PA use in Brazil could be better exploited, and therefore, a more rational use of non-renewable resources.

References

Adrian, A. M., Norwood, S. H., & Mask, P. L. (2005). Producers' perceptions and attitudes toward precision agriculture technologies. *Computers and Electronics in Agriculture, 48*, 256-271. http://dx.doi.org/10.1016/j.compag.2005.04.004

Batte, M. T., & Arnholt, M. W. (2003). Precision farming adoption and use in Ohio: Case studies of six leading-edge adopters. *Computers and Electronics in Agriculture, 38*, 125-139. http://dx.doi.org/10.1016/S0168-1699(02)00143-6

Bora, G. C., Nowatzki, J. F., & Roberts, D. C. (2012). Energy savings by adopting precision agriculture in rural USA. *Energy, Sustainability and Society, 2*, 22. http://dx.doi.org/10.1186/2192-0567-2-22

Bramley, B. G. V., Hill, P. A., Thorburn, P. J., Kroon, F. J., & Panten, K. (2008). Precision agriculture for improved environmental outcomes: Some Australian perspectives. *Agriculture and Forestry Research, 3*, 161-178.

Costa, C. C., & Guilhoto, J. J. M. (2011). Impactos da agricultura de precisão na economia brasileira. In R. Y. Inamasu, J. M. Naime, A. V. Resende, L. H. Bassoi, & A. C. Bernardi (Eds.), *Agricultura de precisão: Um novo olhar* (pp. 307-313). São Carlos: Embrapa.

Costa, C. C., & Guilhoto, J. J. M. (2013). Impactos potenciais da agricultura de precisão sobre a economia brasileira. *Revista de Economia e Agronegócio, 2*, 177-204.

Diekmann, F., & Batte, M. T. (2010). 2010 Ohio Farming Practices Survey: Adoption and Use of Precision Farming Technology in Ohio. *Ohio State University Extension, 2010* (p. 19, Report Series, AEDE-RP-0129-10).

Holland, K. J., Erickson, B., & Widmar, D. A. (2013). *2013 Precision Agricultural Services Dealership Survey Results*. Department of Agricultural Economics, Purdue University, West Lafayette, IN. Retrieved September 7, 2015, from http://agribusiness.purdue.edu/files/resources/rs-11-2013-holland-erickson-wid mar-d-croplife.pdf

INCRA (Instituto Nacional de Colonização e Reforma Agrária). (2012). *Total Brasil*. Retrieved September 16, 2015, from http://www.incra.gov.br/media/politica_fundiaria/regularizacao_fundiaria/estatisitcas_cadastrai s/imoveis_total_brasil.pdf

Larson, J. A., Roberts, R. K., English, B. C., Larkin, S. L., Marra, M. C., Martin, S. W., ... Reeves, J. M. (2008). Factors affecting farmer adoption of remotely sensed imagery for precision management in cotton production. *Precision Agriculture, 9*, 195-208. http://dx.doi.org/10.1007/s11119-008-9065-1

Mandel, R., Lawes, R., & Robertson, M. (2010). Farmer perspectives of precision agriculture in Western Australia: Issues and the way forward. In H. Dove & R. A. Culvenor (Eds.), *Food Security from Sustainable Agriculture, Proceedings of 15th Agronomy Conference 2010, November 15-18, 2010, Lincoln, New Zealand.*

Pierpaoli, E., Carli, G., Pignatti, E., & Canavari, M. (2013). Drivers of precision agriculture technologies adoption: a literature review. *Procedia Technology, 8*, 61-69. http://dx.doi.org/10.1016/j.protcy.2013.11.010

Reichardt, M., & Jürgens, C. (2009). Adoption and future perspective of precision farming in Germany: Results of several surveys among different agricultural target groups. *Precision Agriculture, 10*, 73-94. http://dx.doi.org/10.1007/s11119-008-9101-1

Silva, C. B., Moraes, M. A. F. D., & Molin, J. P. (2011). Adoption and use of precision agriculture technologies in the sugarcane industry of São Paulo State, Brazil. *Precision Agriculture, 12*, 67-81. http://dx.doi.org/ 10.1007/s11119-009-9155-8

Silva, C. B., Vale, S. M. L. R., Pinto, F. A. C., Müller, C. A. S., & Moura, A. D. (2007). The economic feasibility on precision agriculture in Mato Grosso do Sul State, Brazil: a case study. *Precision Agriculture, 8*, 255-265. http://dx.doi.org/10.1007/s11119-007-9040-2

Swinton, S. M., & Lowenberg-Deboer, J. (2001). Global adoption of precision agriculture technologies: who, when and why? In G. Grenier & S. Blackmore (Eds.), *Proceedings of the 3rd European Conference on Precision Agriculture* (pp. 557-562). Agro Montpellier, Montpellier, France.

Watcharaanantapong, P., Roberts, R. K., Lambert, D. M., Larson, J. A., Velandia, M., English, B. C., ... Wang, C. (2014). Timing of precision agriculture technology adoption in US cotton production. *Precision Agriculture, 15*, 427-446. http://dx.doi.org/10.1007/s11119-013-9338-1

Whipker, L. D., & Akridge, J. T. (2007). *2007 Precision Agricultural Services Dealership Survey Results*. Working Paper #07-13, Department of Agricultural Economics, Purdue University, West Lafayette, IN.

Whipker, L. D., & Akridge, J. T. (2008). *2008 Precision Agricultural Services Dealership Survey Results*. Working Paper #08-09, Department of Agricultural Economics, Purdue University, West Lafayette, IN.

Whipker, L. D., & Akridge, J. T. (2009). *2009 Precision Agricultural Services Dealership Survey Results*. Working Paper #09-16, Department of Agricultural Economics, Purdue University, West Lafayette, IN.

Whipker, L. D., & Erickson, B. (2011). *The state of precision agriculture 2011*. Retrieved May 10, 2014, from http://www.croplife.com/article/print/23009

Whipker, L. D., & Erickson, B. (2013). *2011 Precision Agricultural Services: Dealership Survey Results*. Working Paper #13-2, Department of Agricultural Economics, Purdue University, West Lafayette, IN.

Widmar, D., & Erickson, B. (2015). *2015 precision agricultural dealer survey: top four trends to watch.* Retrieved September 29, 2015, from http://www.croplife.com/equipment/precision-ag/2015-precision-agriculture-dealer-survey-top-four-trends-to-watch

Winstead, M. T., Norwood, S. H., Griffin, T. W., Runge, M., Adrian, A. M., Fulton, J., & Kelton, J. (2010). Adoption and use of precision agriculture technologies by practitioners. *Proceedings of the 10th International Conference on Precision Agriculture (ICPA), Denver, CO, USA, CD-ROM* (pp. 18-21).

17

Light Quality Effect on Corn Growth as Influenced by Weed Species and Nitrogen Rate

Thomas R. Butts[1], Joshua J. Miller[2], J. Derek Pruitt[3], Bruno C. Vieira[1], Maxwel C. Oliveira[4], Salvador Ramirez II[3], & John L. Lindquist[3]

[1] Department of Agronomy & Horticulture, University of Nebraska-Lincoln, North Platte, NE, USA

[2] Department of Plant Pathology, University of Nebraska-Lincoln, Lincoln, NE, USA

[3] Department of Agronomy & Horticulture, University of Nebraska-Lincoln, Lincoln, NE, USA

[4] Department of Agronomy & Horticulture, University of Nebraska-Lincoln, Concord, NE, USA

Correspondence: Thomas R. Butts, Department of Agronomy & Horticulture, University of Nebraska-Lincoln, North Platte, NE 69101, USA. E-mail: tbutts@huskers.unl.edu

Abstract

Corn-weed competition has often been characterized as the competition for limited resources such as light quantity, water, and nutrients. However, growing evidence suggests that light quality, specifically the red:far red ratio (R:FR), is a crucial component to corn-weed interactions. Additionally, a reduction in the R:FR has shown to down-regulate plant genes similarly to a nitrogen (N) deficient environment. A greenhouse study was conducted to evaluate the effect of N stress and R:FR from common waterhemp, velvetleaf, and volunteer corn on corn growth and development. The R:FR for all three weed species tended to be similar but lower than a weed-free treatment. However, observations from the spectral response curves demonstrated significant changes in the patterns of light reflected from each weed species. In the N-sufficient environment, early-season (V5 corn growth stage) R:FR from all three weed species reduced corn height, leaf chlorophyll content, and shoot biomass while increasing fibrous root biomass. However, in the N-deficient environment, no effects were observed on corn growth from changes in light quality, indicating N stress was a greater limiting factor. These results highlight the importance of the critical weed-free period and the need for proper early-season weed management.

Keywords: chlorophyll, competition, light quality, red:far red ratio (R:FR)

1. Introduction

Competition between corn (*Zea mays* L.) and weeds presents a significant barrier to crop production worldwide. Traditionally, competition has been viewed from the understanding that weeds compete for limited resources - namely light, water, and nutrients (Kropff & Laar, 1993). Competition for these resources can influence yield, plant height, and leaf area development (Bonifas et al., 2005; McCullough et al., 1994; Zhou et al., 1997). The timing of weed interference also influences the level of competition on corn. Competitive effects of weeds are more important early in the growing season, so early-season weed control is essential to minimize yield loss (Zimdahl, 1988; Swanton et al., 1999). Several studies have identified critical weed free periods for corn that evaluate both the length of time weeds can remain in a crop before interference occurs, and the length of time weed control efforts must be maintained (Hall et al., 1992; Knezevic et al., 1994; Bosnic & Swanton, 1997). One factor that was found to influence the critical weed free period was the application of nitrogen (N) (Evans et al., 2003).

A growing body of evidence suggests that another mechanism for competition is a reduction in the red (670 nm) to far red (730 nm) ratio (R:FR) intercepted by the corn plant due to the presence of neighboring weeds (Rajcan et al., 2004). This mechanism results from two signal-transducing photoreceptors, phytochrome (600-800 nm) and blue-absorbing photoreceptors (300-500 nm), that acquire information on the light environment to modulate cellular processes (Smith, 1982; Smith & Holmes, 1977). Phytochrome has been the primary focus of research investigating the influence of light quality on crop growth and development. The phytochrome molecule can detect both the proximity and distribution of neighboring plants because of reductions in the R:FR ratio (Ballaré, 2009). Particularly in shade-intolerant crops, this can result in a shade avoidance response (Page et al., 2010).

This response can result in several physiological changes in the plant, including stem elongation, reduction in stem diameter, and a reduction in the root and shoot biomass (Afifi & Swanton, 2011). Additionally, Afifi and Swanton (2012) found that in corn subjected to low R:FR light from both biological and nonbiological sources, phytochrome was involved in decreased anthocyanin and increased lignin, increased H_2O_2 content in the first true leaf and crown, increased stomatal closure on the first and second leaves, and changed expression levels of genes involved in auxin transport, ethylene biosynthesis, scavenging enzymes and anthocyanin and lignin biosynthesis pathways.

Three weeds of concern for crop production in the United States are velvetleaf (*Abutilon theophrasti* Medik), common waterhemp (*Amaranthus rudis* Sauer), and volunteer corn. Velvetleaf has been ranked as one of the most troublesome weeds in the United States (Stoller et al., 1993) and is commonly a competitive annual weed in summer crops globally (Loddo et al., 2013). Although yield loss in corn due to velvetleaf competition can be highly variable (Lindquist et al., 1996), research has suggested that grain yields can be decreased by up to 80% depending on field conditions and velvetleaf density (Lindquist et al., 1998). This loss is primarily due to competition for light (Lindquist & Mortensen, 1999). Common waterhemp is a dioecious, indigenous species native to the Great Plains region of the United States (Sauer, 1957). It has spread recently due to several factors, including the adoption of reduced tillage practices, the reduction in soil applied residual herbicides, and the evolution of herbicide-resistant biotypes (Steckel & Sprague, 2004). Previous research has shown when common waterhemp emerges with corn through the V6 growth stage, grain yield can be dramatically reduced (Steckel & Sprague, 2004). Volunteer corn tends to be less of a concern in hybrid corn production, but can still impact yield of neighboring corn plants. Marquardt et al. (2012) observed reductions in hybrid corn grain yield above 20%; however, no reduction in total grain yield was observed when the volunteer corn grain yield was combined with the hybrid corn grain yield.

In addition to weeds, N stress can significantly affect corn grain yields. Adequate N supply during vegetative stages is essential for achieving optimum yields (Rajcan & Swanton, 2001). Roughly 65-80% of total N uptake by corn plants occurs during the vegetative stages of crop growth (Rajcan & Tollenaar, 1999). Adequate N during the vegetative stages is also important because 50% or more of N found in grain comes from mobilization rather than increased uptake (Pearson & Jacobs, 1987; Ta & Weiland, 1992). Moriles et al. (2012) determined that weed competition and N stress both resulted in the down regulation of genes involved with photosynthesis, auxin signaling, H_2O_2 removal and stomatal movement.

In the current study, changes in R:FR interception by corn due to the presence of velvetleaf, common waterhemp, and volunteer corn plants were evaluated to determine the influence light quality has on corn growth and development. Additionally, because N stress is known to regulate gene expression similarly to weed competition, responses to R:FR changes were evaluated in corn under two N management regimes. The objectives of this study were to 1) determine the differences in reflected R:FR of velvetleaf, common waterhemp, and volunteer corn, 2) evaluate the impact of R:FR from different weed species on corn height, stem diameter, leaf chlorophyll content, and plant biomass, and 3) determine if N stress changes the response of corn to changes in R:FR. From these objectives, we hypothesized the R:FR would vary for each weed species observed and this variance in R:FR would impact the singular corn plant differently: a lower R:FR would decrease corn height, stem diameter, leaf chlorophyll content, and plant biomass more than a higher R:FR. Furthermore, we hypothesized N stress paired with low R:FR would produce an additive negative effect on the corn plant growth characteristics compared with N stressed corn plants alone.

2. Method

2.1 Experimental Design and Maintenance

A greenhouse experiment was conducted in the spring of 2016 at the University of Nebraska-Lincoln East Campus Greenhouse facility. Four weed species (none, common waterhemp, velvetleaf, and volunteer corn) and two N rates [366.5 ppmw (full) and 209.0 ppmw (reduced)] were arranged in a randomized complete block design as a four by two factorial. Treatments were spatially replicated four times. The greenhouse was maintained at 26-30 °C with 14 hours of light.

The experiment was designed to provide light quality competition between the corn plant and weed species, while eliminating light quantity, nutrient, and root architecture competition. To accomplish this, weed species were sown in 7.6-cm diameter pots, and four of these pots were placed equidistantly apart on the outside rim of a 30.5-cm pot in which the corn was planted directly in the center (Figure 1) (Liu et al., 2009). The growing medium was a 1:1 mixture of perlite:vermiculite specifically chosen to allow for complete control of the nutrient amendments (E. T. Paparozzi, University of Nebraska-Lincoln, personal communication). Weed species were sown and allowed to emerge prior to the corn plant to establish a weedy environment in which the corn plant would emerge. Common waterhemp and velvetleaf were planted on February 10 and were thinned to four plants per 7.6-cm pot after emergence. Volunteer corn was planted on February 17 and thinned to two plants per 7.6-cm pot after emergence to better simulate a corn cropping system scenario. DeKalb DKC62-98VT2RR2 corn variety, a typical hybrid corn used in Nebraska, was planted on February 20 and was thinned to one viable plant per 30.5-cm pot. As the plants grew, weed species were trimmed once per week to maintain no competition for light quantity and avoid shading of the centered corn plant.

Figure 1. Experiment set-up with weed species sown in 7.6-cm pots placed equidistantly apart on the outside of a 30.5-cm pot in which a corn plant was planted directly in the center and forced to emerge into a weedy environment

Weeds and corn plants were watered and fertilized with Modified Hoagland nutrient solutions (Table 1) developed using a protocol by Clark (Clark, 1982). To determine the amount of nutrient solution needed each fertigation timing, the 7.6- and 30-cm pots were filled to saturation prior to planting to determine the holding capacity of each size pot. Weeds were initially watered by overhead irrigation with tap water until plants emerged. After emergence, the full N solution was used to fertigate the weed species with 100 mL, and the corn plants were fertigated with 1 L of the respective N solution treatment three times per week.

Table 1. Nutrient concentrations and nitrogen (N) treatments supplied at each fertigation

Nutrient	Full N (ppmw)[a]	Low N (ppmw)
Nitrogen (N)	366.5	209.0
Phosphorus (P)	20.4	20.4
Potassium (K)	282.9	282.9
Calcium (Ca)	238.0	238.0
Magnesium (Mg)	35.7	35.7
Sulfur (S)	62.6	62.6
Chlorine (Cl)	71.3	71.3

Note. [a]Abbreviation: ppmw, parts per million by weight.

2.2 Data Collection

2.2.1 Plant Characteristics

Plant measurements were collected weekly starting seven days after planting (DAP) for six weeks. Plant height was measured from the base of the potting mix to the extended tip of the last collared leaf. Stem diameter was measured with a caliper near the base of the plant. Growth stage of corn plants were recorded each week according to the collar method (Abendroth et al., 2011).

Chlorophyll content was measured using a CCM-300 Chlorophyll Content Meter (Opti-Sciences, Hudson, NH) at 25, 32, and 39 DAP. The device uses a fiber optic probe to detect the emission ratio of red fluorescence (700 nm) to far red fluorescence (735 nm) according to Gitelson et al. (1999). The probe was connected to a leaf clip that was used to secure the probe to the leaf during measurement. Six measurements were recorded on the fifth leaf down from the uppermost collared leaf on each corn plant. The average of the six measurements was calculated and used for further analysis.

Plants were destructively harvested at 45 DAP. Corn plants were cut directly above the uppermost brace roots and the above ground biomass was placed in a paper bag and oven-dried at 55 °C to constant mass. The remaining plant parts were removed from the potting mix and separated into the crown (remaining aboveground portion of plant and first five cm of roots) and the fibrous roots (roots below five cm). Root components were cleaned and placed in separate paper bags and oven-dried at 55 °C to constant mass. Once plants reached constant mass, the dry biomass of the shoot and fibrous roots were recorded.

2.2.2 Light Quality (R:FR)

Spectral measurements were taken at the V5, V11, and V14 corn growth stages. Plants were removed from the greenhouse and brought to a staging room for measurements. Four halogen bulbs were used as a light source that was maintained at a constant distance and angle from the corn plants for all measurements. The potting medium was covered with black felt and the plants were placed on a black table with a black backdrop. An Ocean Optics (Dunedin, FL) USB2000+ radiometer was used to collect data in the range of 363 to 1000 nm with a spectral resolution of about 0.35 nm. The radiometer was equipped with a 25° field-of-view optical fiber that was placed at the base of the corn plant approximately 15 cm above the potting mix and pointed into the canopy of the neighboring weeds. Four measurements were collected per pot, one directed at each of the four smaller pots of weeds, and were treated as subsamples. The four subsamples from within the weed species treatment from each N rate were pooled as weeds were only supplied with the full N rate. This provided a total of eight subsamples for each weed species treatment to be averaged for further data analysis. Data were averaged over every 10 nm to construct spectral response curves for every treatment.

2.3 Statistical Analyses

Data were subjected to ANOVA using a mixed effect model in SAS (SAS v9.4, SAS Institute Inc., Cary, NC). Weed species and N rate were designated as fixed effects. Corn stem diameter, height, leaf chlorophyll content, the R:FR light spectrum ratio, and end-of-season corn shoot biomass were the response variables measured. The R:FR light spectrum ratio was calculated by the summation of the reflected light within the 660-680 nm range divided by the summation of the reflected light within the 720-740 nm range (660-680 nm:720-740 nm) as this compared to values measured in previous research that used a R:FR sensor (Liu et al., 2009). Linear regression correlations were developed to determine the influence of the R:FR and timing on corn growth characteristics. All data except for corn shoot biomass were initially analyzed using a repeated measures design. The time factor was significant in all analyses; therefore, response variables were each analyzed separately by date of data collected. Variance and normality assumptions were found to hold true for all data collected. When fixed main effects were significant ($P \leq 0.05$), means were separated using Fisher's protected LSD. When fixed effect interactions were significant ($P \leq 0.1$), means of simple effects were separated using Fisher's protected LSD.

3. Results and Discussion

3.1 Light Quality (R:FR)

Measurements of R:FR decreased from the initial measurement at V5 to the last measurement at V14 within each weed treatment. The weed-free treatment consistently had a greater R:FR than the weed treatments at every measurement (Figure 2). The dicot species (common waterhemp and velvetleaf) R:FR estimates were lower than the monocot species (volunteer corn) at the V5 and V14 corn growth stages, which supports results observed by Cressman et al. (2011); however, the estimates were only statistically different at V14.

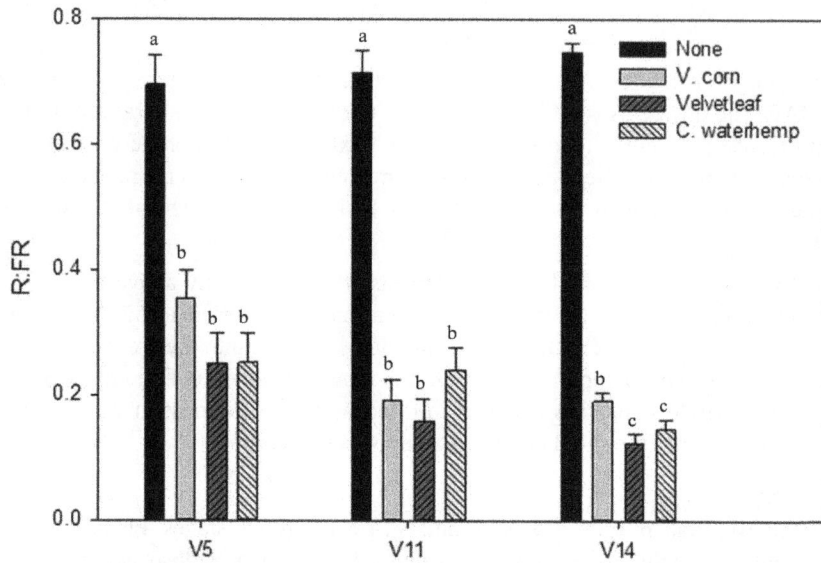

Figure 2. Red to far-red ratio (R:FR) of weed treatments at three corn growth stages

Note. Means within corn growth stage followed by the same letter are not different according to Fisher's protected LSD at $\alpha = 0.05$.

Differences in the spectral response curves of common waterhemp, velvetleaf, and volunteer corn are shown in Figure 3. Although R:FR was similar for velvetleaf and common waterhemp, the spectral properties of the weeds are markedly different. Reflectance in the red wavebands (660-680 nm) are lowest for velvetleaf, whereas common waterhemp and volunteer corn are similar. Reflectance in the far-red wavebands (720-740 nm) were highest for common waterhemp, whereas velvetleaf and volunteer corn are similar.

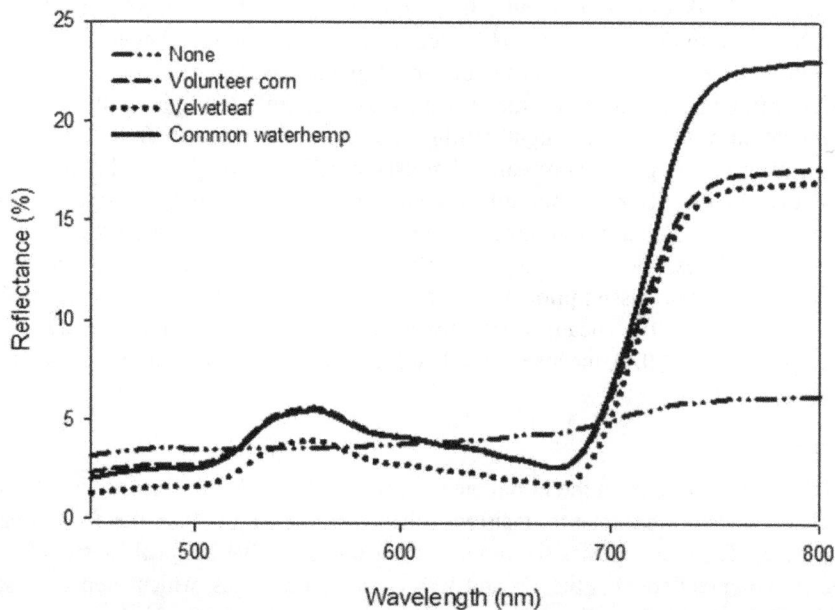

Figure 3. Spectral response curves for weed species at the V14 corn growth stage

3.2 Corn Height and Stem Diameter

Measurements of corn height and stem diameter were recorded and analyzed for six dates (Table 2). The relationship over time is shown for corn stem diameter (Figure 4) and height (Figure 5) in response to weed species and N rate. Both corn growth measurements were not different across treatments when compared at the V1 growth stage. At the V3 and V5 growth stages, the full N rate increased corn height by 6.4 (P < 0.0001) and 14.8 cm (P < 0.0001), respectively, and increased corn stem diameter by 0.1 (P = 0.0002) and 0.5 cm (P < 0.0001), respectively, compared to the reduced N rate. Data collected from the V8 and V11 growth stages resulted in a significant weed*N rate interaction for both corn height (P = 0.0194 and P = 0.0210, respectively) and stem diameter (P = 0.0557 and P = 0.0918, respectively). The full N rate and weed-free treatment produced the tallest corn and largest stem diameters at the V8, V11, and V14 growth stages. This leads to the conclusion that N rate was more critical for influencing early season growth of corn than changes in light quality; however, with full season exposure to weed presence, light quality additionally impacted late-season growth of corn in corroboration with N rate. Furthermore, two general trends appear for each response variable. In the full N rate treatments, all three weed species decreased corn height and stem diameter compared to the weed-free treatment at the V8, V11, and V14 corn growth stages. However, in the reduced N rate treatments, weed species buffered the impact of the reduced N rate and the corn plant tended to be taller with larger stem diameters than the weed-free treatment. We hypothesize this can be explained through two means. First, there is potential for experimental error as the full N rate supplied to weed species may have leached from the pots and provided extra N compared to the weed-free treatment. However, this effect is likely minor as the saturation capacity of the pots measured prior to the experiment was used to determine the total amount of solution added at each fertigation. Secondly, the N-deficient environment may be a greater limiting factor on corn growth than changes in light quality by neighboring weeds. At the V14 growth stage, corn height was impacted similarly as the previous significant weed species*N rate interaction. However, corn stem diameter was no longer impacted by weed species, but the full N rate increased the corn stem diameter by 0.3 cm compared with the reduced N rate.

Table 2. Corn growth stage corresponding to date of data collected for each response variable[a]

Date	Corn growth stage[b]	Corn height	Corn diameter	Corn leaf chlorophyll content	R:FR[c]	Corn shoot biomass
1	V1	Y	Y	N	N	N
2	V3	Y	Y	N	N	N
3	V5	Y	Y	N	Y	N
4	V8	Y	Y	Y	N	N
5	V11	Y	Y	Y	Y	N
6	V14	Y	Y	Y	Y	Y

Note. [a] Y = Data were collected; N = No data collected; [b] Average vegetative growth stage of corn plants based on number of collared leaves; [c] Red:far red light spectrum ratio (660-680 nm:720-740 nm).

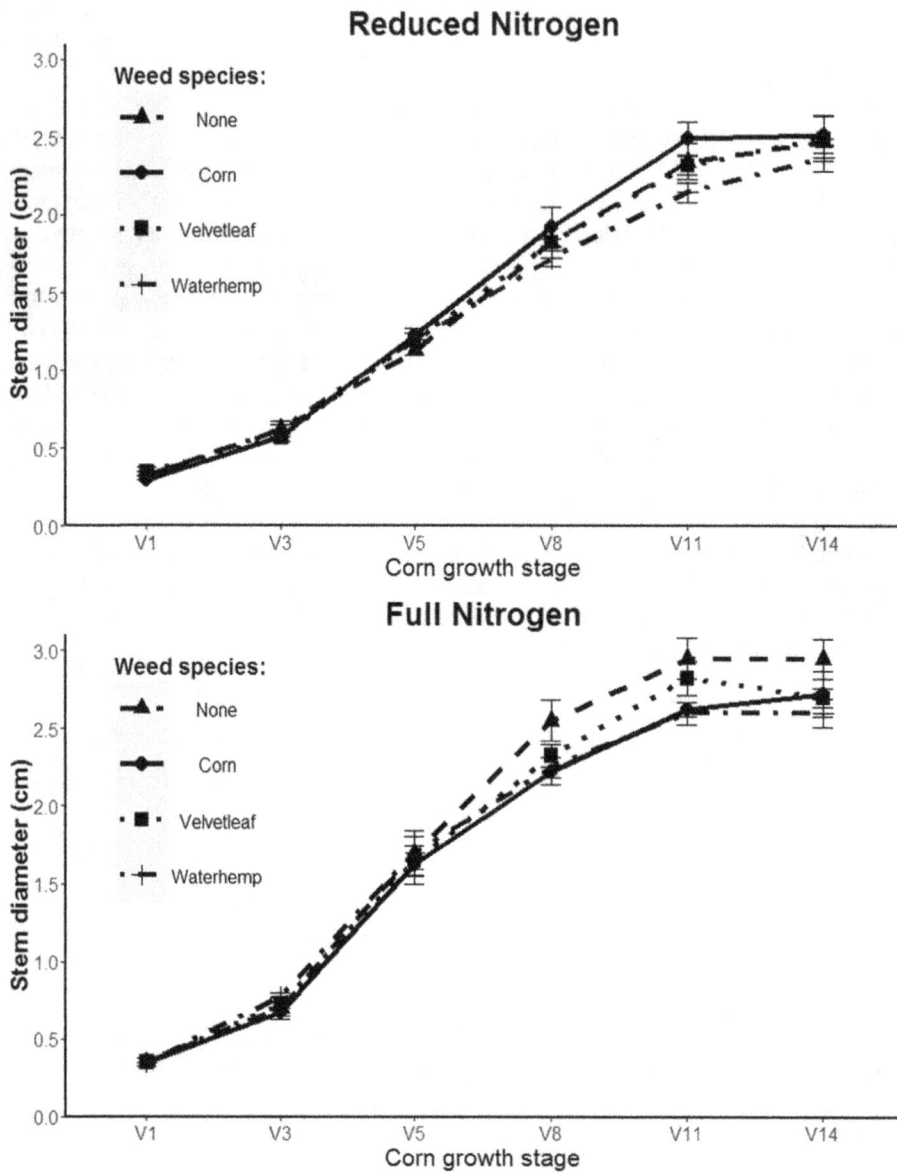

Figure 4. Measured corn stem diameter (cm) for four weed species treatments across six dates with reduced nitrogen (N) (top) or full N (bottom)

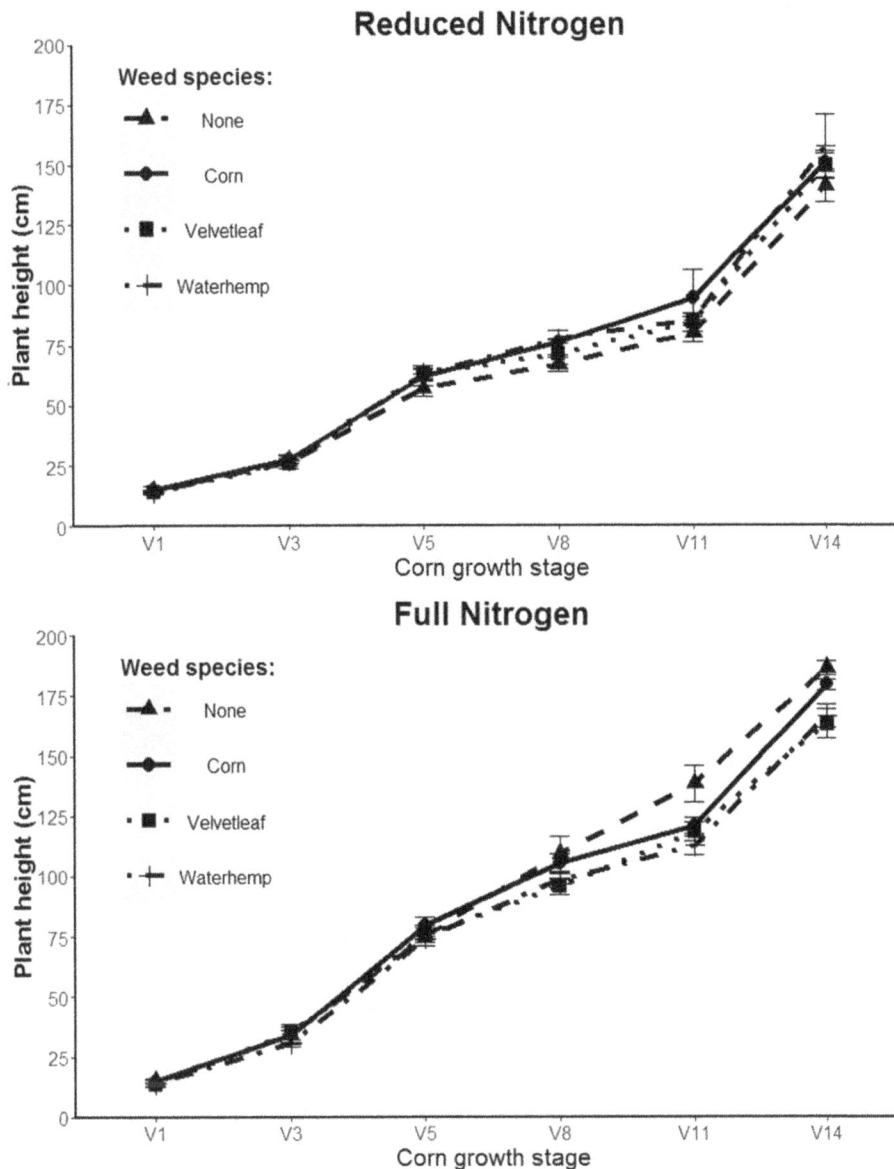

Figure 5. Measured corn height (cm) for four weed species treatments across six dates with reduced nitrogen (N) (top) or full N (bottom)

From the linear regression correlations, significant relationships were observed between late-season corn height and early-season R:FR measurements (data not shown). Lower R:FR at the V5 growth stage was correlated with reduced corn height at the V8, V11, and V14 growth stages in the full N rate treatments. No correlations were observed between R:FR and corn height in the reduced N rate treatments (data not shown), which further illustrated the greater influence of N rate as the limiting factor when compared to light quality.

3.3 Corn Leaf Chlorophyll Content

The weed species*N rate interaction influenced chlorophyll content in corn leaves at the V8 growth stage (Table 3). The full N rate and weed-free treatment resulted in the highest chlorophyll content (770.0 mg m^{-2}); however, it was not different than the common waterhemp (740.8 mg m^{-2}) and volunteer corn (740.0 mg m^{-2}) treatments. The chlorophyll content of the velvetleaf (720.0 mg m^{-2}) treatment was 6.5% less than the weed-free treatment but was similar to the common waterhemp and volunteer corn treatments. The influence of light quality on leaf chlorophyll content is not well understood. There is evidence that *PIF1*, a transcription factor in the phytochrome interaction factor protein family, inhibits protochlorophyllide, a precursor of chlorophyll (Jiao et al., 2007). Therefore, further studies investigating the molecular mechanisms involved are needed.

Table 3. The weed species*nitrogen (N) rate interaction effect on leaf chlorophyll content of corn plants at the V8 growth stage[a]

Weed species*N rate	Chlorophyll (mg m^{-2})
None*Full	770.0 a
Volunteer corn*Full	740.0 ab
Velvetleaf*Full	720.0 b
Common waterhemp*Full	740.8 ab
None*Reduced	668.0 c
Volunteer corn*Reduced	712.5 bc
Velvetleaf*Reduced	704.3 bc
Common waterhemp*Reduced	721.0 b
ANOVA	P-value
N rate	0.001
Weed species	0.666
Weed species*N rate	0.044

Note. [a] Means within a column with the same letter are not significantly different ($P \le 0.1$).

At the reduced N rate, the corn leaf chlorophyll content of the weed-free treatment was 53.0 mg m^{-2} lower than the common waterhemp treatment (Table 3). These results demonstrate that neighboring weeds variably affect corn leaf chlorophyll content. In a resource dependent weed removal study in corn, Cordes et al. (2004) showed that corn leaf chlorophyll content was similar in weed-free and weedy treatments at low common waterhemp densities (35-82 plants m^{-2}), however, at higher common waterhemp densities (> 369 plants m^{-2}), chlorophyll content was reduced by direct competition. Therefore, it appears that corn leaf chlorophyll content may be weed density dependent.

The weed species*N rate interaction was not significant for later corn growth stages (Table 4). The full N rate resulted in higher corn leaf chlorophyll content than the reduced N rate. The full N rate resulted in a 56.5 and 45.6 mg m^{-2} increase in corn leaf chlorophyll content over the reduced N rate at the V11 and V14 growth stages, respectively. Our results are consistent with previous studies that have demonstrated the positive correlation between corn leaf chlorophyll content and N application rates (Lindquist et al., 2010; Ziadi et al., 2008; Scharf et al., 2006). Additionally, lower R:FR at the V5 growth stage correlated with lower corn leaf chlorophyll content at the V8 and V14 growth stages in the full N rate treatments (data not shown).

Table 4. Nitrogen (N) rate effect on leaf chlorophyll contents of corn plants at the V11 and V14 growth stages[a]

N rate	V11	V14
	------------------------------ mg m^{-2} ------------------------------	
Full	743.4 a	627.4 a
Reduced	686.9 b	581.8 b
ANOVA	P-value	P-value
N rate	0.027	0.014
Weed species	0.576	0.845
Weed species*N rate	0.970	0.365

Note. [a] Means within a column with the same letter are not significantly different ($P \le 0.05$).

3.4 Corn Biomass

Corn shoot biomass varied as a function of weed species ($P = 0.0491$) and nitrogen rate ($P < 0.0001$), although no significant interaction was detected between factors ($P = 0.1365$). Therefore, nitrogen rate and weed species main effects were analyzed individually. As expected, corn plants that were supplied with the full N rate had greater shoot biomass when compared to the reduced N rate treatments (Table 5). Corn shoot biomass for the common waterhemp treatment was lower than the volunteer corn and velvetleaf treatments by 9.0 and 6.6 g, respectively. No differences were detected among the velvetleaf, volunteer corn, and weed-free treatments.

Table 5. Influence of weed species and nitrogen (N) rate on corn shoot biomass and root:shoot ratio[a]

Factor	Shoot biomass (g)	Root:shoot (g:g)
N rate		
Full	90.18 a	0.84 a
Reduced	70.29 b	0.89 a
Weed species		
None	80.23 ab	0.64 b
Volunteer corn	84.03 a	0.83 ab
Velvetleaf	81.65 a	0.92 ab
Common waterhemp	75.03 b	1.06 a
ANOVA	P-value	P-value
N rate	< 0.0001	0.5823
Weed species	0.0491	0.0468
Weed species*N rate	0.1365	0.1636

Note. [a] Means within a column and factor with the same letter are not significantly different (P ≤ 0.05).

A significant weed species*N rate interaction (P = 0.0937) influencing corn root biomass was observed, thus simple effects of both factors were investigated. For the full N rate, the common waterhemp treatment resulted in a greater corn root biomass when compared with the corn and weed-free treatments (Table 6). No differences were detected among the velvetleaf, volunteer corn, and weed-free treatments. Corn root biomass did not vary across weed species at the reduced N rate.

Table 6. Influence of weed species and nitrogen (N) rate interaction on corn root biomass[a]

Factor	Root biomass (g)
Weed species*N rate	
None*Full	60.41 b
Volunteer corn*Full	60.42 b
Velvetleaf*Full	72.73 ab
Common waterhemp*Full	104.74 a
None*Reduced	41.96 b
Volunteer corn*Reduced	76.57 ab
Velvetleaf*Reduced	75.04 ab
Common waterhemp*Reduced	60.79 b
ANOVA	P-value
N rate	0.2047
Weed species	0.0863
Weed species*N rate	0.0937

Note. [a] Means within a column with the same letter are not significantly different (P ≤ 0.1).

Corn root:shoot ratio varied as a function of weed species (P = 0.0468), whereas N rate had no effect (P = 0.5823) (Table 5). The common waterhemp treatment had a greater root:shoot ratio when compared to the weed-free treatment (P = 0.0069), but no differences were observed among velvetleaf, volunteer corn, and weed-free treatments.

The results suggest that the presence of common waterhemp influenced biomass partitioning of corn plants, where plants tended to accumulate more biomass in the roots. Similar results were reported by Liu et al. (2009), where corn plants in the presence of neighboring redroot pigweed had an increase in the root:shoot biomass partitioning at early growth stages. The authors associated that result with lower R:FR interception by the corn plants. However, the same study reported a decrease in the root:shoot ratio at the 9-leaf tip stage, whereas no differences were observed at later growth stages. Croster et al. (2003) reported that the total biomass production of nightshade species (*Solanum* spp.) was not influenced by the R:FR ratio, although the stem biomass was greater at the lower R:FR treatment. Contrary to the results reported in this research, Rajcan et al. (2004) reported that total corn biomass was not affected by upwardly reflected R:FR radiation treatments, whereas root:shoot ratio was greater in the higher R:FR radiation treatments. Kasperbauer and Hunt (1992) reported an increase in soybean root:shoot ratio when they were exposed to higher R:FR radiation when compared to plants exposed to lower R:FR radiation. Similar results were reported in corn seedlings, where lower R:FR radiation resulted in lower root:shoot ratios (Kasperbauer & Karlen, 1994). In the current study, lower early-season R:FR (V5 growth stage) correlated with lower shoot biomass and increased root biomass and root:shoot ratio in the full N rate treatments.

Spectral characteristics of common waterhemp, velvetleaf, and volunteer corn varied in the red and far-red waveband regions and the R:FR was consistently lower in the dicot species than volunteer corn. The effect of reduced early-season R:FR from neighboring weeds impacted corn growth characteristics under an optimal N regime. When the corn plants were not stressed from inadequate N, lower R:FR at the V5 growth stage reduced corn height and corn leaf chlorophyll content and increased fibrous root biomass. Changes in R:FR did not have an impact on these growth characteristics under reduced N treatments, indicating N stress was a greater limiting factor to corn growth than light quality. Therefore, this supports that competition for resources is still more important than the effect of light quality on the outcome of interplant competition. Results highlight the importance of the critical weed-free period and the need for proper early-season weed management.

Acknowledgements

The authors would like to thank Dr. Ellen Paparozzi, Dr. Timothy Arkebauer, and Dr. Bryan Leavitt for their technical expertise with experimental design, maintenance, and data collection techniques.

References

Abendroth, L. J., Elmore, R. W., Boyer, M. J., & Marlay, S. K. (2011). *Corn growth and development*. Iowa State Univ. Extension Publication #PMR-1009. Retrieved April 15, 2016, from https://store.extension.iastate.edu/Product/Corn-Growth-and-Development

Afifi, M., & Swanton, C. J. (2011). Maize seed and stem roots differ in response to neighboring weeds. *Weed Research, 51*, 442-450. http://dx.doi.org/10.1111/j.1365-3180.2011.00865.x

Afifi, M., & Swanton, C. J. (2012). Early physiological mechanisms of weed competition. *Weed Science, 60*, 542-551. http://dx.doi.org/10.1614/WS-D-12-00013.1

Ballaré, C. L. (2009). Illuminated behavior: Phytochrome as a key regulator of light forging and plant anti-herbivore defence. *Plant Cell Environment, 32*, 713-725. http://dx.doi.org/10.1111/j.1365-3040.2009.01958.x

Bonifas, K. D., Walters, D. T., Cassman, K. G., & Lindquist, J. L. (2005). Nitrogen supply affects root:shoot ratio in corn and velvetleaf (*Abutilon theophrasti*). *Weed Science, 53*, 670-675. http://dx.doi.org/10.1614/WS-05-002R.1

Bosnic, C. A., & Swanton, C. J. (1997). Influence of barnyardgrass (*Echinochloa crus-galli*) time of emergence and density on corn (*Zea mays*). *Weed Science, 45*, 276-282.

Clark, R. B. (1982). Nutrient solution growth of sorghum and corn in mineral nutrition studies. *Journal of Plant Nutrition, 5*, 1039-1057. http://dx.doi.org/10.1080/01904168209363037

Cordes, J. C., Johnson, W. G., Scharf, P., & Smeda, R. J. (2004). Late-emerging common waterhemp (*Amaranthus rudis*) interference in conventional tillage corn. *Weed Technology, 18*, 999-1005. http://dx.doi.org/10.1614/WT-03-185R

Cressman, S. T., Page, E. R., & Swanton, C. J. (2011). Weeds and the red to far-red ratio of reflected light: Characterizing the Influence of herbicide selection, dose, and weed species. *Weed Science, 59*, 424-430. http://dx.doi.org/10.1614/WS-D-10-00166.1

Croster, M. P., Witt, W. W., & Spomer, L. A. (2003). Neutral density shading and far-red radiation influence black nightshade (*Solanum nigrum*) and eastern black nightshade (*Solanum ptycanthum*) growth. *Weed Science, 51*, 208-213. http://dx.doi.org/10.1614/0043-1745(2003)051[0208:NDSAFR]2.0.CO;2

Evans, S. P., Knezevic, S. Z., Lindquist, J. L., Shapiro, C. A., & Blankenship, E. E. (2003). Nitrogen application influences the critical period for weed control in corn. *Weed Science, 51*, 408-417. http://dx.doi.org/10.1614/0043-1745(2003)051[0408:NAITCP]2.0.CO;2

Gitelson, A. A., Buschmann, C., & Lichtenhaler, H. K. (1999). The chlorophyll fluorescence ratio F735/F700 as an accurate measure of chlorophyll content in plants. *Remote Sensing of Environment, 69*, 296-302. http://dx.doi.org/10.1016/S0034-4257(99)00023-1

Hall, M. R., Swanton, C. J., & Anderson, G. W. (1992). The critical period of weed control in grain corn (*Zea mays*). *Weed Science, 40*, 441-447.

Jiao, Y., Lau, S., & Deng, X. W. (2007). Light-regulated transcriptional networks in higher plants. *Nature Reviews Genetics, 8*, 217-230. http://dx.doi.org/10.1038/nrg2049

Kasperbauer, M. J., & Hunt, P. G. (1992). Root size and shoot root ratio as influenced by light environment of the shoot. *Journal of Plant Nutrition, 15*, 685-697. http://dx.doi.org/10.1080/01904169209364355

Kasperbauer, M. J., & Karlen, D. L. (1994). Plant spacing and reflected far-red light effects on phytochrome-regulated photosynthate allocation in corn seedlings. *Crop Science, 34*, 1564-1569. http://dx.doi.org/10.2135/cropsci1994.0011183X003400060027x

Knezevic, S. Z., Weise, S. F., & Swanton, C. J. (1994). Interference of redroot pigweed (*Amaranthus retroflexus* L.) in corn (*Zea mays* L.). *Weed Science, 42*, 568-573.

Kropff, M. J., & van Laar, H. H. (1993). *Modelling crop-weed interactions*. Wallingford, U.K.: CAB International and Manila, Philippines: International Rice Research Institute.

Lindquist, J. L., & Mortensen, D. A. (1999). Ecophysiological characteristics of four maize hybrids and Abutilon theophrasti. *Weed Research, 39*, 271-285. http://dx.doi.org/10.1046/j.1365-3180.1999.00143.x

Lindquist, J. L., Evans, S. P., Shapiro, C. A., & Knezevic, S. Z. (2010). Effect of nitrogen addition and weed management interference on soil nitrogen and corn nitrogen status. *Weed Technology, 24*, 50-58. http://dx.doi.org/10.1614/WT-09-070.1

Lindquist, J. L., Mortensen, D. A., & Johnson, B. E. (1998). Mechanisms of corn tolerance and velvetleaf suppressive ability. *Agronomy Journal, 90*, 787-792. http://dx.doi.org/10.2134/agronj1998.00021962009000060012x

Lindquist, J. L., Mortensen, D. A., Clay, S. A., Schmenk, R., Kells, J. J., Howatt, K., & Westra, P. (1996). Stability of corn (*Zea mays*)-velvetleaf (*Abutilon theophrasti*) interference relationships. *Weed Science, 44*, 309-313.

Liu, J. G., Mahoney, K. J., Sikkema, P. H., & Swanton, C. J. (2009). The importance of light quality in crop-weed competition. *Weed Research, 49*, 217-224. http://dx.doi.org/10.1111/j.1365-3180.2008.00687.x

Loddo, D., Sousa, E., Masin, R., Calha, I., Zanin, G., Fernández-Quintanilla, C., & Dorado, J. (2013). Estimation and comparison of base temperatures for germination of European populations of velvetleaf (Abutilon theophrasti) and jimsonweed (*Datura stramonium*). *Weed Science, 61*, 443-451. http://dx.doi.org/10.1614/WS-D-12-00162.1

Marquardt, P. T., Terry, R., Krupke, C. H., & Johnson, W. G. (2012). Competitive effects of volunteer corn on hybrid corn growth and yield. *Weed Science, 60*, 537-541. http://dx.doi.org/10.1614/WS-D-11-00219.1

McCullough, D. E., Aguilera, A., & Tollenaar, M. (1994). N uptake, N partitioning, and photosynthetic N-use efficiency of an old and a new maize hybrid. *Canadian Journal of Plant Science, 74*, 479-484. http://dx.doi.org/10.4141/cjps94-088

Moriles, J., Hansen, S., Horvath, D. P., Reicks, G., Clay, D. E., & Clay, S. A. (2012). Microarray and growth analyses identify differences and similarities of early corn response to weeds, shade, and nitrogen stress. *Weed Science, 60*, 158-166. http://dx.doi.org/10.1614/WS-D-11-00090.1

Page, E. R., Tollenaar, M., Lee, E. A., Lukens, L., & Swanton, C. J. (2010). Shade avoidance: An integral component of crop weed competition. *Weed Research, 50*, 281-288. http://dx.doi.org/10.1111/j.1365-3180.2010.00781.x

Pearson, C., & Jacobs, B. (1987). Yield components and nitrogen partitioning of maize in response to nitrogen before and after anthesis. *Australian Journal of Agricultural Research, 38*, 1001-1009. http://dx.doi.org/10.1071/AR9871001

Rajcan, I., & Swanton, C. J. (2001). Understanding maize-weed competition: Resource competition, light quality and the whole plant. *Field Crop Research, 71*, 139-150. http://dx.doi.org/10.1016/S0378-4290(01)00159-9

Rajcan, I., & Tollenaar, M. (1999). Source : Sink ratio and leaf senescence in maize: II. Nitrogen metabolism during grain filling. *Field Crop Research, 60*, 255-265. http://dx.doi.org/10.1016/S0378-4290(98)00143-9

Rajcan, I., Chandler, K., & Swanton, C. J. (2004). Red-far-red ratio of reflected light: A hypothesis of why early-season weed control is important in corn. *Weed Science, 52*, 774-778. http://dx.doi.org/10.1614/WS-03-158R

Sauer, J. (1957). Recent migration and evolution of the dioecious Amaranths. *Evolution, 11*, 11-31. http://dx.doi.org/10.2307/2405808

Scharf, P. C., Brouder, S. M., & Hoeft, R. G. (2006). Chlorophyll meter readings can predict nitrogen need and yield response of corn in the North-Central USA. *Agronomy Journal, 98*, 655-665. http://dx.doi.org/10.2134/agronj2005.0070

Smith, H. (1982). Light quality, photoperception, and plant strategy. *Annual Review of Plant Physiology and Plant Molecular Biology, 33*, 481-518. http://dx.doi.org/10.1146/annurev.pp.33.060182.002405

Smith, H., & Holmes, M. G. (1977). The function of phytochrome in the natural environment. III. Measurement and calculation of phytochrome photoequilibrium. *Photochemistry and Photobiology, 25*, 547-550. http://dx.doi.org/10.1111/j.1751-1097.1977.tb09126.x

Steckel, L. E., & Sprague, C. L. (2004). Common waterhemp (*Amaranthus rudis*) interference in corn. *Weed Science, 52*, 359-364. http://dx.doi.org/10.1614/WS-03-066R1

Stoller, E. W., Wax, L. M., & Alm, D. M. (1993). Survey results on environmental issues and weed science research priorities within the corn belt. *Weed Technology, 7*, 763-770.

Swanton, C. J., Shrestha, A., Roy, R. C., Ball-Coelho, B. R., & Knezevic, S. Z. (1999). Effect of tillage systems N, and cover crop on the composition of weed flora. *Weed Science, 47*, 454-461.

Ta, C. T., & Weiland, R. T. (1992). Nitrogen partitioning in maize during ear development. *Crop Science, 32*, 443-451. http://dx.doi.org/10.2135/cropsci1992.0011183X003200020032x

Zhou, X. M., Madramootoo, C. A., MacKenzie, A. F., & Smith, D. L. (1997). Biomass production and nitrogen uptake in corn-ryegrass systems. *Agronomy Journal, 89*, 749-756. http://dx.doi.org/10.2134/agronj1997.00021962008900050007x

Ziadi, N., Brassard, B. M., Bélanger, G., Claessens, A., Tremblay, N., Cambourins, A. N., ... Parent, L. (2008). Chlorophyll measurements and nitrogen nutrition index for the evaluation of corn nitrogen status. *Agronomy Journal, 100*, 1264-1273. http://dx.doi.org/10.2134/agronj2008.0016

Zimdahl, R. L. (1988). The concept and application of the critical weed-free period. In M. A. Altieri & M. Liebman (Eds.), *Weed Management in Agroecosystems: Ecological Approaches* (pp. 145-155). CRC Press, Boca Raton, FL.

18

The Influence Factors Decomposition of Grain Output Increase in China: 2003-2014

Xiaoquan Hua[1]

[1] School of Economics and Management, Huainan Normal University, Huainan, Anhui, China

Correspondence: Xiaoquan Hua, School of Economics and Management, Huainan Normal University, Quanshan Road, Huainan, Anhui 232038, China. E-mail: huaxq2007@126.com

The research is financed by "The Humanities and Social Science Foundation of Anhui Educational Department" (SK2016A0846).

Abstract

As the world's most populous country taking up 18.84% of the world's total population, China's grain supply has attracted high concerns in the world. This article takes the grain output increase in China from 2003 to 2014 as the research object, and wants to explore the influence factors of China's grain growth and spatial effect analysis. It employs LMDI method to decompose China's total grain output increase during 2003-2014. The increase of OFPPU paid maximum contribution to China's total grain output increase among ACL, MCI and GPP. The conclusion of China's grain output provincial spatial analysis is that the six provinces including Heilongjiang, Henan, Inner Mongolia, Jilin, Anhui and Shandong, contributed 62.2% share to China's increase. The results of predicted OFPPU by Exponential Smoothing Method of Hoter-Winter No-Seasonal are that grain growth in China can meet requirements of grain security for China's population growth and living improvement. The author wishes China can increase investment on land consolidation and renovate facilities for farmland and water conservancy projects.

Keywords: index decomposition, grain output increase, spatial analysis

1. Introduction

As the world's most populous country of 18.84% world's total population, China's grain supply has attracted global attentions of the world in 1990s after a article titled 'who will feed China?' published by Lester Brown, director of the World Watch Institute in the United States in 1994. In the article, he claimed that China's population would reach 1.6 billion in early twenty-first Century, and food supply would fall behind its demand. The food price would rise if China imports food from the world market, which would cause the world's food crisis. Over the past 20 years, China's population had reached 1.375 billion by the end of 2015, China's grain output reached 621.44 million tons, 445.5 kg per capita, and this year witnessed grain output growth in 12 consecutive years.

Scholars have maintained high enthusiasm for the topic of food supply and demand in China despite China's grains production improved rapidly since 1980s because of the household-contract-responsibility-system extension in rural. Hong and Xiubin (2000) studied connection between cultivated land and food supply in China; Peter et al. (2000) analyzed the spatial exploration characteristics of land use change and grain production in China; Jie (2006) focused on the challenge of soil protection and food security along with rapid urbanization in China; Conflict between food security and conservation has never ceased in recent years in China (Zhigang et al., 2005); environmental change and resilience to drought of three major food crops in China became more and more outstanding in recently decades (Elisabeth et al., 2005); Hong et al. (2009) worried about negative effect of biofuel large-scale implications for food supply. The existing quantitative analysis of China's grain production was lack, and researcher paied little attention to the space unbalanced between supply and demand of broad land China.

The single-child policy (Note 1), which has been implemented since 1980, was ended in China in 2016. Now a couple is allowed to have two children, while a couple is even allowed to have three children in remote areas such as some districts in Heilongjiang Province. Will this policy set off an explosion of China's population? Can limited food production in China meet a rising demand of China's growing population and increasing demand from poor people after shaking off poverty? This paper will analyze the influence factors of China's grain output in the past, and predict the imbalance between grain supply and demand in the future in China.

This article is organized as follows: the first part is the introduction, explaining why China's grain growth is chosen as the research topic; the second part describes the characteristics of food production and consumption in China in nearly 30 years; the third part introduces the research methods on China's grain growth, and analyzes different contribution factors to China's grain growth; the fourth part studies the spatial distribution to China's grain growth and describes the spatial characteristics of grain supply and demand gap with province as a unit; the fifth part predicts grain gap between supply and demand in China in next 15 years by using Hoter-Winter No-Seasonal method; the last part is the conclusions and suggestions of this article.

2. Descriptions of Grain Output in China

In the future the challenge of China's grain supply comes from three aspects: one is the China's accelerated industrialization and urbanization. In this stage, a large population transfer from rural areas to urban areas, from agriculture to industry and service. In China as a large number of rural young labors transfer from agriculture to industry; the elders, women, and children are engaged in agricultural production to supply grain crops for 1.375 billion Chinese with random planting, which pays attention to grain quantity instead of quality and safety.

The second factor is that farmland area decreased rapidly with city expansion in China. It is estimated that about 500 thousand acres (34 thousand hectares) farmland disappeared annually. China's Ministry of Land and Resources demanded that the total cultivated land should keep balance through cultivating new land to make up the loss. In fact, the new land is not ready for agriculture production, which is only a response to inspection of the Ministry of Land and Resources.

The third factor is low profit of China's agriculture. The China's government promoted agricultural modernization through developing large-scale farmlands, in order to solve the problem of low-efficient land use caused by land fragmentation. In the early 1980's, the transition of agricultural production form from collective organization to family tenant management system promoted grain production rapidly in China, while rural land was divided into small pieces to ensure fair distribution of better farmlands. The lack of agricultural inputs by government have cause economic loss of small farming. A large number of small-scale falmlands were no longer under cultivation in China's rural area. In 2016, grain price fell while the average commodity price has beening rising. So, few capitalists are engaged in agricultural production in China.

Figure 1 shows the line diagram of grain total output and net import volume (import-export) after 1978 in China. According to Figure 1, the period of China's grain production from 1978 to now can be divided into four stages: the first stage is from 1978 to 1990. In this stage, grain output has maintained a faster growth because of institutional bonus from extension of family tenant management system in China; The second stage is from 1991 to 1998. In this stage, grain production kept growing slowly with fluctuations. The rural factories boomed and peasants entered into factory from fields because of high income. Prior to this, city factories only accept urban labors and refuse rural labors in China; The third stage is from 1999 to 2003. In this stage, grain output declined steadily. From the late 1990s, the tax burden of China's rural areas increased rapidly. So a large number of peasants were forced to go out for living with farmland abandoned. Two books described the current situation in Chinese rural. One is 'I Tell Prime Minister Truth' written by Changping Li, a town's mayor of Hubei province. Another is 'The Survey of China's Peasants' written by Guidi Chen and Chuntao, a couple scholars in Hefei city, Anhui province; The fourth stage is from 2004 to now. In this stage, grain output has maintained steady growth. In 2006, government abolished agricultural tax nationwide which had been implemented for 2600 years in China. Refering to the provisions of WTO, the central government would offer agricultural subsidies to peasants. In 2015, peasants were subsidized by 250 yuan RMB (about $40) per mu (666.7 square meters) from government. The policy of agricultural subsidy increases the enthusiasm of peasants, and farm production became profitable.

But some Chinese citizens especially rich people don't trust Chinese grain products. They think that the taste of China's grain grain is not good, and worry about grain food safety because of excessive use of chemical fertilizers and pesticides in the process of grain production. So they choose imported grain products. Shown by Figure 1, China's grain net import total has increased significantly since 2003, and reached 1.03 hundred million tons in 2014, accounting for about 1/6 of China's total grain output in this year.

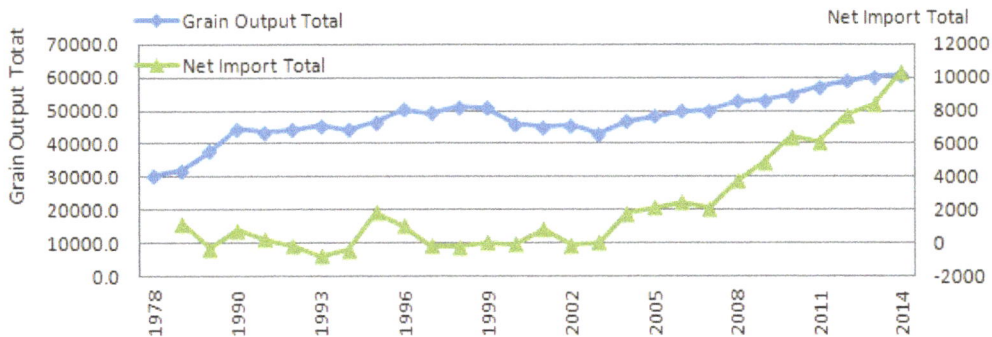

Figure 1. The grain output total and net import total in China: 1978-2014

Note. Unit: million tons.

Source: China Statistical Yearbook (2015). Beijing: China Statistical Press.

Comparing grain supply in several countries in East Asia from Figure 2, we found that the situation of East Asia's grain supply is similar to China's, and China's grain supply almost determines the trend of East Asian's. The capacity of China's grain supply is better than that of India and Japan, but worse than that of South Korea, while these countries have similar food structure with grain consumption as mainly.

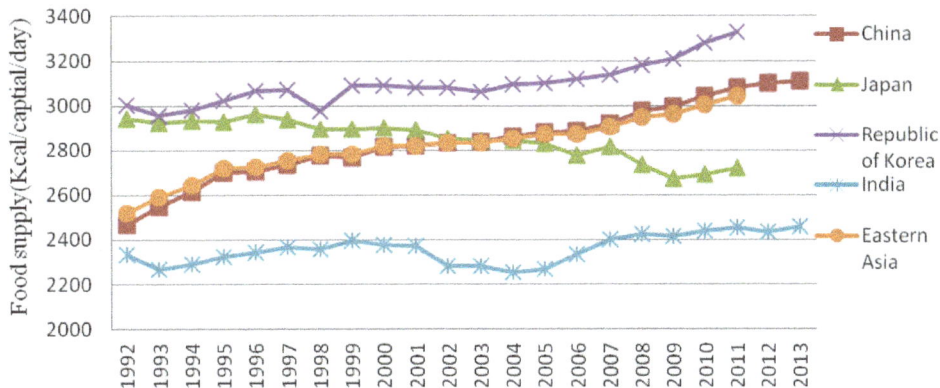

Figure 2. Food supply in selected country (Kcal/captial/day): 1992-2013

Source: FAOSTAT. Retrieved from http://faostat3.fao.org/browse/FB/*/E

3. LMDI Decomposition Method and Application to Grain Output Increase in China

3.1 Method of Grain Output LMDI Decomposition

Quantitative analysis of influence factors of grain output in China is usually used by two methods. The first method is production function. By establishing input-output function of grain production, we get the contribution of input-factors change to grain output. In China, Cobb-Douglass production function and transcendental logarithmic production function model is used frequently. But the input factors of production function and the selection method of production function is doubted, so the reliability of the conclusion obtained from production function method is regard as poor. The second method is the factor decomposition method of Logarithmic Mean Divisia Index-LMDI, which was first used by Ang B. W. in 1998. Benefiting from residual error limination of the traditional regression equation, LMDI method was often used in decomposition of carbon emissions growth for achieving complete decomposition of influence factors (Zhang et al., 2001; Ang, 2005). This approach is rather effective in solving a number of variable contributions to the growth total in a country.

In this paper, the LMDI method applied to China's grain output growth is as follows:

$$Total\ quantity\ of\ grain\ production = Area\ of\ cltivated\ land \times \frac{Total\ sown\ area}{Area\ of\ cultivated\ land} \tag{1}$$

$$\times \frac{Sown\ area\ of\ grain\ crops}{Total\ sown\ area} \times \frac{Total\ quantity\ of\ grain\ production}{Sown\ area\ of\ grain\ crops}$$

$$\frac{Total\ sown\ area}{Area\ of\ cultivated\ land} = Multiple\ crop\ index \tag{2}$$

$$\frac{Sown\ area\ of\ grain\ crops}{Total\ sown\ area} = Grain\ planting\ proportion \tag{3}$$

$$\frac{Total\ quantity\ of\ grain\ production}{Sown\ area\ of\ grain\ crops} = Output\ of\ farm\ product\ per\ unit \tag{4}$$

Therefore, in one year total quantity of grain production-TQGN is equal to multiply continuously of four factors: area of cultivated land-ACL, multiple crop index-MCI, grain planting proportion-GPP and output of farm product per unit-OFPPU. The identity in Equation (1) may be written as:

$$TQGN = \sum TQGN_i = \sum ACL_i \times MCI_i \times GPP_i \times OFPPU_i \tag{5}$$

In Equation (5), the total quantity of grain production is divided into $TQGN$ sum of i district. $DTQGN$ or $\Delta TQGN$ represent the aggregate change from $TQGN_0$ in time 0 to $TQGN_t$ in time t. In multiplicative decomposition, the ratio is decomposed as follows:

$$D_{TQGN} = TQGN^t/TQGN^0 = D_{ACL} \times D_{MCI} \times D_{GPP} \times D_{OFPPU} \tag{6}$$

In additive decomposition, the difference is decomposed as follows:

$$\Delta TQGN = TQGN^t - TQGN^0 = \Delta ACL + \Delta MCI + \Delta GPP + \Delta OFPPU \tag{7}$$

The formulation process of Equations (6) and (7) are summarized in Table 1.

Table 1. LMDI formulae for decomposing in TQGN

Identity	$TQGN = \sum TQGN_i = \sum ACL_i \times MCI_i \times GPP_i \times OFPPU_i$	
Change scheme	Multiplicative Decomposition $DTQGN=TQGN_t/TQGN_0=DAC \times DMCI \times DGPP \times DOFPPU$	Additive Decomposition $\Delta TQGN=TQGN_t-TQGN_0=\Delta ACL+\Delta MCI+\Delta GPP+\Delta OFPPU$
LMDI fomulae	$D_{ACL} = Exp\left(\sum_i \frac{(TQGN_i^t - TQGN_i^0)/(\ln TQGN_i^t - \ln TQGN_i^0)}{(TQGN^t - TQGN^0)/(\ln TQGN^t - \ln TQGN^0)} \times \ln(\frac{ACL_i^t}{ACL_i^0}) \right)$	$\Delta ACL = \sum_i \frac{TQGN_i^t - TQGN_i^0}{\ln TQGN_i^t - \ln TQGN_i^0} \times \ln(\frac{ACL_i^t}{ACL_i^0})$
	$D_{MCI} = Exp\left(\sum_i \frac{(TQGN_i^t - TQGN_i^0)/(\ln TQGN_i^t - \ln TQGN_i^0)}{(TQGN^t - TQGN^0)/(\ln TQGN^t - \ln TQGN^0)} \times \ln(\frac{MCI_i^t}{MCI_i^0}) \right)$	$\Delta MCI = \sum_i \frac{TQGN_i^t - TQGN_i^0}{\ln TQGN_i^t - \ln TQGN_i^0} \times \ln(\frac{MCI_i^t}{MCI_i^0})$
	$D_{GPP} = Exp\left(\sum_i \frac{(TQGN_i^t - TQGN_i^0)/(\ln TQGN_i^t - \ln TQGN_i^0)}{(TQGN^t - TQGN^0)/(\ln TQGN^t - \ln TQGN^0)} \times \ln(\frac{GPP_i^t}{GPP_i^0}) \right)$	$\Delta GPP = \sum_i \frac{TQGN_i^t - TQGN_i^0}{\ln TQGN_i^t - \ln TQGN_i^0} \times \ln(\frac{GPP_i^t}{GPP_i^0})$
	$D_{OFPPU} = Exp\left(\sum_i \frac{(TQGN_i^t - TQGN_i^0)/(\ln TQGN_i^t - \ln TQGN_i^0)}{(TQGN^t - TQGN^0)/(\ln TQGN^t - \ln TQGN^0)} \times \ln(\frac{OFPPU_i^t}{OFPPU_i^0}) \right)$	$\Delta OFPPU = \sum_i \frac{TQGN_i^t - TQGN_i^0}{\ln TQGN_i^t - \ln TQGN_i^0} \times \ln(\frac{OFPPU_i^t}{OFPPU_i^0})$

3.2 Results of LMDI Decomposition

Through 12-year consecutive growth, the total grain output rose from 430.695 million tons, or 333.3 kg per capita in 2003 to 621.44 million tons, or 445.5 kg per capita in 2015 in China. The growth rate of China's grain output per capita has averaged 2.90% annually, which is rare in the grain crops production history in the World. China's grain production capacity shows that China is capable of feeding 1.6 billion people in the future, which has taken up for nearly 22% of world's total population.

Next, we will use the LMDI method to decompose China's total grain output increase from 2003 to 2014 (the detail data of 2015 in China hasn't published openly up to now). The results are shown in Tables 2 and 3.

Table 2. Results of total grain output decomposition for China, 2003-2014: multiplicative decomposition

DTQGN	DACL	DMCI	DGPP	DOFPPU
1.4094	1.0311	1.0522	1.0457	1.2424

Table 3. Results of total grain output decomposition for China, 2003-2014: additive decomposition (million tons)

ΔTQGN	ΔACL	ΔMCI	ΔGPP	ΔOFPPU
1763.3	157.35	261.38	229.38	1115.21

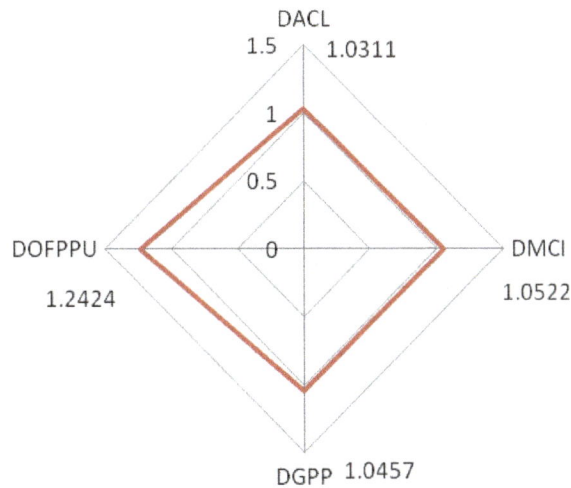

Figure 3. Presentation of multiplication decomsition results in Table 2

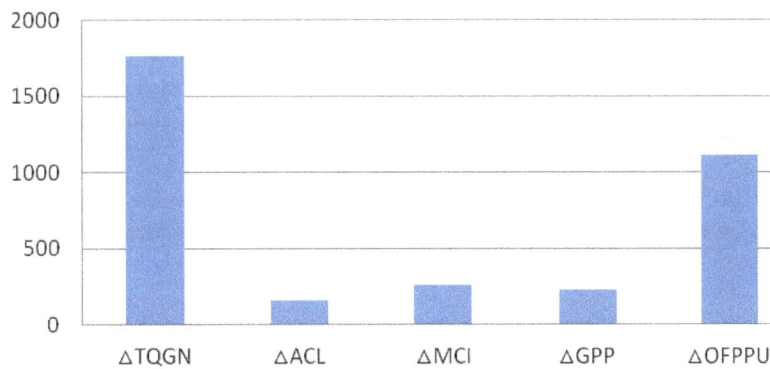

Figure 4. Presentation of additive decomsition results in Table 3 (million tons)

Figures 3 and 4 drawn according to Tables 2 and 3 more clearly show the comparable decomposition results of various factors. The increase of OFPPU made maximum contribution to China's total grain output increase during 2003-2014. From 2003 to 2014, China's total grain output has increased by 1763.3 million tons or 40.94%. The LMDI decomposition results showed that the increase of OFPPU led to total grain output increase of 1115.21 million tons or 24.24%. The increase of DMI led to total grain output increase of 261.38 million tons or 5.22%, and the increase of GPP led to an increase of 229.38 million tons or 4.57%. The increase of ACL made minimum contribution to China's total grain output increase, only leading to an increase of 157.35 million tons or 3.11%.

4. The Spatial Analysis of Grain Output in China

China's total population has reached 1.374 billion at the end of 2015, the largest in the world, distributed in a wide area of 31 provinces with Hongkong, Macao Special Administrative Region and Taiwan Province excluded. Especially since the 1990s, a large amount of rural population coming from middle and west provinces flows to southeastern regions of China to find jobs to earn money, attracted by strong economy growth in this districts such as Guangdong, Zhejiang, Jiangsu and Fujian Provinces. According to statistics published by China's National Bureau, the migrating population in 2014 reached 252 million in China. However, the affluent provinces of land area per capita, including Heilongjiang, Inner Mongolia, Jilin and Xinjiang, has relatively fewer population. Thus, the space distribution between grain output and demand in China remains imbalance.

In ancient China, the approach for space coordination between grain output and demand is river transportation. The Grand Canal, which was built in the Sui dynasty, has served as the channel of grain crops transportation in ancient China until railway became popular.

In China, due to poor profit of grain production, few provinces are willing to engage in it. Regional leaders are willing to develop industries for more fiscal revenue. So over a long time after the foundation of P.R.C. independence, grain crops was always in short supply. The implementation policy of the central government is "the governor of province should be responsible for grain output and supply of their provincal residents".

In order to investigate the spatial characteristics for China's grain output increase during 2003-2014, we build a matrix of grain production and growth in China. Table 4 shows that there are 14 provinces belong to Dog style with both low base of grain output and low growth rate in China from 2003 to 2014. There are only 6 provinces matching Star style with both high base of grain output and high growth rate, whose four are located in the northeast China and the remaining two are in the middle. So now the traditional grain transportation from south to north has changed into the one from north to south in China.

Table 4. The spatial matrix of grain output increase in china, 2003-2014

	Low base of grain output per capita	High base of grain output per capita
Low increase rate	Dog style:	Taurus style:
	Beijing, Hebei, Shanxi, Shanghai, Zhejiang, Fujian, Guangdong, Guangxi, Hainan, Guizhou, Yunnan, Tibet, Shaanxi, Qinghai	Liaoning, Shandong, Hubei, Hunan, Chongqing, Sichuan, Ningxia
High increase rate	Infant style:	Star style:
	Tianjin, Jiangsu, Gansu, Jiangxi	Inner Mongolia, Jilin, Xinjiang, Heilongjiang, Henan, Anhui

Using the LMDI method to decompose China's province total grain output increase from 2003 to 2014, we obtained the contribution of provincal influence factors to total grain output increase in China. The results are shown in Table 5.

Table 5. Results of total grain output LDMI decomposition for China province, 2003-2014

District	Total contribution		Contribution of aclchange		Contribution of MIC change		Contribution of GPP change		Contribution of OFPPU change	
	Output	Percentage	Output	Percentage	Output	Percentage	Output	Percentage	Output	Percentage
Beijing	0.59	0.03	-2.69	-1.71	-0.08	-0.03	1.78	0.78	1.58	0.14
Tianjin	5.67	0.32	-1.49	-0.95	0.83	0.32	4.93	2.15	1.40	0.13
Hebei	97.24	5.51	-14.08	-8.95	16.52	6.32	15.55	6.78	79.24	7.11
Shanxi	37.19	2.11	-13.83	-8.79	16.12	6.17	14.53	6.34	20.37	1.83
Inner Mongolia	139.23	7.90	22.69	14.42	25.88	9.90	17.17	7.49	73.48	6.59
Liaoning	25.56	1.45	28.93	18.39	-10.60	-4.05	8.42	3.67	-1.20	-0.11
Jilin	127.32	7.22	64.94	41.27	-15.26	-5.84	12.96	5.65	64.69	5.80
Heilongjiang	372.99	21.15	122.23	77.68	-31.70	-12.13	59.30	25.85	223.16	20.01
Shanghai	1.38	0.08	-5.45	-3.46	3.75	1.44	2.81	1.22	0.26	0.02
Jiangsu	101.88	5.78	-29.42	-18.70	29.31	11.21	42.34	18.46	59.65	5.35
Zhejiang	-3.60	-0.20	-5.55	-3.53	-11.53	-4.41	7.80	3.40	5.68	0.51
Anhui	120.10	6.81	-4.14	-2.63	-1.35	-0.52	25.96	11.32	99.64	8.93
Fujian	-4.61	-0.26	-4.78	-3.03	-1.34	-0.51	-8.06	-3.52	9.57	0.86
Jiangxi	69.32	3.93	5.48	3.48	13.79	5.27	14.82	6.46	35.23	3.16
Shandong	116.11	6.58	-2.90	-1.84	8.46	3.24	53.54	23.34	57.02	5.11
Henan	220.28	12.49	1.71	1.09	20.96	8.02	39.05	17.03	158.56	14.22
Hubei	66.32	3.76	14.53	9.24	14.07	5.38	17.40	7.59	20.31	1.82
Hunan	55.85	3.17	13.16	8.36	20.86	7.98	-8.59	-3.74	30.42	2.73
Guangdong	-7.31	-0.41	-30.88	-19.62	26.87	10.28	-9.99	-4.35	6.69	0.60
Guangxi	6.93	0.39	0.39	0.25	-8.97	-3.43	-9.90	-4.32	25.41	2.28
Hainan	-1.80	-0.10	-0.93	-0.59	-0.11	-0.04	-5.18	-2.26	4.42	0.40
Chongqing	5.74	0.33	-21.92	-13.93	27.56	10.54	-16.37	-7.14	16.48	1.48
Sichuan	32.08	1.82	27.61	17.55	-18.03	-6.90	-5.58	-2.43	28.08	2.52
Guizhou	3.42	0.19	-8.44	-5.36	27.98	10.70	-15.28	-6.66	-0.84	-0.08
Yunnan	38.97	2.21	-5.29	-3.36	42.28	16.18	-19.97	-8.71	21.95	1.97
Tibet	0.14	0.01	1.92	1.22	-1.23	-0.47	-1.21	-0.53	0.65	0.06
Shaanxi	22.94	1.30	-27.28	-17.34	32.64	12.49	-6.97	-3.04	24.55	2.20
Gansu	36.93	2.09	6.55	4.16	7.66	2.93	-1.84	-0.80	24.56	2.20
Qinghai	1.80	0.10	-1.50	-0.95	3.13	1.20	-0.47	-0.20	0.64	0.06
Ningxia	10.77	0.61	0.31	0.20	3.03	1.16	-4.72	-2.06	12.16	1.09
Xinjiang	63.90	3.62	27.46	17.45	19.88	7.60	5.13	2.24	11.43	1.03

Note. The unit of output is million tons.

From Table 5 we found that the six provinces including Heilongjiang, Henan, Inner Mongolia, Jilin, Anhui and Shandong, contributed 62.2% share to China's total grain output increase from 2003 to 2014. As China's major grain producing provinces, the increase of OFPPU in these six provinces contributed 38.4% share to China's total grain output increase from 2003 to 2014. In these six main grain production provinces, only Shandong province is the relatively developed region while the other five provinces are less developed regions in China. This brings challenge to China's more and more big regional gap, because developed region believe they can solve the problem of food supply through increasing income while the less developed region get confined because of low-profit of grain production and slim possibilities of rapid industrial development. As a result, only central government pay enough attention to grain production in China.

The gap of per capita grain reflects the gap between grain output and demand. Figure 5 is four sub bitmap of provincial grain gap per capita in China in 2003 and 2014. Comparing four sub bitmap of provincial grain gap per capita in 2003 with that in 2014, Figure 5 shows that the gap between grain supply and demand in China has little provincial space change. The 8 provinces of maximum grain gap per capita in 2014 have remained the same compared with that in 2003. But 8 provinces of the minimum grain gap per capita (i.e. maximum grain surplus per capita) in 2014 have changed. Henan, Anhui and Jiangxi provinces have maximum grain surplus per capita replacing Chongqing, Sichuan and Shandong province. As for the degree of maximum grain gap per capita include Beijing, Tianjin, Shanghai, Zhejiang, Fujian, Guangdong and Qinghai provinces have not changed from 2003 to 2014.

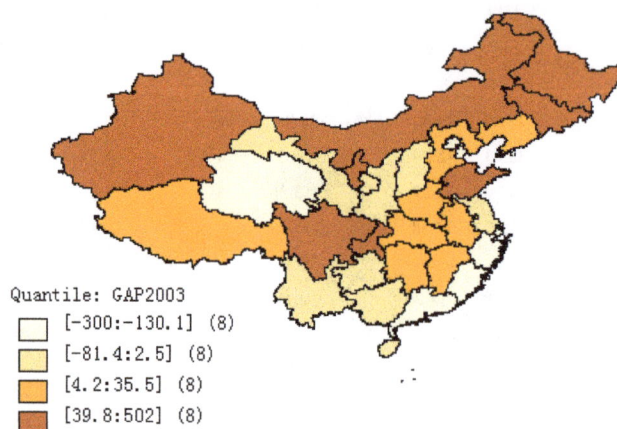

Quantile: GAP2003
[-300:-130.1] (8)
[-81.4:2.5] (8)
[4.2:35.5] (8)
[39.8:502] (8)

Figure 5(a). Four sub bitmap of provincial grain gap per capita in 2003

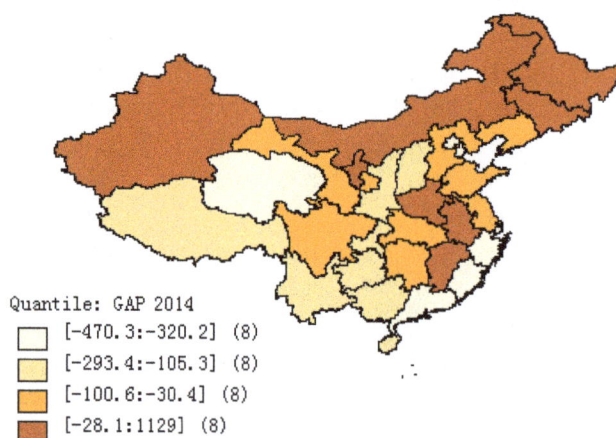

Quantile: GAP 2014
[-470.3:-320.2] (8)
[-293.4:-105.3] (8)
[-100.6:-30.4] (8)
[-28.1:1129] (8)

Figure 5(b). Four sub bitmap of provincial grain gap per capita in 2014

Note. China's provincial map has 32 units including the South Sea Islands, excluding Taiwan, Hongkong and Macao.

5. The Prediction of China's Grain Gap between Supply and Demand in the Future

The growth of food demand in China comes from two aspects: one cause is the slow growth of the population. The total population of China will maintain a low growth rate of about 5/1000 until 2030 and reached maximum peak of 1.6 billion. Then China population will begin to decline afterwards. The second cause is the growth demand of grain by poverty, which brings about a 1/1000 increase in total grain demand. By the end of 2015, there are still 80 million people in China living under food insecurity. The China's government puts forward the policy of "Accurate Poverty Alleviation" in 2016. Subsequently, local governments ensure to achieve the goal of poverty alleviation in the next five years to their superior government.

Therefore, China's total grain output in the next 15 years will still need to maintain a growth rate of up to 6/1000 in order to ensure the grain target of over 90% self-sufficiency, which has been proposed by China central government. Can China achieve this goal? Or is it that China can feed itself in the next 15 years? The stable growth of grain output per units is the answer to this goal in China.

In the four factors to grain output growth in China, there will be a little change of ACL, MCI and GPP in the future. Next, we use the Exponential Smoothing Method of Hoter-Winter No-Seasonal method to predict the variation trend of OFPPU.

Using Stata 14.0 software, we establish the Hoter-Winter No-Seasonal model of OFPPU from 1995 to 2014. The prediction formula used in the forecast period is as follows:

$$\text{OFPPUF}_{T+K} = a_T + b_T \times k = 5389.05 + 40.19 \times K \quad (\text{Kg/ha}) \ (k = 1, 2, 3, \ldots) \tag{8}$$

In the formula, $T = 20$ (end of sample). The predicted results are found in Table 6.

Table 6. Results of the predicted OFPPU (Unit: Kg/ha)

year	2015	2016	2017	2018	2019	2020	2021	2022
OFPPUF	5429.24	5469.43	5509.63	5549.82	5590.01	5630.21	5670.40	5710.59
year	2023	2024	2025	2026	2027	2028	2029	2030
OFPPUF	5750.78	5790.98	5831.17	5871.36	5911.56	5951.75	5991.94	6032.14

The results of predicted OFPPU by Exponential Smoothing Method of Hoter-Winter No -Seasonal in Table 6 shows that China's grain production per unit area will increase stably from 5429.24 Kg/ha in 2015 to 5750. 78 Kg/ha in 2030, while the growth rate of grain output per unit area will be reduced from 0.7403% in 2016 to 0.6847% in 2030. Fortunately, in the next 15 years, China's will keep average 0.7045% growth rate of grain output per unit area, which is higher than 0.6%, the demand rate of grain growth for China's population growth and living improvement.

6. Conclusion

This article focuses on China's grain security, one of top world concern. It describes the general situation of China's grain output and import after 1978, and compares the China's grain supply with its neighboring countries. LMDI decomposition from 2003 to 2014 shows that the increase of OFPPU paied biger contribution than ACL, MCI or GPP to China's total grain output increase during 2003-2014, which led to total grain output increase of 1115.21 million tons or 24.24%. The conclusion of China's grain output provincial spatial analysis is that the six provinces including Heilongjiang, Henan, Inner Mongolia, Jilin, Anhui and Shandong, contributed 62.2% share to China's total grain output increase from 2003 to 2014. The results of predicted OFPPU by Exponential Smoothing Method of Hoter-Winter No-Seasonal is that grain growth in China can meet grain security requirement for China's population growth and living improvement.

The biggest challenge of China's grain security outlook is deterioration of the ecological environment. Greenhouse gases emission and discharge of 'The Three Wastes' (industrial wastewater, waste gases and residues) increased rapidly along with the acceleration of urbanization and industrialization in China. The extreme natural disasters occured frequently in past ten years. The water shortage became quite frequent in Northern China. So China should increase investment for land consolidation, renovate facilities for farmland and water conservancy projects to ensure the basic cultivation conditions of arable land.

References

Ang, B. W. (2005). The LMDI approach to decomposition analysis: A practical guide. *Energy Policy, 33*, 867-871. http://dx.doi.org/10.1016/j.enpol.2003.10.010

Ang, B. W., & Liu, F. L. (2001). A new energy decomposition method perfect in decomposition and consistent in aggregation. *Energy, 26*, 537-495. http://dx.doi.org/10.1016/S0360-5442(01)00022-6

Ang, B. W., Zhang, F. Q., & Choi, K. H. (1998). Factorizing changes in energy and environmental indictors through decomposition. *Energy, 23*, 489-495. http://dx.doi.org/10.1016/S0360-5442(98)00016-4

Elisabeth, S., Evan, D. G. F., Mette, T., Piers, M. F., & Andrew, J. D. (2008). Typologies of crop-drought vulnerability: an empirical analysis of the socio-economic factors that influence the sensitivity and resilience to drought of three major food crops in China (1961-2001). *Environmental Science and Policy, 4*, 438-452.

Hong, Y., & Xiubin, L. (2000). Cultivated land and food supply in China. *Land Use Policy, 2*, 73-88.

Hong, Y., Yuan, Z., & Junguo, L. (2009). Land and water requirements of biofuel and implications for food supply and the environment in China. *Energy Policy, 5*, 1876-1885.

Jie, C. (2006). Rapid urbanization in China: A real challenge to soil protection and food security. *Catena, 1*, 1-15.

Lester, R. B. (1995). *Who Will Feed China? Wake-Up Call for a Small Planet*. W. W. Norton & Company, New York.

Peter, H. V., Youqi, C., & Tom, A. V. (2000). Spatial explorations of land use change and grain production in China. *Agriculture, Ecosystems and Environment, 1*, 333-354.

Zhigang, X., Jintao, X., Xiangzheng, D., & Jikun, H. (2005). Grain for Green versus Grain: Conflict between Food Security and Conservation Set-Aside in China. *World Development, 1*, 130-148.

Notes

Note 1. September 1980, the CCCC opened a letter named "How to control the growth of China's population problem" to all Party members and Communist Youth League, the main content of the open letter is "promoting a couple only birth single child". Family planning has become a basic national policy for China since then. In the city, couples only allow single child birth, except couple can have two children birth if they are both single children. In the countryside, couple allow second child if their first child is a girl except for farmers living inYunnan, Qinghai, Ningxia, Xinjiang and Hainan province; Farmers in Tibetan have no restriction numbers of children.

19

Facile Green Synthetic Route to the Zinc Oxide (ZnONPs) Nanoparticles: Effect on Green Peach Aphid and Antibacterial Activity

Alaa Y. Ghidan[1], Tawfiq M. Al-Antary[1], Nidá M. Salem[1] & Akl M. Awwad[2]

[1] School of Agriculture, the University of Jordan, Amman, Jordan

[2] Department of Materials Science, Royal Scientific Society, Amman, Jordan

Correspondence: Akl M. Awwad, Royal Scientific Society, Amman, Jordan.
E-mail: akl.awwad@yahoo.com; akl.awwad@rss.jo

The research is financed by SRF, Jordan (Agr/2/13/2014).

Abstract

In this study, zinc oxide nanoparticles (ZnONPs) were synthesized using *Punica granatum* peel extract in one–step reaction at room temperature. Zinc oxide nanoparticles were characterized by Fourier transform infrared spectroscopy (FT-IR), X-ray diffraction (XRD), ultraviolet visible spectroscopy (UV-vis) and scanning electron microscopy (SEM). The UV-vis absorption spectrum shows an absorption band at 278 nm due to ZnO nanoparticles. XRD characterized the final product as highly crystalline ZnO with sizes in the range 10-40 nm. The SEM results reveal a presence of network of randomly oriented ZnO nanoplatelets with an average size of 40 nm and thicknesses of about 8 nm. This study determined the effect of zinc oxide nanoparticles on green peach aphid and antibacterial activity.

Keywords: green synthesis, zinc oxide nanoparticles, *Punica granatum* peel, antibacterial activity, green peach aphid

1. Introduction

Zinc oxide has a potential application in sensors, solar cells, UV lasers, semiconductors, hydrogen storage devices, and catalysts. These application led many researchers to develop different routes to synthesis zinc oxide nanoparticles such as chemical route (Singh & Gopal, 2008; Abbasi et al., 2017), hydrothermal route (Ipeksac et al., 2013; Peng et al., 2016), sol-gel template process (Kumari et al., 2010; El Ghoul et al., 2012), photoluminescence emission technique (Rocha et al., 2014), microwave-assisted hydrothermal and decomposition (Tseng et al., 2012; Mousa et al., 2013), aerosol process (Ozcelik & Ergun, 2014), sonochemical synthesis (Zak et al., 2013), laser ablation (Thareja & Shukla, 2007), microemulsion method (Yıldırım & Durucan, 2010), precipitation method (An et al., 2014), hydrolyzed in polar organic solvents (Ehlert et al., 2014), solid-state thermal decomposition (Soofivand et al., 2013), microwave synthesis (Sutradhar et al., 2016). These routes have many disadvantages due to difficulty of scale up the synthesis process, separation and purification of nanoparticles from surfactants, co-surfactants, organic solvents, high energy consumption, and toxic by-products. Developing green methods for synthesis ZnONPs are of importance and still a challenge for materials researchers. Recently, plant extracts have been suggested as possible eco-friendly, alternative to chemicals in synthesis of nanoparticles. Different plants extracts have been reported in the open literature for green synthesis of zinc oxide nanoparticles, such as *Olea europea* (Awwad et al., 2014), *Eichhornia crassipes* leaf (Vanathi et al., 2014), *Aloe barbadensis* Miller leaf (Sangeetha et al., 2011), *Hibiscus subdariffa* leaf (Bala et al., 2015), *Solanum nigrum* leaf (Ramesh et al., 2015), *Camelia sinesis* leaf (Shah et al., 2015), *Hibiscus rosa-sinensis* (Devi & Gayathri, 2014), *Cassia fistula* plant extract (Suresh et al., 2015), *Ocimum tenuiflorum* leaves (Raut et al., 2015), *Trifoliumpratense* flower extract (Dobrucka & Dugaszewska, 2016), *Mimosa Pudica* leaves and coffee powder (Fatimah et al., 2016), *Rambutan* peel extract (Karnan & Selvakumar, 2016) *Jacaranda mimosifolia* flowers (Sharma et al., 2016), *Terminalia chebula* fruits (Rana et al., 2016) and *Azadirachia indica* (Madan et al., 2016).

In this research work, we developed a green and fast route for synthesis zinc oxide nanoparticles using *Punica granatum* peel aqueous extract. Also this research work determined the effect of synthesized ZnONPs on green peach aphid (GPA) and antibacterial efficacy against standard strains of Gram positive *Staphylococcus aureus* and Gram negative *Escherichia coli*.

2. Materials and Methods

2.1 Materials

Zinc acetate dihydrate ($ZnC_4H_6O_2 \cdot 2H_2O$) and potassium hydroxide (KOH) are pure grade purchased from SIGMA-Aldrich and used without further purification. Distilled water was used in all experimental work.

2.2 Preparation of Punica granatum Peel Extract

Fresh fruits of healthy *Punica granatum* were collected from local market, Amman, Jordan. Peels were washed with water to remove dust particles and then dried in shade for two weeks to remove the residual moisture. Peels aqueous extract was prepared by placing 20 g of dried fine powder in 500 ml glass beaker along with 400 ml of sterile distilled water. The mixture was boiled for 10 min until the color of aqueous solution changed from watery to brown. Then the mixture was cooled to room temperature and filtered with Whatman No. 1 filter paper before centrifuging at 1,200 rpm for 5 minutes to remove biomaterials. The aqueous extract was stored at room temperature in order to be used for further experiments.

2.3 Green Synthesis of Zinc Oxide Nanoparticles (ZnONPs)

In a typical reaction mixture, 1.2 g of zinc acetate dihydrate was dissolved in 100 ml of the distilled water in 250 mL conical flask and stirred magnetically at room temperature for 5 min. *Punica granatum* peel aqueous extract (pH 5) was adjusted by potassium hydroxide to pH > 12. Afterwards, the alkaline peels aqueous extract was added drop wise under stirring, as soon as, the peels extract comes in contact zinc ions spontaneous change the colorless of zinc ions to yellow color. The obtained yellow mixture was left under stirring at room temperature. After one min, the yellow mixture started changing to a yellow-white suspended mixture, indicating the formation of water soluble monodispersed zinc oxide nanoparticles.

2.4 Characterization Techniques

Scanning electron microscopy (SEM) analysis of synthesized zinc oxide nanoparticles was done using a Hitachi S-4500 SEM machine. Powder X-ray diffraction (XRD) was performed using a X-ray diffractometer, Shimadzu, XRD-6000 with CuKα radiation λ = 1.5405 Å over a wide range of Bragg angles ($3° \leq 2\theta \leq 80°$). Fourier transform infrared (FT-IR) spectroscopic measurements were done using Shimadzu, IR-Prestige-21 spectrophotometer. UV-vis spectrum of zinc oxide nanoparticles was recorded, by taking 0.1 ml of the sample and diluting it with 2 ml deionized water, as a function of time of reaction using a Schimadzu 1601 spectrophotometer in the wave length region 200 to 700 nm operated at a resolution of 1 nm.

2.5 Green Peach Aphid

Zinc oxide nanoparticles synthesized using aqueous *Punica granatum* peels was tested against the green peach aphid. Five concentrations of ZnONPs were used to find their effect on this global pest. The testing method is similar to that adopted by Ghidan et al. (2016). Two ways analysis of variance was used to compare between mortalities caused by the ZnONPs concentrations.

2.6 Antibacterial Activity

Zinc oxide nanoparticles synthesized using aqueous *Punica granatum* peel extract was tested for its potential antimicrobial activity against some selected microbes. To analyze the antimicrobial activity of the sample, the samples were subjected to Agar well Diffusion method. Diameter of the zone of inhibition was measured in mm and expressed as Mean ± Standard Deviation.

3. Results and Discussion

3.1 XRD Analysis

The X-ray diffraction (XRD) pattern of synthesized ZnO nanoparticles is illustrated in Figure 1. The XRD pattern revealed the orientation and crystalline nature of zinc oxide nanoparticles. The peak position with 2θ values of 31.76°, 34.47°, 36.25°, 47.54°, 56.58°, 62.82°, 65.55°, 67.94°, 68.88°, 72.33, and 76.67° are indexed as (100), (002), (101), (102), (110), (103), (200), (112), (201), (004), and (202) planes, which are in good agreement with those of powder ZnO obtained from the International Center of Diffraction Data card (JCPDS-36-1451) confirming the formation of a crystalline monoclinic structure. No extra diffraction peaks of other phases are detected, indicating the phase purity of ZnO nanoparticles. The average crystallite size of the synthesized zinc oxide nanoparticles was calculated to be 20 nm using Debye-Scherrer equation (Awwad et al., 2014; Raut et al., 2015).

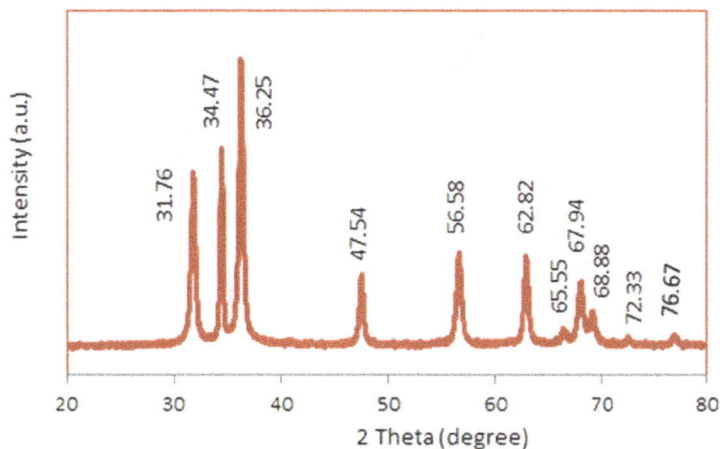

Figure 1. XRD pattern of synthesize zinc oxide nanoparticles

3.2 FT-IR Analysis

FT-IR Spectrum of *Punica granatum* peels, Figure 2 display strong and abroad absorption bands at 3518-3232 cm^{-1}, which could be ascribed to the stretching absorption bands of amino (-NH) and hydroxyl (-OH) stretching H bonded alcohols and phenols. The strong absorption peak at 2928 cm^{-1} could be assigned to the asymmetric and symmetric stretching of -CH_2 and -CH_3 functional groups of aliphatic. Peak at 1728 cm^{-1} corresponds to stretching carboxyl groups. The band at 1616 cm^{-1} is characteristic of amide carbonyl group in amide I and amide II. The bands at 1520 cm^{-1} and 1446 cm^{-1} is assigned to the methylene scissoring vibrations from the proteins. C-N stretch of aromatic amines and carboxylic acids gives rise to band at 1350 cm^{-1}. The band at 1234 $cm^{-1}$is due to CO vibrations of alcohols, phenols and C-N stretching vibrations of amine. The band at 1026 cm^{-1} assigned to the C-O stretching vibrations of alcohols. The peaks at 879 cm^{-1}, 768 cm^{-1}, and 509 cm^{-1} can be assigned to aromatic compounds. These functional groups act as dispersing, capping and stabilizing agents for ZnONPs during the process of synthesis. FT-IR spectrum of the synthesized ZnONPs, Figure 3 indicated a new chemistry linkage on the surface of ZnO nanoparticles. This suggests that *Punica granatum* peel extract can bind to zinc oxide nanoparticles through carbonyl of the amino acid residues in the protein of the extract, therefore acting as stabilizer and dispersing agent for synthesized zinc oxide nanoparticles and prevent agglomeration of nanoparticles. The main characteristic peaks of *Punica granatum* peel extract were observed in FT-IR spectrum of zinc oxide nanoparticles; the FT-IR spectrum of the ZnO nanoparticles show strong and sharp peaks at 567 cm^{-1} and 440 cm^{-1}.

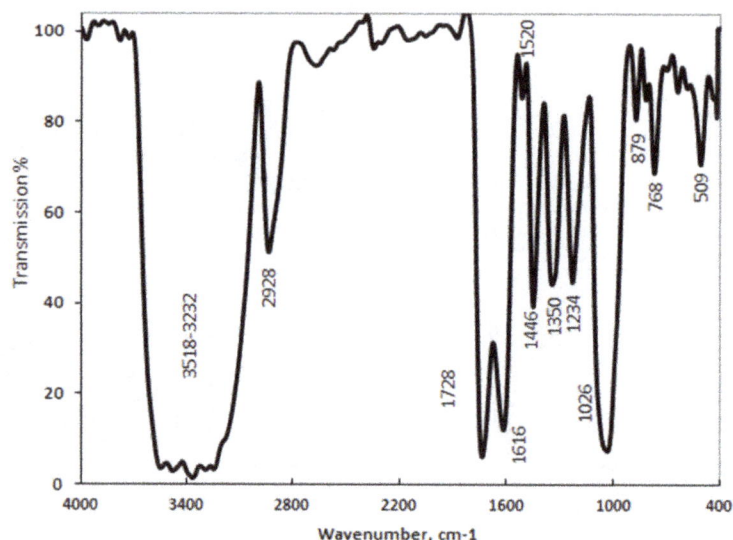

Figure 2. FT-IR spectrum of *Punica granatum* peel extract

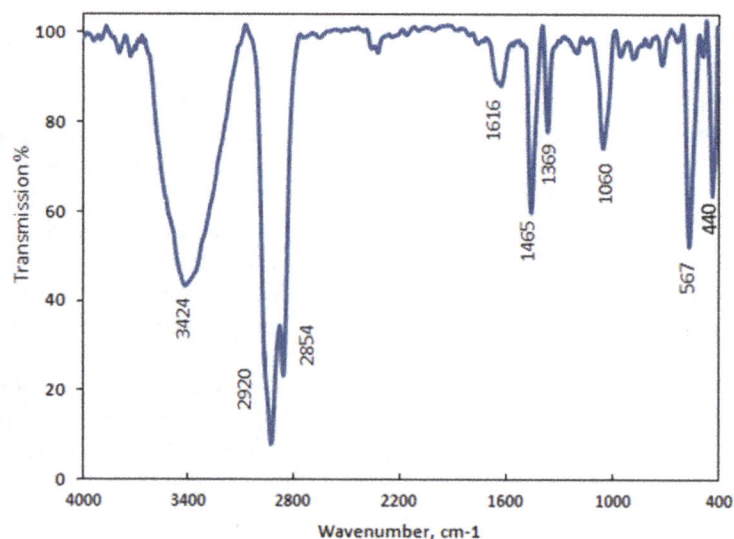

Figure 3. FT-IR spectrum of synthesized ZnONPS

3.3 UV-vis Spectroscopy

It is known that UV-vis spectroscopy is the most widely used technique for the structural characterization of nanoparticles. The absorbance of the ZnO reaction mixture was monitored in the range of 200-700 nm. Typical excitation absorption at 278 nm was observed at room temperature.

3.4 SEM Images Analysis

Scanning electron microscopy (SEM) images of synthesized zinc oxide nanoparticles by this green method are shown in Figure 4. It was found that the *Punica granatum* peel extract strongly influences the morphologies of the resultant zinc oxide nanoparticles. As-synthesized zinc oxide were mainly composed of Nano-flowers and Nano-platelets sizes ranges with the average size in the range of 40 nm and primary particles coalesced together to form larger-sized secondary particles,. While using larger quantities of *Punica granatum* peel extract in synthesis process, zinc oxide Nano-flowers were obtained (Figure 4A). Smaller quantities with quite surface were synthesized with an average size of ca. 40 nm and average thickness of 8 nm. Figure 4B showed a quite dense morphology compromised of randomly oriented overlapping, thin sheets of ZnONPs where the individual sheets appear to have a lateral dimension of less than 300 nm. A good estimate of the Nano-sheets thickness is in the range of 10 to 30 nm. However, it is possible that the thicker sheets may consist of several thinner sheets aggregated to form Nano-platelets networks. Such ZnONPs platelet format form a useful baseline in the sense of demonstrating that ZnO of very high crystallinity and near-perfect stoichiometry which agrees well with the XRD results.

Figure 4. SEM images of synthesized ZnONPs: (A) flowers shape and (B) platelets shape

3.5 Antibacterial Activity

The green synthesized zinc oxide nanoparticles showed a significant antibacterial activity against Gram positive and Gram negative bacteria. The antibacterial activity results against Gram positive *Staphylococcus aureus* and Gram negative *Escherichia coli*. Chloramphenicol was used as a control antibacterial agent. Maximum zone of inhibition (MZI) are listed in Table 1. It was observed that an increase in ZnONPs concentration increases the MZI of *Staphylococcus aureus* and *Escherichia coli*. The results indicated that green synthesized zinc oxide nanoparticles showed effective antibacterial activity against *S. aureus* and *E. coli*. Also our results indicated that the inhibitory effect increase with an increase in the concentrations of ZnONPs.

Table 1. Antibacterial activity of ZnONPs against selected bacteria

ZnONPs (μg/L)	S. aureus	E. coli
200	16±0.21	11.8±0.05
100	14±0.06	10.7±0.06
50	12±0.07	9±0.07
Reference drug	20±2.44	20±2.28

3.6 Effect on Green Peach Aphid

The effect of different concentrations of ZnONPs on GPA is shown in Tables 2 and 3. Means of mortality % of the 1st and 2nd nymphal instars caused by the five concentrations were different significantly, Table 2. Means of mortality % of 3rd and 4th nymphal instars of the same aphid caused by the same concentrations were also different significantly. However, mortality % of both aphid categories in all concentrations was greater significantly than the control treatment.

The highest mortality % of 1st and 2nd nymphal instars was in 8000 μg/ml concentration. It was 75.5% after 24 h, then reached 100% after 48 h and 72 h, while mortality increased at 4000 μg/ml concentration reached 90 and 100 after 48 h and 72 h respectively. The lowest mortality was in case of using 250 μg/ml after 24 h then increased to 43 and 75% after 48 h and 72 h, respectively. The mortalities in case of control were significantly the least after 24 h, 48 h and 72 h compared with the other treatments.

The highest mortality % of 3rd and 4th nymphal instars was in 8000 μg/ml concentration, Table 3. It was 74.0% after 24 h, then reached 65%, 100% after 48 h and 72 h, respectively, while mortality increased at 4000 μg/ml concentration reached 55.0% after 48 h and 72 h respectively. The lowest mortality was in case of using 250 µg/ml after 24 h then increased to 35.0% and 58.5% after 48 h and 72 h, respectively. The mortalities in case of control were significantly the least after 24 h, 48 h and 72 h compared with the other treatments.

Table 2. Means of mortality percent (%) of 1st and 2nd nymphal instars of the green peach aphid by five different concentrations of ZnONPs

Concentration (µg/ml)	Mortality % of 1st and 2nd nymphal instars		
	After 24 h	After 48 h	After 72 h
Control	4.0d	5.0e	5.0d
250	29.5c	43.0d	75.0c
1000	34.5c	55.0c	83.0bc
2000	60.5b	92.0b	92.0ab
4000	66.0ab	90.0b	100a
8000	75.5a	100a	100a

Note. * Means within the same colum of the same period of time sharing the same letter do not differ significantly at 5% level using Frisher protected LSD test.

Table 3. Means of mortality percent (%) of 3rd and 4th nymphal instars of the green peach aphid by five different concentrations of ZnONPs

Concentration (µg/ml)	Mortality % of 3rd and 4th nymphal instars		
	After 24 h	After 48 h	After 72 h
Control	3.0e	4.0c	5.0c
250	17.5d	35.0ab	58.5b
1000	25.0d	31.0ab	61.0b
2000	38.0c	65.0a	55.0b
4000	58.0b	55.0a	55.0b
8000	74.0a	65.0a	100a

Note. * Means within the same colum of the same period of time sharing the same letter do not differ significantly at 5% level using Frisher protected LSD test.

4. Conclusion

In the present work, we first report an eco-friendly and simple method for the synthesis of zinc oxide nanoparticles using *Punica granatum* peel extract. FTIR analysis of aqueous *Punica granatum* peel extract indicated the presence of phyto-constituents such as amines, aldehydes, phenols, and alcohols which were the surface active molecules stabilized the zinc oxide nanoparticles. XRD analysis reveals that the average size of the nanoparticles was found to be 20 nm which was calculated by Debye-Scherrer equation. FT-IR and XRD results corroborated the purity of the synthesized ZnONPs. Green synthesized ZnONPs was evaluated against green peach aphid, the results showed significant effect on this aphid compared with the control treatment. Gram positive *Staphylococcus aureus* and Gram negative *Escherichia coli* showing significant effective activity. The method of the present study offers several important advantageous features. First, the synthesis route is economical and environmentally friendly, because it involves inexpensive and non-toxic materials for second, large scale synthesis.

References

Abbasi, H. Y., Habib, A., & Tanveer, M. J. (2017). Synthesis and characterization of nanostructures of ZnO and ZnO/Graphene composites for the application in hybrid solar cells. *J. Alloys and Compounds, 690*, 21-26. http://dx.doi.org/10.1016/j.jallcom.2016.08.161

An, D., Li, Y., Lian, X., Zou, Y., & Deng, D. (2014). Synthesis of porous ZnO structure for gas sensor and photocatalytic applications. *Colloids and Surfaces A: Physicochem. Eng. Aspects., 447*, 81-87. http://dx.doi.org/10.1016/j.colsurfa.2014.01.060

Awwad, A. M., Albiss, B., & Ahmad, A. L. (2014). Green Synthesis, Characterization and Optical Properties of Zinc Oxide Nanosheets Using Olea Europea Leaf Extract. *Adv. Mater. Lett., 5*, 520-524. http://dx.doi.org/10.5185/amlett.2014.5575

Bala, N., Saha, S., Chakraborty, M., Maiti, M., Das, S., Basu, R., & Nandy, P. (2015). Green synthesis of zinc oxide nanoparticles using Hibiscus subdariffa leaf extract: effect of temperature on synthesis, anti-bacterial activity and anti-diabetic activity. *RSC Adv., 5*, 4993-5003. http://dx.doi.org/10.1039/C4RA12784F

Devi, R. S., & Gayathri, R. (2014). Green synthesis of zinc oxide nanoparticles by using Hibiscus rosa-sinensis. *Int. J. Curr. Eng. Technol., 4*, 2444-2446. Retrieved from http://inpressco.com/category/ijcet

Dobrucka, R., & Dugaszewska, J. (2016). Biosynthesis and antibacterial activity of ZnO nanoparticles using *Trifolium pratense* flower extract. *Saudi J. Biolog. Sci., 23*, 517-523. http://dx.doi.org/10.1016/j.sjbs.2015.05.016

Ehlert, S., Lunkenbein, T., Breu, J., & Förster, S. (2014). Facile large-scale synthetic route to monodisperse ZnO nanocrystals. *Colloids and Surfaces A: Physicochem. Eng. Aspects, 444*, 76-80. http://dx.doi.org/10.1016/j.colsurfa.2013.12.034

El Ghoul, J., Barthou, C., & El Mir, L. (2012). Synthesis by sol-gel process, structural and optical properties of nanoparticles of zinc oxide doped vanadium. *Superlattices and Microstructures, 51*, 942-951. http://dx.doi.org/10.1016/j.spmi.2012.03.013

Fatimah, I., Pradita, R. Y., & Nurfalinda, A. (2016). Plant extract mediated of ZnO nanoparticles by using ethanol extract of *Mimosa pudica* Leaves and coffee powder. *Procedia Eng., 148*, 43-48. http://dx.doi.org/10.1016/j.proeng.2016.06.483

Ghidan, A. Y., Al-Antarya, T. M., & Awwad, A. M. (2016). Green synthesis of copper oxide nanoparticles using Punica granatumpeels extract: Effect on green peach Aphid. *Environmental Nanotechnology, Monitoring & Management, 6*, 95-98. http://dx.doi.org/10.1016/j.enmm.2016.08.002

Ipeksac, T., Kaya, F., & Kaya, C. (2013). Hydrothermal synthesis of zinc oxide (ZnO) nanotubes and its electrophoretic deposition on nickel filter. *Mater. Lett., 100*, 11-14. http://dx.doi.org/10.1016/j.matlet.2013.02.099

Karnan, T., & Selvakumar, S. A. S. (2016). Biosynthesis of ZnO nanoparticles using rambutan (*Nephelium lappaceum* L.) peel extract and their photocatalytic activity on methyl orange dye. *J. Molecular Structure, 1125*, 358-365. http://dx.doi.org/10.1016/j.molstruc.2016.07.029

Kumari, L., Li, W. Z., Vannoy, C. H., Leblanc, R. M., & Wang, D. Z. (2010). Zinc oxide micro- and nanoparticles: Synthesis, structure and optical properties. *Mater. Res. Bulletin, 45*, 190-196. http://dx.doi.org/10.1016/j.materresbull.2009.09.021

Madan, H. R., Sharma, S. C., Udayabhanu, Suresh, D., Vidyad, Y. S., Nagabhushanae, H., ... Maiya, P. S. (2016). Facile green fabrication of nanostructure ZnO plates, bullets, flower, prismatic tip, closed pine cone: Their antibacterial, antioxidant, photoluminescent and photocatalytic properties. *Spectrochimica Acta Part A: Molecular and Biomolecular Spectroscopy, 152*, 404-416. http://dx.doi.org/10.1016/j.saa.2015.07.067

Mousa, M. A., Bayoumy, W. A. A., & Khairy, M. (2013). Characterization and photo-chemical applications of nano-ZnO prepared by wet chemical and thermal decomposition methods. *Mater. Res. Bulletin, 48*, 4576-4582. http://dx.doi.org/10.1016/j.materresbull.2013.07.050

Ozcelik, B. K., & Ergun, C. (2014). Synthesis of ZnO nanoparticles by an aerosol process. *Ceramics Inter., 40*, 7107-7116. http://dx.doi.org/10.1016/j.ceramint.2013.12.044

Peng, J., Guo, W., Yang, W., Shi, C., Liu, M., Zheng, Y., ... Yuqian, Y. (2016). Synthesis of three-dimensional flower-like hierarchical ZnO nanostructure and its enhanced acetone gas sensing properties. *J. Alloys and Compounds, 654*, 371-378. http://dx.doi.org/10.1016/j.jallcom.2015.09.120

Ramesh, M., Anbuvannan, M., & Viruthagiri, G. (2015). Green synthesis of ZnO nanoparticles using *Solanum nigrum* leaf extract and their antibacterial activity. *Spectrochimica Acta Part A: Molecular and Biomolecular Spectroscopy, 136, Part B*, 864-870. http://dx.doi.org/10.1016/j.saa.2014.09.105

Rana, N., Chand, S., & Gathania, A. K. (2016). Green synthesis of zinc oxide nanosized spherical particles using *Terminalia chebula* fruits extract for their photocatalytic applications. *Inter. Nano Lett., 6*, 91-98. http://dx.doi.org/10.1007/s40089-015-0171-6

Raut, S., Thorat, P. V., & Thakre, R. (2015). Green synthesis of zinc oxide (ZnO) nanoparticles using *Ocimum tenuiflorum* leaves. *Inter. J. Sci. and Res., 4*, 1225-1228. Retrieved from http://www.ijsr.net/SUB154428

Rocha, L. S. R., Deus, R. C., Foschini, C. R., Moura, F., Gonzalez Garcia, F., & Simões, A. Z. (2014). Photoluminescence emission at room temperature in zinc oxide nano-columns. *Mater. Res. Bulletin, 50*, 12-17. http://dx.doi.org/10.1016/j.materresbull.2013.09.049

Sangeetha, G., Rajeshwari, S., & Venckatesh, R. (2011). Green synthesis of zinc oxide nanoparticles by *aloe barbadensis* miller leaf extract: Structure and optical properties. *Mater. Res. Bulletin, 46*, 2560-2566. http://dx.doi.org/10.1016/j.materresbull.2011.07.046

Shah, R. K., Boruah, F., & Parween, N. (2015). Synthesis and Characterization of ZnO Nanoparticles using Leaf Extract of *Camelia sinesis* and evaluation of their antimicrobial efficacy. *Int. J. Curr. Microbiol. App. Sci., 4*, 444-450. Retrieved from http://www.ijcmas.com

Sharma, D., Myalowenkosi, I., Sabela, M. I., Kanchi, S., Mdluli, P. S., Singh, G., ... Bisetty, K. (2016). Biosynthesis of ZnO nanoparticles using *Jacaranda mimosifolia* flowers extract: Synergistic antibacterial activity and molecular simulated facet specific adsorption studies. *J. Photochem. & Photobiol. B: Biology, 162*, 199-207. http://dx.doi.org/10.1016/j.jphotobiol.2016.06.043

Singh, S. C., & Gopal, R. (2008). Synthesis of colloidal zinc oxide nanoparticles by pulsed laser ablation in aqueous media. *Physica E, 40*, 724-730. http://dx.doi.org/10.1016/j.physe.2007.08.155

Soofivand, F., Salavati-Niasari, M., & Mohandes, F. (2013). Novel precursor assisted synthesis and characterization of zinc oxide nanoparticles/nanofibers. *Mater. Lett., 98*, 55-58. http://dx.doi.org/10.1016/j.matlet.2013.01.129

Suresh, D., Nethravathi, P. C., Udayabhanu, Rajanaika, H., & Sharma, S. C. (2015). Green synthesis of multifunctional zinc oxide (ZnO) nanoparticles using *Cassia fistula* plant extract and their photodegradative, antioxidant and antibacterial activities. *Mater. Sci. Semiconductor Process, 31*, 446-454. http://dx.doi.org/10.1016/j.mssp.2014.12.023

Sutradhar, P., Debbarma, M., & Saha, M. (2016). Microwave synthesis of zinc oxide nanoparticles using coffee powder extract and its application for solar cell, Synthesis and Reactivity in Inorganic. *Metal-Organic and Nano-Metal Chemistry, 46*, 1622-1627. http://dx.doi.org/10.1080/15533174.2015.1137035

Thareja, R. K., & Shukla, S. (2007). Synthesis and characterization of zinc oxide nanoparticles by laser ablation of zinc in liquid. *Appl. Surf. Sci., 253*, 8889-8895. http://dx.doi.org/10.1016/j.apsusc.2007.04.088

Tseng, C.-C., Chou, Y.-H., Liu, C.-M., Liu, Y.-M., & Ger, M.-D. (2012). Microwave assisted hydrothermal synthesis of zinc oxide particles starting from chloride precursor. *Mater. Res. Bulletin., 47*, 96-100. http://dx.doi.org/10.1016/j.materresbull.2011.09.027

Vanathi, P., Rajiva, P., Narendhran, S., Rajeshwari, S., Rahman, P. K. S. M., & Venckatesh, R. (2014). Biosynthesis and characterization of phyto mediated zinc oxide nanoparticles: A green chemistry approach. *Mater. Lett., 134*, 13-15. http://dx.doi.org/10.1016/j.matlet.2014.07.029

Yıldırım, O. A., & Durucan, C. (2010). Synthesis of zinc oxide nanoparticles elaborated by microemulsion method. *J. Alloys and Compounds, 506*, 944-949. http://dx.doi.org/10.1016/j.jallcom.2010.07.125

Zak, A. K., Majid, W. H., Wang, H. Z., Yousefi, R., Golsheikh, A. M., & Ren, Z. F. (2013). Sonochemical synthesis of hierarchical ZnO nanostructures. *Ultrasonics Sonochemistry, 20*, 395-400. http://dx.doi.org/10.1016/j.ultsonch.2012.07.001

20

Toxicity and Efficacy of Chlorantraniliprole on *Pieris rapae* (Linnaeus) (Lepidoptera: Pieridae) on Cabbage

Qi Su[1], Hong Tong[1], Jiaxu Cheng[1], Guohui Zhang[1], Caihua Shi[1], Chuanren Li[1] & Wenkai Wang[1]

[1] Institute of Insect Sciences, College of Agriculture, Yangtze University, Jingzhou, Hubei, China

Correspondence: Qi Su, Institute of Insect Sciences, College of Agriculture, Yangtze University, Jingzhou, Hubei 434025, China. E-mail: qsu@yangtzeu.edu.cn

This work was supported by the National Natural Science Foundation of China (31572010, 31501641) and Hubei Provincial Natural Science Foundation of China (2016CFB304).

Abstract

Toxicity of chlorantraniliprole was assayed against young (first and second instars) and older larvae (third and fourth instars) of cabbage *Pieris rapae* (Lepidoptera: Pieridae) on cabbage (*Brassicae oleracea*), and persistence of field–aged leaf residue of chlorantraniliprole was assayed with 5-old-day larvae of *P. rapae* on cabbage. Efficacies of chlorantraniliprole and other newer insecticides to *P. rapae* were tested under field conditions for two seasons in Hubei province in China. The LC_{50} value of chlorantraniliprole for early and later *P. rapae* larvae were 7.92 and 11.34 mg/L by contact toxicity, respectively. The LC_{50} value of chlorantraniliprole for early and later *P. rapae* larvae were 0.95 and 4.32 mg/L through ingestion, respectively. The toxicity of field-aged leaf residues of chlorantraniliprole (0-, 3-, 5-, 7-, 10-, 14-, 21-, 25-, and 28-day-old residues) declined gradually under the field conditions. Almost all larvae died on day 5 after feeding on the leaves with 0-21-day residue, and the mortalities were as high as 83.3% and 72.5% for the 21- and 25-day-old leaf residues. Chlorantraniliprole application suppressed *P. rapae* larvae below the economic threshold for 21-28 days. The field efficacy trials show that chlorantraniliprole at 52 mg a.i /L rate was effective against *P. rapae* larvae on cabbage, providing marketable cabbage with three applications per season. In addition, chlorantraniliprole was as effective as indoxacarb and spinosad and significantly more effective than emamectin benzoate.

Keywords: *Pieris rapae*, chlorantraniliprole, reduced-risk insecticides

1. Introduction

The imported cabbage worm, *Pieris rapae* (L.) (Lepidoptera: Pieridae), is one of the most serious insect crop pest in China, costing more than $RMB371 million ($US45 million) in insecticide treatment in 1997 (Wu, 2000). Control of the pest is becoming increasingly difficult due to its resistance to many conventional synthetic insecticides in many areas (Mu et al., 1984; Han et al., 1987; Armes et al., 1997; Li et al., 1999). With the imposed quality restrictions on fresh market vegetables, management of Lepidopteran pests on cabbage has been based on either a low threshold (one larva per three plants) or on scheduled weekly sprays (Cartwright et al., 1987). Thus, some newer kinds of insecticides with a chemistry of the anthranilic diamide, have been introduced as substitutes to control *P. rapae*.

Chlorantraniliprole is a novel anthranilic diamide insecticide discovered by DuPont, also known as Rynaxypyr and DPX-E2Y45, which belongs to a new chemical class of selective ryanodine receptor (RyR) agonists (Lahm et al., 2005; Cordova et al., 2006). Upon ingestion, chlorantraniliprole activates the release and depletion of internal calcium stores in muscles (Bassi et al., 2007). In the target organism, this causes impaired regulation of muscle contraction and leads to feeding cessation, lethargy, paralysis, and death. Differential selectivity towards insect RyRs explains chlorantraniliprole's outstanding profile of mammalian toxicity. It is being developed worldwide by DuPont in a broad range of crop to control a range of pests belonging to the order Lepidoptera and some Coleoptera, Diptera and Isoptera species. It is primarily active on chewing pests by ingestion and secondarily by contact, showing good larvicidal active.

In this study, we tested chlorantraniliprole and several other newer insecticides under laboratory and field conditions for several years in Hubei province in China. We reported the toxicities of chlorantraniliprole on *P. rapae* larvae, toxicities of field-aged leaf residues against *P. rapae* larvae, and the result of field trials compared with other reduced-risk and commonly used insecticides.

2. Materials and Methods

2.1 Pieris Rapae

The *P. rapae* lab colony used in this study were derived from cultures maintained on plotted cabbage (Bailey) without exposure to any insecticide in the greenhouse for > 5 years at 24-30 °C and 60-80% relative humidity (RH) under a photoperiod of 14:10 h light:dark at the Laboratory of Insect Toxicology, Yangtze University, Jingzhou, China. The emergent butterflies were fed on a 10% aqueous honey solution daily during the experiment. Adults were allowed to mate and oviposit on four small cabbage plants that each had 4-5 fully expanded leaves. The leaves bearing *P. rapae* eggs were detached from the plants, and placed in plastic rearing trays (20 × 40 × 15 cm, F1020-No Holes plastic Flat Tray, T. O. Plastic Inc., Minneapolis, MN). After hatching, the larvae were fed with fresh cabbage leaves as described above.

2.2 Cabbage

We used cabbage as the host plants for both field tests and laboratory bioassay. Cabbage plants for laboratory bioassays were grown Metro-Mix 300 growing medium (Grace Sierra, Horticultural Products, Milpitas, CA) in a greenhouse with 24-30 °C and 60-80%RH and natural lighting. At the time of planting, a slow-release fertilizer (N:P:K = 12:8:6) was applied to cabbage. Plants used in the experiments were at the 30-35 cm in the height with 10-12 leaves.

2.3 Insecticides

Chlorantraniliprole (Rynaxypyr, 20% purity, recommended field rate 26-52 mg a.i/l, DuPont, USA), indoxacarb (Avatar, 15% purity, recommended field rate 47-63 mg a.i/l, DuPont, USA), spinosad (Tracer, 2.5% purity, recommended field rate 2.0-43 mg a.i/l, Dow AgroSciences Corporation, USA), emamectin benzoate (Banleptm, 1% purity, recommended field rate 2.6-2.9 mg a.i/l, Pharmaceutical Group Corporation, Hebei, China).

2.4 Toxicity Bioassays

2.4.1 Topical Toxicity Test

Topical toxicity test were performed to determine the effective concentration of chlorantraniliprole that would kill larvae *P. rapae* by contact. Larvae were inoculated with 1μl droplet of chlorantraniliprole using a microsyringe at delivered dosages of 1.5, 3, 6, 12, 24 mg/l. Treated larvae were transferred to petri dishes (9 cm diameter and 1.5 cm depth) with four layers of filter paper and held in a chamber maintained at 26±2 °C, 70±5% RH for 48 h. Each petri dish containing ten larvae were set up for each dose. All treatment had five replications. Mortality, defined as the inability to move when prodded, was recorded by counting the number of dead and moribund larvae after 24 h of exposure.

2.4.2 Ingestion Toxicity to Larvae

Leaf-dip assay was used to evaluate the ingestion toxicity effect of the chlorantraniliprole according to the method of Insecticide Resistance Action Committee (2000). Leaf discs (2.5 cm diameter) cut from cabbage leaves with a cork borer, were dipped for 5s in the chlorantraniliprole or control solutions (sterile distilled water) and dried in the air for 1h at room temperature. Four leaf discs were transferred to Petri dishes (9 cm diameter and 1.5 cm depth) with four layers of filter paper. Ten *P. rapae* larvae were placed alongside the leaf discs at the centre of the Petri dish. Mortality was defined as described above. All experiments were assessed after 24 h. All treatment had five replications.

2.4.3 Toxicity of Field-Aged Leaf Residue to Larvae

Chlorantraniliprole was sprayed at 52 mg a.i/l and untreated plants were used as the control. The insecticides were applied at nine different times to have the residue ages of 1, 3, 7, 10, 14, 21, 25 and 28 days after treatment (DAT). Bioassays were conducted in a laboratory at 26± 2 °C, 50±5% RH, and a photoperiod of 14 h:10 h (L:D). Bioassays were initiated 2 h after treatment at 0, 3, 7, 10, 14, 17, 25 and 28 DAT. For each date, a single leaf disk (8-9 cm diameter) was cut from one leaf in each plot. The leaves were selected based on location on the plant to increase likelihood of good spray coverage. Each leaf disk was placed in a large clear plastic Petri dish and ten third instars were placed on the leaf disk. Each treatment had six replications. Larvae were fed with freshly cut leaf disks from the treated plants 2 days after initial exposure. Larval mortality was recorded daily for 5 days. Experiments with control mortality more than 20% were discarded and repeated.

2.4.4 Field Efficacy Trials

Two field trials were conducted to evaluate efficacies of four insecticides: chlorantraniliprole (52 mg a.i/l), indoxacarb (63 mg a.i/l), emamectin benzoate (3.9 mg a.i/l) and spinosad (43 mg a.i/l) during 2015-2016 at the research center of pesticide in Yangtze University, Jingzhou, China. For field trials in 2015, cabbage was planted on 10 March 2015. Insecticides were applied four times on 12, 19, 26 May, and 1 June, 2015. For field trials in 2016, cabbage was planted on 10 March 2016. Insecticides were applied on 12 May when *P. rapae* population in some plots reached the economic threshold. Thereafter, the insecticides were applied once per week on 19 and 26 May and 1 June. At termination on 5 July in 2015 and 1 June in 2016, a damage-quality evaluation was made on 10 plants per plot as described below. In general, field plots consisted of two rows of cabbage on 1-m bed with 30-cm within-row plant spacing and 10-m length. All plots were separated with *sorghum* windbreaks and a 1.3-m alleyway. All treatments were arranged in a randomized complete block design with four replications. Insecticide applications were initiated as larval densities exceeded the threshold level of 0.3 larvae per plant as determined in Hubei (Cartwright et al., 1987). Insecticides were applied using a tractor-mounted sprayer. The tractor-mounted sprayer was equipped with three ceramic hollow cone nozzles per row (TX6-red, one over the plant, and one on each side of the row directed into the plant) with a spray pressure of 689.5 kPa and a delivery rate of 280 l ha^{-1} at 3.2 km h^{-1}. To monitor *P. rapae* larval populations, cabbage plants were scouted weekly. Number of *P. rapae* larvae and eggs per plant were counted by checking both the upper and lower surfaces of every leaf on each plant, and 10 plants were examined from each plot. A damage-quality evaluation was made on plants per plot based on the six categories as described by Greene et al. (1969): 0, no apparent damage; 1, minor feeding damage on wrapper outer leaves, or 1% leaf area eaten; 2, minor-moderate feeding damage, or 2-5% leaf area eaten; 3, moderate damage, or 6-10% leaf area eaten, but no head damage; 4, moderate-heavy damage on wrapper and outer leaves with minor damage on head, or 11-30% leaf area eaten; and 5, heavy damage on wrapper and head, or > 30% leaf area eaten.

2.5 Data Analysis

Toxicity of chlorantraniliprole to *P. rapae* eggs and larvae, including LC_{50} and LC_{90} with related parameters, were analyzed using POLO (Robertson & Preisler, 1991). Field-aged leaf residue bioassay data, including percentage of larvae mortality, were transformed to the arcsine square root before analysis to stabilize error variance (K. A. Gomez & A. A. Gomez, 1984). Mean percentage of numbers of *P. rapae* larvae and damages among the treatments were analyzed using analysis of variance (ANOVA) and were separated using the least significant different test least significant different (LSD) following a significant *F* test at *P* = 0.05. Although all tests of significance were based on transformed data, the untransformed percentages of mortalities are presented. All data were analyzed using the SPSS software package (ver.17, SPSS Inc., Chicago, IL, USA).

3. Results

3.1 Toxicity Bioassays

Based on dose-mortality responses, the LC_{50} value of chlorantraniliprole for early and later larvae ranged from 5.07 to 12.13 mg/l and 7.12 to 19.71 mg/l by contact, respectively (Table 1). The LC_{90} value of chlorantraniliprole for early and later larvae ranged from 22.58 to 122.98 mg/l and 33.10 to 156.09 mg/l by contact, respectively (Table 1). Chlorantraniliprole exhibit high level contact toxicity on *P. rapae* larvae. Similarly, chlorantraniliprole are highly toxic to *P. rapae* larvae through ingestion. The LC_{50} and LC_{90} value for early and later larvae by ingestion were 0.95, 4.32, 7.01 and 19.65 mg/l, respectively (Table 1). The results indicated that chlorantraniliprole is primarily active on chewing pests by ingestion and secondarily by contact, showing good and larvicidal activity. The LC_{50} and LC_{90} values of chlorantraniliprole to young larvae (first and second instars) and older larvae (third and fourth instars) were significantly different, and younger larvae were significantly more susceptible to chlorantraniliprole than the older larvae (Figures 1 and 2).

Table 1. Comparison of toxicity of chlorantraniliprole at 24 h to early and late larvae of *P. rapae* as mean of LC$_{50}$ and LC$_{90}$ through exposure and leaf-dip assay at 95% of fiducial limit

Larval age	Methods	n	LC$_{50}$ (mg/L)[a]	LC$_{90}$ (mg/L)	Slope ± SE	χ^2(df)
Early instars	Exposure	300	7.92(5.07-12.13)	42.92(22.58-122.98)	1.74±0.33	0.76(3)
	Leaf-dip	300	0.95(0.41-2.57)	7.01(2.53-20.34)	1.47±0.63	0.79(3)
Late instars	Exposure	300	11.34(7.12-19.71)	66.42(33.10-156.09)	1.67±0.61	1.13(3)
	Leaf-dip	300	4.32(2.79-6.00)	19.65(12.7-44.40)	1.95±0.37	0.53(3)

Note. [a] indicates that the LC values differ significantly between the exposure and Leaf-dip method; LC values in parenthesis indicate the range of toxicities measured.

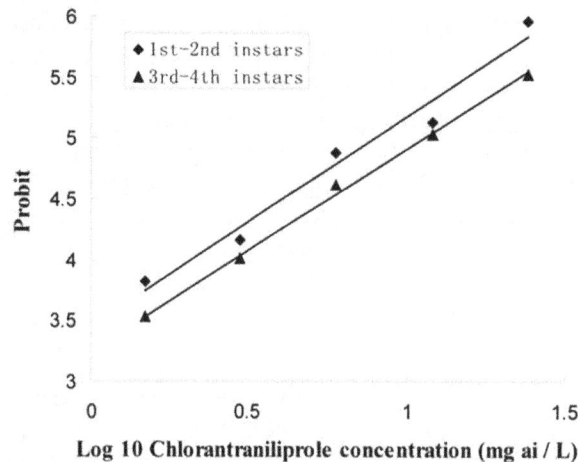

Figure 1. Log dose-probit response lines and slopes for *P. rapae* larvae exposed topically to chlorantraniliprole concentrations

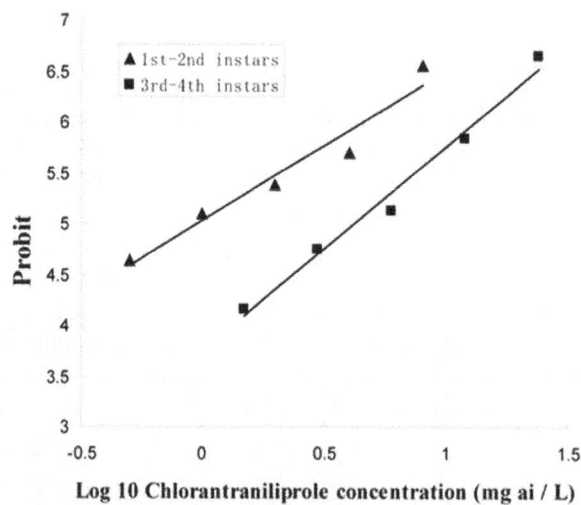

Figure 2. Log dose-probit response lines and slopes for *P. rapae* larvae through ingestion to chlorantraniliprole concentrations

3.2 Field-Aged Leaf Residue Bioassay

Field-aged leaf residue of chlorantraniliprole on cabbage was highly toxic to 5-day-old larvae of *P. rapae*. Percentage mortalities of *P. rapae* larvae were closely related to the ages of leaf residues and the durations of exposure (Figure 3). Field-aged leaf residues were highly toxic to larvae, resulting in 100%, 100%, 95.0%, 92.3%, 82.5 and 83.3% mortalities at 6 day after exposure after feeding on the treated leaves of 0, 3, 7, 10, 14 and 17 d, respectively. Mortality still reached 72.5% for the residue of 25 d, but it dropped to 52.5% for the residue of 28 d. These results give us a hypothesis about the mortality-time after treatment dependence and show that chlorantraniliprole could be effective against *P. rapae* for at least 17-25 days on cabbage under the field conditions of Hubei province, in China.

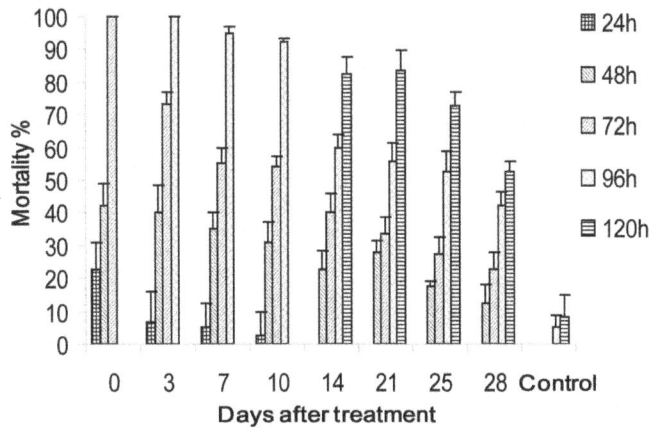

Figure 3. Persistence of filed-aged leaf residue of chlorantraniliprole on cabbage on 5-day-old *P. rapae* larvae

3.3 Field Efficacy Trials

3.3.1 Field Trial in 2015

P. rapae larval populations were low at the beginning of the season, and increased rapidly 2 weeks after the first application and then exceeded the economic threshold in the untreated check throughout the remaining period (Figure 4). Numbers of larvae per plant among the five treatments were significantly different throughout the season after the first application (Figure 4). Among the insecticides, chlorantraniliprole, indoxacard and spinosad significantly reduced the larval population after the first application and maintained the populations below the economic threshold throughout the season. Larval densities in the treatment of chlorantraniliprole maintained at levels below the economic threshold for 28 days. At harvest time, untreated plants were totally unmarketable with a damage rating of 3.6, whereas the plants treated with chlorantraniliprole, indoxacard and spinosad had no or little damage with ratings of 0.4, 0.6, and 0.5, respectively, which were all marketable. However, plants treated with emamectin benzoate were not marketable with a damage rating of 3.2.

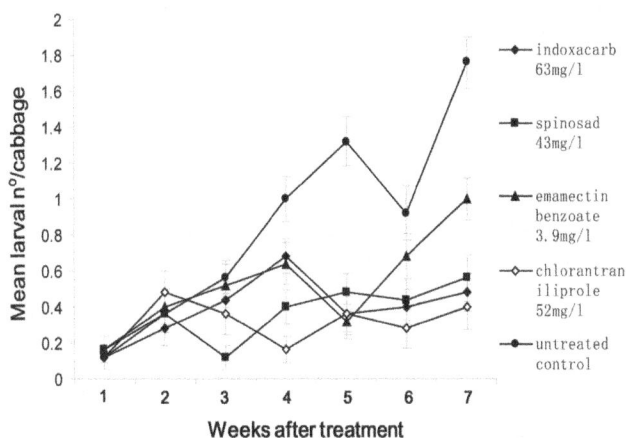

Figure 4. The mean larval number on cabbage plants each week following 4 insecticide application schedules and one control treatment in 2015

3.3.2 Field Trial in 2016

P. rapae populations were slightly higher than those in spring of 2015. All insecticides used were effective against *P. rapae* larva (Figure 5). There were no significant differences in *P. rapae* larval densities on cabbage plants in the first three sampling dates. Although *P. rapae* larval densities on cabbage plants were high with the applications of these insecticides, the overall efficacies of the insecticides were significantly different, and larval density was the lowest on the plants treated with chlorantraniliprole on the last four sampling dates (Figure 5). In two of the four sampling dates, larvae on the plants treated with spinosad were not significantly different from those on the untreated plants. In one of the four sampling dates, larval densities on the plants tested with indoxacard were not significantly different from those on the untreated plants. At harvest time, untreated plants were totally unmarketable with damage rating of 3.9 whereas plants treated with chlorantraniliprole, indoxacard and spinosad had no or little damage with ratings if 0.3, 0.5 and 0.7 respectively, which were all marketable. In contrast, plants treated with emamectin benzoate were rated 2.0 which were marginally marketable depending the demand.

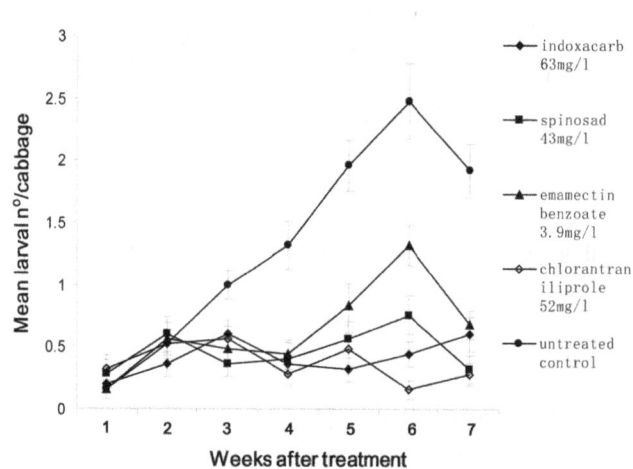

Figure 5. The mean larval number on cabbage plants each week following 4 insecticide application schedules and one control treatment in 2016

4. Discussion

The present study showed that the knockdown activities of chlorantraniliprole to *P. rapae* larvae were very quick, with high mortality within 1 d after exposure. We also observed inhibition of insect feeding occurs rapidly (minutes to a few hours after ingestion) and death normally occurs within 24-72 hours. In addition to larvicidal activity, chlorantraniliprole has been found to have significant ovicidal activity among other Lepidopteran pests (Lahm et al., 2009). Similarly, field-aged leaf residues of chlorantraniliprole knocked down the larvae within 72 h, causing mortality quicker to young larvae than to older ones, indicating that older larvae are more tolerant to chlorantraniliprole than younger ones. The similar result has been reported in Liu et al. (2003) that older *Plutella xyllostella* larvae were more tolerant to indoxacarb and λ-cyhalothrin than younger ones. To achieve control, chlorantraniliprole should be applied with proper timing when most larvae are relatively young (first and second instars). However, chlorantraniliprole will provide good control of older larvae given adequate time. In addition, chlorantraniliprole should be applied thoroughly to ensure good coverage on all plant surfaces, including the underside of the leaf surface where the larvae are located and feed.

We also observed the larvae rapidly stops feeding, becomes paralyzed, subsequently developed curved or discolored bodies and ultimately dies after they were exposed to treated cabbage leaf disks. However, less intoxicated larvae stopped feeding and remained alive for several weeks before they died. When late-instar larvae were fed with chlorantraniliprole-treated cabbage leaves, some *P. rapae* completed their larval development but did not pupate properly. Some larvae could not make cocoon before pupation, whereas some could spin a cocoon, but it was very loose. Some larvae, when pupating, could not molt properly, with exuvia remaining attached to the body surface, or only a portion of body becoming a "pupa". Some adults were able to emerge, but almost all were abnormal with wings that were either twisted, or not well developed. Some adults even lacked appendages, missing antennae or mouthparts, and these adults could not fly. These results clearly indicated that chlorantraniliprole has significant growth regulating effects on *P. rapae* larvae, pupae and adult formation.

The data from the field-aged leaf residues of chlorantraniliprole showed that 21-day-old residue was still highly toxic to *P. rapae* larvae. In addition to its quick knockdown, within 1-3 days after exposure and feeding with treated leaf material, it provide excellent cabbage protection up to 3 weeks or longer under field conditions in Hubei. Our data were consistent with Knight and Flexner (2007), who reported that exposure to chlorantranilipole residues applied to sleeve cages and apple foliage effectively disrupted mating for at least 3weeks under field conditions. Chlorantranilipole showed similar effectiveness in bioassays against field-collected larval populations of codling moth exhibiting a five-fold range of tolerance to azinphos-methyl, and season-long field trials have demonstrated that it can be used effectively to manage codling moth with applications timed every 21 days. It has been reported that chlorantraniliprole has no or little effects on birds, fish, invertebrate, earthworm, honeybee, wasp parasitoid, and predatory mite (Larson et al., 2012). In addition to the indicated species above, several field tests have confirmed minimal to no impact upon beneficial arthropods (Bassi, 2007). Therefore, chlorantraniliprole provides a much safer alternative to currently registered organophosphates, pyrethroids, carbamates, and other high risk conventional insecticides (Bassi, 2007). Although chlorantraniliprole has no cross-resistance with other insecticides, the risk of resistance development has been considered from beginning. We recommended for chlorantraniliprole using with a restricted number of applications per season, within spray programmers that include other effective insecticides with different modes of action.

5. Conclusions

Based on both larval densities and plants damage evaluations, chlorantraniliprole, indoxacarb and spinosad were the most effective insecticides against *P. rapae* on cabbage. Although emamectin benzonate significantly reduced larval densities below the untreated check, they often did not perform as well as the other new products being evaluated. In conclusion, chlorantraniliprole is highly toxic to *P. rapae* larvae not only through ingestion but also through the cuticle. Its effectiveness under field conditions persisted up to 25 d after treated, and the residue will likely last longer than 25 d. Although biological and other bio-rational methods could play important roles in managing *P. rapae*, it is normally difficult to produce cabbage and other leaf vegetable for fresh market, with the necessary cosmetic quality and low cull rate, without using insecticide to control pests. Chlorantraniliprole and the newer insecticide evaluated in our field trials represent valuable new chemical control tools that provide growers with alternative to currently used insecticides.

References

Armes, N. J., Wightman, J. A., Jadhav, D. R., & Ranga, R. G. V. (1997). Status of insecticide resistance in *Spodoptera litura* in Andhra Pradesh, India. *Pesticide Science, 50*, 240-248. https://doi.org/10.1002/(SICI) 1096-9063(199707)50:3%3C240::AID-PS579%3E3.0.CO;2-9

Bassi, A., Alber, R., Wiles, J. A., Rison, J. L., Frost, N. M., Marmor, F. W., & Marcon, P. C. (2007). *Chlorantraniliprole: A novel anthranilic diamide insecticide* (pp. 52-59). XVI International Plant Protection Congress.

Cartwright, B. J., Edelson, V., & Chambers, C. (1987). Composite action thresholds for the control of lepidopterous pests on fresh-market cabbage in the Lower Rio Grande Valley of Texas. *Journal of Economic Entomology, 80*, 175-181. https://doi.org/10.1093/jee/80.1.175

Cordova, D., Benner, E. A., Sacher, M. D., Rauh, J. J., Sopa, J. S., Lahm, G. P., ... Tao, Y. (2006). Anthranilic diamides: A new class of insecticides with a novel mode of action, ryanodine receptor activation. *Pesticide Biochemistry and Physiology, 84*, 196-214. https://doi.org/10.1016/j.pestbp.2005.07.005

Gomez, K. A., & Gomez, A. A. (1984). *Statistical procedures for agricultural research* (2nd ed.). Wiley New York.

Han, X. L., Zhang, W. J., Chen, N. C., & Luo, J. T. (1987). Studies on resistance of imported cabbage worm (*Artogeia rapae* L) to insecticides II: The monitoring and evaluation of imported cabbage worm to insecticides in Beijing. *Journal of China Agriculture University, 13*, 193-198.

Insecticide Resistance Action Committee. (2000). *Proposed Insecticide/Acaricide susceptibility Tests*. IRAC Method No. 3.

Knight, A. L., & Flexner, L. (2007). Disruption of mating in codling moth (Lepidoptera: Tortricidae) by chlorantranilipole, an anthranilic diamide insecticide. *Pest Management Science, 63*, 180-189. https://doi.org/10.1002/ps.1318

Lahm, G. P., Cordova, D., & Barry, J. D. (2009). New and selective ryanodine receptor activators for insect control. *Bioorganic & Medicinal Chemistry, 17*, 4127-4133. https://doi.org/10.1016/j.bmc.2009.01.018

Lahm, G. P., Selby, T. P., Freudenberger, J. H., Stevenson, T. M., Myers, B. J., Seburyamo, G., … Cordova, D. (2005). Insecticidal anthranilic diamides: A new class of potent ryanodine receptor activators. *Bioorganic & Medicinal Chemistry Letters, 15*, 4898-4906. https://doi.org/10.1016/j.bmcl.2005.08.034

Larson, J. L., Redmond, C. T., & Potter, D. A. (2012). Comparative impact of an anthranilic diamide and other insecticidal chemistries on beneficial invertebrates and ecosystem services in turfgrass. *Pest Management Science, 68*, 740-748. https://doi.org/10.1002/ps.2321

Li, X. F., Zhang, W. J., & Wang, C. J. (1999). The sensitivity measurement of different instars cabbageworm to different insecticides. *Chinese Journal of Pesticide Science, 1*, 84-86.

Liu, T. X., Hutchison, W. D., Chen, W., & Burkness, E. C. (2003). Comparative susceptibilities of diamondback moth (Lepidoptera: Plutellidae) and cabbage looper (Lepidoptera: Noctuidae) from Minnesota and South Texas to λ-cyhalothrin and indoxacarb. *Journal of Economic Entomology, 94*, 1230-1236. https://doi.org/10.1603/0022-0493-96.4.1230

Mu, L. Y., Wang, K. Y., Luo, W. C., & Zhao, Y. (1984). A study of resistance of cabbage worm (*Pieris rapae* L) to insecticides. *Acta Phytophylacica Sinica, 11*, 267-273.

Robertson, J. L., & Preisler, H. K. (1991). *Pesticide bioassays with arthropods*. CRC, Boca Raton, FL.

Ruberson, J. L., & Tillman, P. G. (1999). Effect of selected insecticides on natural enemies in cotton: Laboratory studies. In P. Dugger & D. Richer (Eds.), *Proceedings, Beltwide Cotton Conference, Orlando, FL, 3-7 January* (Vol. 2, pp. 1210-1213). National Cotton Council, Memphis, TN.

Tao, Y. (2006). Anthranilic diamides: A new class of insecticides with a novel mode of action, ryanodine receptor activation. *Pesticide Biochemistry and Physiology, 84*, 196-214. https://doi.org/10.1016/j.pestbp.2005.07.005

Wu, S. X. (2000). Pesticide market and target control in China. *Pesticides, 39*, 7-10.

21

Morphological Variation in Selected Accessions of Bambara Groundnut (*Vigna subterranea* L. Verdc) in South Africa

Amara Evangeline Unigwe[1], Abe Shegro Gerrano[2], Patrick Adebola[3] & Michael Pillay[1]

[1] Department of Biotechnology, Vaal University of Technology, Vanderbijlpark, South Africa

[2] Agricultural Research Council-Vegetable and Ornamental Plant Institute (ARC-VOPI), Pretoria, South Africa

[3] Africa Rice Center (AfricaRice), C/o Central Agricultural Research Institute (CARI) Suakoko, Bong County, Monrovia, Liberia

Correspondence: Michael Pillay, Department of Biotechnology, Vaal University of Technology, Private Bag X021, Vanderbijlpark, South Africa. E-mail: mpillay@vut.ac.za

This research has been funded by the Vaal University of Technology.

Abstract

Bambara groundnut (*Vigna subterranea* L. Verdc) is an underutilized crop in the African continent. It is a drought tolerant crop and fixes atmospheric nitrogen. Bambara groundnut is primarily grown for the protein content of its seeds and is mainly produced by small scale farmers at subsistence level. The objective of the study was to assess the morphological variation of landraces of bambara groundnut in South Africa. Thirty accessions of bambara groundnut were evaluated for their variability in agronomic and morphological traits. The field experiment was conducted at ARC-VOPI in Roodeplaat research farm during the 2014/2015 summer cropping season. The field trial was arranged as a complete randomized block design with three replications. Eighteen quantitative traits were recorded to estimate the level of genetic variability among accessions. The analysis of variance revealed significant differences among the phenotypic traits evaluated. The UPGMA cluster analysis based on the quantitative traits produced four distinct groups of genotypes and a singleton. Genotypes SB11-1A, SB19-1A, SB12-3B and Bambara-12 were found to possess good vegetative characters and are recommended for use as suitable parents when breeding cultivars for fodder production. Desirable yield and yield-related traits were identified in B7-1, SB4-4C, SB19-1A, Bambara-12 and SB16-5A and are recommended as suitable parental lines for bambara groundnut grain production improvement. The phenotypic characters therefore provide a useful measure of genetic variability among bambara genotypes and will enable the identification of potential parental materials for future breeding programs in South Africa.

Keywords: morphological variability, multivariate, quantitative traits, *Vigna subterranea*

1. Introduction

Bambara groundnut (*Vigna subterranea* L. Verdc) is an indigenous African legume primarily grown for its seeds. It is becoming increasingly popular as a food crop in rural areas of many countries in Africa (Vurayai et al., 2011). Bambara groundnut has been ranked as the third most important grain legume after groundnut (*Arachis hypogaea* L.) and cowpea (*Vigna unguiculata*) (Howell et al., 1994) in Africa. The crop has been cultivated in the tropical regions of sub-Saharan Africa and in Madagascar for many centuries (Godwin & Moses, 2013). Bambara groundnut is essentially grown for human consumption and has been described as a complete balanced diet due to the high carbohydrate (65%) and protein (18%) content of its seed (Ouedraogo et al., 2008). The protein content of bambara groundnut is high in lysine (Massawe et al., 2005).

The immature seeds of bambara groundnut can be boiled or grilled before being eaten while the mature seeds can be roasted in oil or ground into flour and then mixed with oil or butter to form a porridge. In South Africa and Swaziland, bambara groundnut is used to add variety to the daily diets and the boiled seeds can also be pounded and mixed with samp or used to make soup (Masindeni, 2006). Traditionally, bambara groundnut is used to cure nausea especially in pregnant women by chewing and swallowing the raw bean (Department of Agriculture, Forestry and Fisheries, 2011).

The leaves can also be used to feed livestock. Bambara groundnut is a drought tolerant crop and is readily adaptable to different environmental conditions and has the ability to be intercropped making it an important economic crop in many developing countries (Rungnoi et al., 2012).

The breeding system of bambara groundnut is not well understood and there are no cultivars. Landraces of the crop are still cultivated. There is a need to develop improved varieties for particular agro-ecological conditions or production systems. Bambara groundnut is still regarded as a poor man's crop grown for subsistence and very little progress has been made in improving the crop germplasm (Ayana & Bekele, 2000). Many researchers have used several morphological traits to characterize bambara groundnut accessions. Goli et al. (1997) characterized and evaluated the collection of bambara groundnut at the International Institute of Tropical Agriculture. The variability between local and exotic bambara groundnut landraces in Botswana was reported by Karikari (2000). Jonah et al. (2012) evaluated the seasonal variation and the correlation between yield and yield components in bambara groundnut accessions in Nigeria. Another study in Nigeria used multivariate analysis and character association for the growth and yield of bambara groundnut (Jonah et al., 2014). Mohammed (2014) did pre-breeding of bambara groundnut accessions in Kano State of Nigeria. Shegro et al. (2013) reported morphological variation in bambara groundnut in South Africa. However, there are a large number of land races that are planted in South Africa and it is important to further assess the variation in this germplasm for use in the breeding programs. Consequently, this study examined the agro-morphological variation in selected accessions of bambara groundnut in South Africa.

2. Materials and Methods

2.1 Plant Material

Thirty accessions of bambara groundnut landraces were obtained from the germplasm bank of Agricultural Research Council-Vegetable and Ornamental Plant Institute (ARC-VOPI), Roodeplaat, South Africa. The list of bambara groundnut genotypes used in this study, their seed morphology, leaf shape and growth habit is given in Table 1. The accessions were planted under open field conditions at the ARC-VOPI Roodeplaat research farm during the 2014/2015 summer cropping season. Roodeplaat lies at 25°59′S latitude and 28°21′E longitudes at an altitude of 1164 meters above sea level. The soil type is a clay loam with the pH of 7.08 (ARC-VOPI, 2012).

Table 1. The list of bambara groundnut accessions used in this study with their seed colours, leaf shape and growth habit

No	Accessions names	Seed colour	Leaf shape	Growth habit
1	SB1-1	Brown/spotted purple	Lanceolate	Erect
2	SB7-1C	Dark red	Oval	Semi-erect
3	SB2-1B	Cream brown	Round	Semi-erect
4	SB4-1	Dark red	Oval	Semi-erect
5	SB4-2	Cream brown	Elliptic	Erect
6	SB4-4	Black	Oval	Spreading
7	SB4-4C	Cream	Oval	Erect
8	SB7-1	Dark red	Round	Spreading
9	SB7-1A	Light red	Oval	Semi-erect
10	SB7-2	Light Red	Lanceolate	Semi-erect
11	SB8-1	Brown	Round	Spreading
12	SB8-3A	Dark red	Round	Semi-erect
13	SB9-1A	Cream brown	Oval	Spreading
14	SB10-1	Dark red	Elliptic	Spreading
15	SB10-1A	Brown	Lanceolate	Spreading
16	SB10-1C	Black	Oval	Spreading
17	SB10-2	Dark red	Elliptic	Semi-erect
18	SB11-1A	Cream	Elliptic	Erect
19	SB11-5	Speckle brown	Elliptic	Erect
20	SB12-3B	Cream brown	Lanceolate	Erect
21	SB16-5A	Speckle brown	Elliptic	Erect
22	SB17-1	Dark red	Lanceolate	Spreading
23	SB17-1A	Light red	Oval	Semi-erect
24	SB19-1A	Brown	Lanceolate	Spreading
25	SB19-3	Black	Lanceolate	Spreading
26	SB19-3B	Black	Lanceolate	Semi-erect
27	BAMBARA-6	Cream	Elliptic	Semi-erect
28	BAMBARA-7	Speckle brown	Lanceolate	Erect
29	BAMBARA-9	Cream	Lanceolate	Semi-erect
30	BAMBARA12	Cream black	Lanceolate	spreading

2.2 Experimental Design

The seeds of the 30 bambara groundnut accessions were evaluated in an open field experiment that was laid out as a randomized complete block design with three replications. Recommended crop management practices (land preparation, land clearing, weeding and irrigation) were carried out during the cropping season. The distance between plants was 0.3 m, the row length was 0.5 m, plot distance was 1.5 m and the distance between replications was 2 m. Two seeds were hand sown per hole and the seedlings thinned to one at two weeks after planting when they were fully established. Five randomly selected plants were selected from each plot to estimate the genetic variability using the morphological traits among the accessions evaluated.

2.3 Parameters Measured and Data Analysis

Eighteen quantitative characters (Table 2) were measured among the thirty bambara groundnut accessions from 14 days after planting until harvest at weekly intervals.

Table 2. List of quantitative morphological characters recorded from 30 bambara groundnut accessions

Quantitative characters	Code	Description	Measurement type
Days to 50% flowering (count)	D50%F	Number of days from emergence to when 50% of the plants have started flowering in the plot	Observation
Leaf length (cm)	LL	Length of the middle leaf	Leaf area meter
Leaf width (cm)	LW	Width of the middle leaf	Leaf area meter
Leaf area (mm²)	LA	Area of the middle leaf	Leaf area meter
Initial plant stand (count)	IS	Number of plant after 50% emergence in the plot	Counting
Panicle length (cm)	PL	Length of panicle from its base to the tip	Tape measure
Plant Height (cm)	PH	Height of main stalk from the ground to the tip of the main panicle	Tape measure
Number of leaves per plant (count)	NLPP	Count of total number of leaves from the plant	Counting
Number of branches per plant (count)	NBPP	Count of total number of branches from the main stem	Counting
Days to 50% maturity (count)	D50%M	Number of days from emergence to when 50% of the plants have matured in the plot	Counting
Days to harvest (count)	DH	Total number of days from planting to harvest	Counting
Final plant stand (count)	FS	Number of plant after 50% maturity in the plot	Counting
Fresh weight (g)	Fwt	Average weight of five harvested fresh seed	Weighing balance
Dry weight (g)	Dwt	Average weight of five harvested dried seed	Weighing balance
Number of seed per plant (count)	NSPp	Total count of number of seed per five plant	Counting
Yield per plant (g)	YPP	Weight of seed per plant (average of five plants)	Weighing balance
Hundred seed weight (g)	Hswt	Weight of hundred seed counts at 12% moisture content	Weighing balance
Yield per plot (g)	YPPlot	Total weight of seed per plot	Weighing balance

The leaf area was measured using a leaf area meter (AM300 ADC BioScientific Limited, Hoddesdon, UK). The quantitative traits for all the accessions of the three replications were computed and subjected to analysis of variance (ANOVA), based on the lattice procedure using Agrobase statistical software (Agrobase, 2008). Means were compared by least significance difference (LSD) at $P \leq 0.01$. Cluster and Principal Component Analyses were conducted to determine similarities and dissimilarities among the genotypes based on the traits recorded using SPSS (IBM SPSS Statistics, 2015).

3. Results

3.1 Analysis of Variance

Analysis of variance (ANOVA) revealed significant differences among the phenotypic traits evaluated (Table 3). The mean number of days to 50% flowering (D50% F) ranged from 42 to 67 days, with a mean value of 59 days. The earliest flowering accession was SB11-1A and was followed by SB7-1 and SB7-1A. Bambara-12 and SB2-1B bloomed late.

The leaves of SB11-1A had the longest length and width (4.34 cm), while SB4-2 had the largest leaf surface area. SB7-1 had the shortest leaf, while SB10-1 had the narrowest leaf. The lowest leaf area was found in SB10-1. SB19-1A was the tallest plant with a height of 27.48 cm and longest petiole length. Furthermore, SB12-3B had the highest number of branches per plant and the highest number of leaves per plant. Accession SB11-1A was the shortest plant with the shortest petiole. The smallest number of leaves and branches per plant was recorded for SB7-1. There was a significant difference ($P \leq 0.05$) between plant height and petiole length in the accessions that were evaluated. The number of days to harvest ranged from 128 to 152 with an overall mean of 137. Early maturing varieties included SB7-1A and SB7-2, while Bamara-9 matured very late. There was a highly significant ($P \leq 0.01$) difference among all accessions for grain yield and other yield related traits.

The highest fresh and dry weight were observed in SB7-1. SB8-3A had the lowest fresh and dry weight. SB4-4C produced the highest number of seeds per plant, while SB16-5A produced the fewest number of seeds. The highest seed yield per plant was obtained from SB19-1A (37.81 g), while SB8-3A (2.97 g) had the lowest yield. The seed weight ranged between 20.45 g and 68.19 g with a mean of 43.03. The highest mean hundred seed weight was observed in SB16-5A while the lowest was found in SB11-1A. Bambara-12 had the highest yield per plot while SB11-1A had the lowest yield.

3.2 Correlational Matrix

The phenotypic correlation matrix among the 18 quantitative traits of 30 bambara groundnut is presented in Table 4. A high significant positive correlation was observed between number of branches and leaves per plant. The days to 50% flowering showed a positive correlation with plant height, days to harvest and yield per plot. Leaf length was negatively correlated with initial plant stand, days to 50% maturity, finial plant stand and hundred seed weight.

Conversely, a significant negative correlation was observed between initial plant stand and days to harvest, fresh weight, dry weight and number of seeds per plants. There was a strong positive correlation of plant height and petiole length (r = 0.85) and a moderate positive correction between petiole length and hundred seed weight. Yield per plot was moderately correlated with days to 50% flowering, petiole length, plant height, days to 50% maturity, days to, final plant stand, number of seed per plant and yield per plot. A very strong significant positive correlation coefficient was obtained between days to 50% maturity and days to harvest (r = 0.75). Days to 50% maturity correlated moderately with number of seeds per plant, yield per plant and yield per plot. A significant positive correlation was observed between days to harvest and fresh weight, dry weight, number of seed per plant and yield per plan). There was a strong significant and positive correlation (r = 0.95) between fresh and dry weight. Similarly, yield per plant had a very strong positive correlation with number of seeds per plant (r = 0.88), fresh (r = 0.71) and dry weight (r = 0.77). Finally, fresh weight was strongly and positively correlated with the number of seed per plants.

Table 3. Mean values of the 18 quantitative traits evaluated on 30 bambara groundnut accessions during the 2014/2015 cropping season

No	Accessions	D50%F	LL	LW	LA	IS	PL	PH	NLPP	NBPP	D50%M	DH	FS	Fwt	Dwt	NSPp	YPP	Hswr	YPPlot
1	SB19-1A	62.33	6.27	3.00	1394.70	16.67	18.00	27.48	407.40	135.80	117.33	136.67	11.33	152.50	54.11	51.70	37.81	52.32	222.87
2	SB8-1	60.00	5.64	2.79	1229.70	14.67	13.58	19.12	248.60	82.87	119.33	139.33	10.33	20.50	6.91	8.60	6.36	41.08	150.50
3	BAMBARA-9	64.00	6.31	2.53	1222.90	5.33	16.42	23.48	307.27	102.42	125.00	151.67	5.00	135.17	43.93	56.60	31.19	42.62	128.30
4	SB7-1C	57.33	5.58	2.44	1164.00	17.00	16.08	22.71	291.80	97.27	109.00	137.33	10.00	58.00	24.15	42.60	15.24	41.20	108.40
5	SB19-3B	63.33	6.05	2.59	1192.00	13.00	12.50	23.40	360.10	126.70	119.00	139.33	4.00	94.00	30.60	51.10	16.61	40.20	104.70
6	SB10-2	59.33	5.70	2.81	1188.30	15.00	15.85	21.61	254.73	87.13	110.67	137.00	12.33	24.50	8.85	13.20	3.23	36.08	185.10
7	BAMBARA-7	65.33	6.35	2.83	1313.30	7.00	16.68	25.80	366.80	122.27	114.00	138.67	7.67	122.00	32.75	34.75	20.90	49.52	131.30
8	SB4-1	59.00	6.05	2.57	1219.30	13.00	15.01	21.15	388.13	114.00	115.00	136.00	12.33	78.50	30.67	43.10	20.27	45.97	138.00
9	SB17-1A	52.33	5.86	2.97	1290.70	17.67	16.93	23.63	264.00	88.00	112.67	131.67	13.67	39.00	26.64	18.00	4.03	45.28	153.40
10	SB16-5A	57.33	5.93	2.23	1177.60	17.00	16.17	22.90	348.00	116.00	112.67	130.67	10.33	43.00	13.70	7.90	4.56	68.19	56.80
11	SB10-1	57.33	5.49	1.70	883.80	17.67	12.38	19.91	340.00	113.33	109.33	130.33	13.00	37.50	17.89	23.20	8.33	37.11	48.75
12	SB4-4C	57.00	5.94	2.20	1106.70	6.67	14.49	20.38	364.17	121.40	119.00	144.33	5.67	101.00	40.43	67.80	27.13	40.98	154.1
13	SB10-1A	59.33	5.91	1.93	1043.40	11.67	12.91	20.83	248.40	82.80	108.33	131.00	7.00	16.50	5.98	33.40	15.55	48.75	138.77
14	SB4-2	55.00	6.70	2.52	1553.10	14.67	14.77	22.13	340.80	113.60	113.67	137.67	13.67	35.50	13.65	26.50	13.46	39.85	65.82
15	SB8-3A	64.00	5.98	2.46	1136.70	15.67	13.67	19.05	251.60	73.87	117.00	134.00	7.67	12.67	5.39	13.20	2.97	43.57	66.98
16	SB11-5	62.33	5.98	2.52	1112.00	12.00	17.11	24.73	338.60	112.87	116.20	140.67	15.67	45.00	14.44	17.70	6.26	49.54	42.34
17	SB7-2	59.00	6.01	2.47	1196.20	17.67	16.59	22.88	325.93	108.60	114.00	129.67	15.00	24.50	5.76	14.60	3.39	37.58	60.06
18	SB2-1B	67.33	6.08	2.52	1240.70	10.00	13.75	21.25	332.90	110.97	114.90	134.00	4.67	53.72	18.40	18.00	6.77	37.45	38.67
19	BAMBARA-6	64.67	5.85	3.09	1345.00	10.67	16.33	13.53	264.80	88.27	120.33	138.67	14.00	48.00	26.86	37.00	17.58	37.64	242.47
20	SB9-1A	58.67	5.53	2.49	1069.30	18.67	15.64	21.83	354.33	118.11	119.00	133.67	15.00	41.17	11.89	18.67	7.88	38.53	83.17
21	SB11-1A	41.67	10.82	4.34	1016.70	1.67	8.83	12.67	360.27	103.15	103.33	132.00	1.33	100.33	31.51	47.00	10.55	20.45	17.48
22	SB19-3	64.33	6.33	2.68	1287.80	3.00	13.80	21.16	352.53	117.50	120.33	145.33	9.33	91.00	35.53	57.25	23.59	45.21	143.77
23	SB12-3B	61.67	5.86	2.75	1254.00	13.67	16.41	24.53	482.40	160.80	114.33	140.00	11.67	51.45	22.11	29.60	15.06	54.48	253.10
24	SB10-1C	64.00	5.76	2.73	1173.00	12.3	12.21	19.28	351.60	117.20	111.67	134.33	13.67	55.50	15.89	18.40	12.17	33.72	121.00
25	SB1-1	62.33	6.24	2.47	1199.70	6.67	14.15	21.64	305.20	101.72	111.00	136.00	6.00	38.00	18.87	16.00	7.61	36.65	38.01
26	BAMBARA-12	66.00	5.75	2.42	1140.30	17.00	16.01	22.32	341.00	113.67	119.33	147.33	17.67	121.00	43.70	50.00	29.30	42.61	333.00
27	SB7-1	51.67	5.25	2.57	1029.00	1.67	14.17	21.33	180.73	60.24	114.00	136.67	1.33	188.77	55.04	30.00	11.78	51.14	40.07
28	SB17-1	64.67	6.07	2.52	1241.40	10.33	14.55	23.14	331.77	110.60	120.33	143.67	8.00	96.50	32.68	61.40	26.64	46.10	124.53
29	SB4-4	53.00	5.83	2.34	1137.70	5.00	10.82	19.82	304.30	131.43	120.00	143.33	4.67	46.88	16.96	27.63	13.66	47.21	54.15
30	SB7-1A	51.67	5.96	2.18	1130.30	17.67	15.71	22.77	304.60	113.53	108.00	128.00	11.67	27.50	9.49	9.90	5.97	41.47	62.03
	Grand Mean	59.53	6.10	2.59	1189.65	12.02	14.72	21.88	324.96	107.20	114.97	137.30	9.79	66.66	23.82	31.56	14.20	43.03	116.92
	Mean squares	92.59	2.65	0.59	47294.58	81.70 **	12.48 *	20.55 *	10122.94	1138.29	69.41	94.51	56.95 **	5870.16 **	597.03 **	926.07 **	262.41 **	201.74 **	16678.30 **
	CV(%)	15.47	25.26	23.00	26.30	11.62	17.51	15.26	29.17	30.96	8.24	6.04	20.15	8.38	18.01	16.52	21.73	15.55	5.00
	LSD	15.05	2.52	0.97	511.29	2.28	4.21	5.46	154.92	54.24	15.49	13.55	3.22	9.13	7.01	8.52	5.04	10.94	9.55

Note. CV = coefficient of variation, LSD = least significant difference. P ≤ 0.05 = significant (*), P≤ 0.01 = significant (**).

Table 4. Pearson correlation coefficients (r) among 18quantitative traits of 30 bambara groundnut accessions.

	D50F	LL	LW	LA	IS	PL	PH	NLPP	NBPP	D50M	DH	FS	Fwt	Dwt	NSPp	YPP	Hswr	YPPlot
D50F	1.00																	
LL	-0.49**	1.00																
LW	-0.30	0.75**	1.00															
LA	0.33	0.03	0.24	1.00														
IS	0.13	-0.41*	-0.34	0.07	1.00													
PL	0.38*	0.47**	0.16	0.45*	0.45*	1.00												
PH	0.51**	0.54**	0.28	0.50**	0.33	0.85**	1.00											
NLPP	0.18	0.23	0.12	0.21	0.10	0.11	0.22	1.00										
NBPP	0.24	0.08	0-.01	0.25	0.17	0.20	0.37*	0.96**	1.00									
D50M	0.57**	-0.38*	-0.14	0.33	-0.13	0.26	0.36	0.06	0.10	1.00								
DH	0.41*	-0.10	0.02	0.25	-0.44*	0.11	0.19	0.14	0.15	0.75**	1.00							
FS	0.24	-0.37	-0.19	0.22	0.74**	0.55**	0.37*	0.20	0.24	0.04	-0.13	1.00						
Fwt	0.03	0.15	0.26	0.04	-0.51**	0.08	0.19	0.14	0.11	0.30	0.49**	-0.39*	1.00					
Dwt	0.04	0.12	0.27	0.12	-0.45*	0.14	0.23	0.17	0.14	0.33	0.53**	-0.30	0.95**	1.00				
NSPp	0.13	0.24	0.17	0.11	-0.43*	-0.05	0.07	0.31	0.26	0.39*	0.65**	-0.29	0.68**	0.76**	1.00			
YPP	0.33	0.04	0.06	0.29	-0.28	0.18	0.33	0.37*	0.37*	0.50**	0.68**	-0.10	0.71**	0.77**	0.88**	1.00		
Hswr	0.21	-0.48**	-0.41*	0.17	0.14	0.49**	0.60**	0.11	0.17	0.26	0.10	0.08	0.15	0.13	0.05	0.16	1.00	
YPPlot	0.42*	-0.25	0.10	0.33	0.22	0.43*	0.38*	0.22	0.25	0.36*	0.44*	0.42*	0.21	0.35	0.39*	0.56**	0.17	1.00

Note. **: Correlatioin is significant at the 0.01 level (2-tailed); *: Correlatioin is significant at the 0.05 level (2-tailed).

3.3 UPGMA Cluster Analysis

The cluster analysis of the 30 bambara groundnut accessions are presented Figure 3. With the exception of SB11-1A, the dendrogram clustered the accessions into four clusters. Cluster I consisted of six accessions including SB7-1, SB4-4, SB16-5A, SB10-1A, SB10-1 and SB4-2. The second cluster comprised of twelve accessions, namely SB7-1A, SB9-1A, SB7-2, SB11-5, SB17-1A, SB7-1C, SB10-1C, SB1-1, SB2-1B, SB10-1C, SB8-3A, andSB8-1. The third cluster included nine accessions including Bambara-12, Bambara-6, Bambara-7, Bambara-9, SB19-3, SB4-4C, SB4-1, SB19-3B and SB17-1. The last cluster contained two accessions, namely, SB12-3B and SB19-1A.

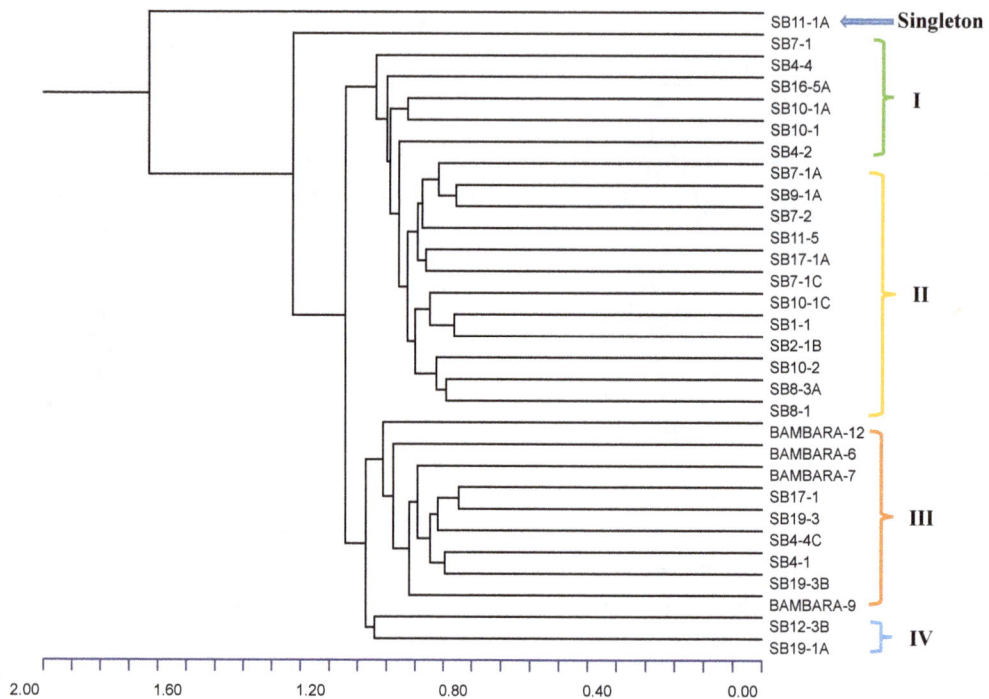

Figure 3. A pair-wise genetic distance matrix of 30 bambara groundnut accessions generated by UPGMA using the data set

4. Discussion

4.1 Morphological Variability

Characterization of bambara groundnut germplasm can provide useful information for a breeding program aiming to genetically improve the crop. The results of this study showed that there was wide genetic variability among the bambara groundnut evaluated for 18 morphological quantitative characters recorded. Days to 50% flowering was quite variable among the bambara groundnut accessions ranging from 42 to 67 days. Similarly, a study by Massawe et al. (2005) showed that days to flowering in bambara groundnut ranged from 64 to 76 days in the Free State Province in South Africa, while Masindeni (2006) recorded 43-80 days in Loughborough, United Kingdom.

The mean days to flowering (60 days) obtained in this study were higher than the values reported in previous studies. Ouedraogo et al. (2008) observed that flowering ranged from 32 to 53 days among bambara groundnut germplasm accessions from Burkina Faso. Similarly, Shegro et al. (2013) reported that flowering occurred between 36-53 days among 20 bambara groundnut accessions assessed under similar conditions to this study in Pretoria, South Africa. A number of environmental factors such as temperature, altitude and soil conditions as well as genotypic factors can affect flowering in bambara groundnut (Shegro et al., 2013). Similar factors may be responsible for the variation in days to flowering in this study. Swanevelder (1997) also reported variation in days to flowering among bambara groundnut accessions in South Africa. Bambara groundnut is a short day plant and when planted during long days there is either delayed or no flowering, a trait which is also cultivar dependent. Generally, first flowering in bambara groundnut occurs about 30 to 45 days after planting (DAP) and might continue until it reaches maturity. It has been reported that 50% flowering may take from 60 up to 80 DAP depending on the cultivar. Early flowering implies early maturity (Shegro et al., 2010) and it is an important trait in bambara groundnut cultivation in South Africa. In this study, the early flowering accessions SB11-1A, SB7-1C and SB7-1A could be selected for early maturity.

Days to physiological maturity was significantly different ($P \leq 0.05$) between genotypes and days to harvest ranged from 128 to 152. A similar maturity range was observed in other studies (Goli et al., 1997; Masindeni, 2006). According to Swanevelder (1998), maturity of the bambara groundnut is dependent on cultivar and climatic conditions, and ranges from three to six months. However, it is also influenced by photoperiod. Linnemann et al. (1995) reported that under long photoperiods maturity is delayed. It was observed that the earlier a genotype flowers, the earlier is the physiological maturity (Shegro et al., 2010). A similar finding was observed in this study. Assessing days to maturity of crops facilitates escape from drought-stressed environmental conditions and may enable selection for adaptation to drought-prone areas of South Africa (Shegro et al., 2010).

No significant difference was observed in the morphological traits such as terminal leaf length and width, leaf area, number of leaves and branches per plant. The mean values obtained for these traits in this study are congruent with those reported by Mohammed (2014) for the same characters. However, there was a significant difference (P ≤ 0.05) between petiole length and plant height among the accessions. This is similar to the results reported by Ntundu et al. (2006) in Tanzania and Shegro et al. (2013) in South Africa.

A highly significance difference (P ≤ 0.01) was obtained between grain-yield and yield related traits such as shoot fresh and dry weight, number of seeds per plant, yield per plant and hundred seed weight indicating high genetic variation among these traits. This is in disagreement with those reported by Ntundu et al. (2006), who reported that there was no significant difference among bambara groundnut accessions for the number of stems per plant, seed number per pod, hundred seed weight and yield per plot over two seasons in Tanzania. However, variation in yield related traits was also reported by Shegro et al. (2013) who suggested that there may be genotype by environmental influence on yield in bambara groundnut.

The yield per plant ranged from 2.97 g to 37.81 g. The mean yield value obtained from this study was significantly higher than those reported by Massawe et al. (2005) and Ouedraogo et al. (2008). Slightly higher yields ranging from 4.0 g to 57.52 g was reported by Shegro et al. (2013) and Mohammed (2014).

Hundred seed weight has been reported as a tool for the assessment of morphological traits (Massawe et al., 2005; Masindeni, 2006; Ntundu et al., 2006; Ouedraogo et al., 2008; Mohammed, 2014; Shegro et al., 2015). The hundred seed weight in this study ranged from 20.45 g to 68.19 g. The phenotypic coefficients of variation for the traits such as number of seeds per plant, plant height, petiole length and hundred seed weight in this study were almost similar, suggesting that the variations for these traits might be due to genetic rather than environmental factors (Kulkarni et al., 2002). In order to assess the real extent of genetic diversity of bambara groundnut accessions, agronomic, physiological, biochemical and molecular evaluation and characterization should be done since morphological genetic markers may be influenced by environmental conditions (Kumar, 1999). Nevertheless, agronomic evaluation and characterization is the first step in assessing the genetic diversity of a crop for breeding programs. The wide genetic variation observed in this study may be useful for breeders planning to enhance the germplasm of bambara groundnut.

4.2 Correlation among the Traits

Days to 50% flowering showed a non-significant positive correlation with all the traits evaluated except for leaf length and leaf width where a negative correlation was observed. It appears that plant height, petiole length and the number of days to 50% maturity were most significant in affecting the days to 50% flowering. The positive correlation between plant height and days to 50% flowering observed in this study was also observed in previous studies (Zongo et al., 1993; Ayana, 2001; Kebede et al., 2001) suggesting that these traits are heritable and can be transferred into desired genotypes. Plant height was positively associated with fresh weight, dry weight, number of seeds per plants, yield per plant, hundred seed weight, yield per plot and the final plant stand was also positively and significantly associated with plant height. This suggests that selection based on these characters may be effective for improving seed yield. As expected, leaf area was positively correlated with leaf length and width. Ayana and Bekele (2000) also reported a functional relationship for these traits in sorghum.

Correlation coefficient is an important parameter in plant breeding since it measures the degree of association, genetic or non-genetic between two or more characters (Jonah et al., 2014). Correlation studies between traits have been of great value for selection of superior genotypes (Adebisi et al., 2004). The highly significant correlation between the number of leaves and branches per plant coupled with the high correlation between fresh and dry weight may suggest that selection based on these traits may be useful for breeding bambara groundnut for fodder production. The highly significant correlation among fresh weight, dry weight and the number of seeds per plant and the moderate positive association among the number of leaves and branches per plant, days to 50% maturity and days to harvest are congruent with the results obtained by Karikari (2000) and Jonah et al. (2014). These characters may be useful to plant breeders for selecting parents that could improve yield in bambara groundnut. However, the hundred seed weight was negatively correlated with the number of seeds per plant. This was due to the variation in seed size among the different accessions.

Panicle length had a very strong positive significant correlation with plant height and a moderate positive correlation with days to 50% flowering, leaf length and width, leaf area and initial plant stand. The positive correlations among and between the various traits recorded in this study clearly indicate that selecting for any of these traits will have a positive effect on selecting for associated traits in a Bambara groundnut improvement program.

4.3 UPGMA Cluster Analysis

The results showed that there was a high level of genetic diversity among the bambara groundnut accessions. The separation of accession SB11-1A as the most divergent is perhaps due to its late flowering, leaf length and width, shortest petiole length and plant height and lowest hundred seed weight and yield per plot. Some of the accessions formed relationships on the basis of shared morphological traits. The separation of accession SB7-1 in cluster 1 is perhaps due to the fact that it had the lowest leaf length, lowest number of branches and leaves per plant. A sister relationship was observed between accessions SB10-1A and SB10-1 and both genotypes were closely associated in the second quadrant in the principle component biplot (data not shown). This sister relationship indicates that these accessions may have a common ancestry. The grouping of the accessions in cluster II may be due to the presence of similar morphological traits in these genotypes. This cluster consisted of accessions with intermediate days to 50% flowering and number of leaves and branches per plant, smallest leaf area, lowest fresh and dry weight, lowest number of seeds and yield per plant. The close grouping of accessions SB8-1 and SB8-3A, SB1-1 and SB2-1B, SB17-1A and SB7-1, and SB9-1A and SB7-2 in this cluster may be due to the similarities in their morphological traits including similar values for plant height and panicle length, lowest fresh and dry weight and lowest yield per plant. The rest of the accessions in this cluster appeared independently.

The clustering of nine accessions in the third cluster appears to be due to shared morphological traits such as days to 50% flowering, smallest leaf length and average leaf width, average panicle length and plant height, late maturing, highest number of seed per plant and medium yield per plant. The uniqueness of Bambara-12 within the cluster may be due to its largest final plant stand and grain yield per plot.

The accessions SB12-3B and SB19-1A were the only accessions contained in cluster IV. This cluster consisted of accession with the longest and widest leaves, largest leaf area, tallest plant, highest number of leaves and branches per plant, medium number of seeds, highest yield and hundred seed weight. Characterization and clustering of accessions on the basis of their morphological traits and genetic similarity can help in the identification and selection of the best parents for hybridisation (Souza & Sorrells, 1991). Therefore, the grouping of accessions by univariate and multivariate methods of analysis based on their similarity in the present study presents important information that will be valuable for bambara groundnut breeders.

5. Conclusions

Characterization and evaluation of bambara groundnut germplasm and identification of the best parents is important for improvement of the crop. The genotypes in this study showed significant variation in phenotypic characters, indicating that the accessions had high genetic diversity which can be exploited for use in a breeding program. The accessions SB19-1A, Bambara-9 and Bambara-12 were associated with desirable grain yield characteristics and may be suitable parental lines for improvement of grain production. Similarly, accessions SB19-1A, SB12-3B and SB7-1 were identified as possessing favorable vegetative traits and these accessions could be used as parents when breeding bambara groundnut for use as fodder production. The genetic potential of the accessions as revealed in this study can assist in choosing suitable parental lines thereby maximizing the efficiency of a bambara groundnut breeding program.

References

Adebisi, M. A., Ariyo, O. J., & Kehinde, O. B. (2004). Variation and correlation studies in quantitative characters in Soyabean. *The Ogun Journal of Agricultural Science, 3*, 134-142.

Agriculture Research Council-Vegetable and Ornamental Plant Institute. (2012). *Soil test analysis*.

Agrobase. (2008). *Generation II*. Agronomix Software Inc., 71 Waterloo St. Winnipeg, Manitoba R3NOS4, Canada.

Ayana, A. (2001). *Genetic diversity in sorghum (Sorghum bicolor (L.) Moench) germplasm from Ethiopia and Eritrea* (Doctoral dissertation). Addis Ababa University, Addis Ababa, Ethiopia.

Ayana, A., & Bekele, E. (2000). Geographical patterns of morphological variation in sorghum (*Sorghum bicolor* (L.) Moench) germplasm from Ethiopia and Eritrea: Quantitative characters. *Euphytica, 115*, 91-104. http://dx.doi.org/10.1023/A:1003998313302

Department of Agriculture, Forestry and Fisheries. (2011). *Production guidelines for Bambara groundnut*. Compiled by Directorate Plant Production in collaboration with the ARC-Grain Crops Institute. Agricultural Information Service Private Bag X144, Pretoria, 0001 South Africa.

Godwin, A. A., & Moses, O. E. (2013). Participatory Rural Appraisal of Bambara Groundnut (*Vigna subterranea* (L.) Verdc.) Production in Southern Guinea Savannah of Nigeria. *Journal of Agricultural Science, 1*, 18-31. http://dx.doi.org/10.12735/as.v1i2p18

Goli, A. E., Begemann, F., & Ng, N. Q. (1997). Characterization and evaluation of IITA's Bambara groundnut collection. In J. B. F. Heller, & J. Mushonga (Eds.), *Conservation and improvement of Bambara groundnut (Vigna subterranea (L.) Verdc.)* (pp. 101-118). International Plant Genetic Resources Institute.

Howell, J. A., Eshbaugh, W. H., Guttman, S., & Rabakonandrianina, E. (1994). Common names given to bambara groundnut (*Vigna subterranea*). *Madagascar Economy Botany, 48*, 217-221.

IBM Corp. (2015). *IBM SPSS Statistics for Windows, Version 22.0*. Armonk, NY: IBM Corp.

Jonah, P. M., Abimiku, O. E., & Adeniji, O. T. (2014). Multivariate Analysis and Character Association on the Growth and Yield of Bambara Groundnut in Mubi, Adamawa State, Nigeria. *International Journal of Management and Social Sciences Research, 3*, 2.

Jonah, P. M., Aliyu, B., Adeniji, T. O., & Bello, D. (2012). Seasonal Variation and Pearson Correlation in Yield and Yield Components in Bambara Groundnut. *World Journal of Agricultural Sciences, 8*, 26-32.

Karikari, S. K. (2000). Variability between local and exotic Bambara groundnut landraces in Botswana. *African Crop Science Journal, 8*, 145-152. http://dx.doi.org/10.4314/acsj.v8i2.27704

Kebede, H., Subudhi, P. K., Rosenow, D. T., & Nguyen, H. T. (2001). Quantitative trait loci influencing drought tolerance in grain sorghum (*Sorghum bicolor* L. Moench). *Theoretical and Applied Genetics, 103*, 266-276. http://dx.doi.org/10.1007/s001220100541

Kulkarni, V. M., Srinivas, L., Satdive, R. K., Bapat, V. A., & Rao, P. S. (2002). Dissection of the genetic variability in elite Indian banana genotypes. *Plant Genetic Resources Newsletter, 132*, 48-52.

Kumar, L. S. (1999). DNA markers in plant improvement: An overview. *Biotechnology Advances, 17*, 143-182. http://dx.doi.org/10.1016/S0734-9750(98)00018-4

Linnemann, A. R., Westphal, E., & Wessel, M. (1995). Photoperiod regulation of development and growth in bambara groundnut (*Vigna subterranea*). *Field Crops Research, 40*, 39-47. http://dx.doi.org/10.1016/03784290

Masindeni, D. R. (2006). *Evaluation of Bambara groundnut (Vigna subterranea) for yield stability and yield related characteristics* (Master's Thesis). University of the Free State, Bloemfontein, South Africa.

Massawe, F. J., Mwale, S. S., Azam-Ali, S. N., & Roberts, J. A. (2005). Breeding in bambara groundnut (*Vigna subterranea* (L.) Verdc.): strategic considerations. *African Journal of Biotechnology, 4*, 463-471.

Mohammed, M. S. (2014). *Pre-breeding of Bambara Groundnut (Vigna subterranea [L.] Verdc.)* (Doctoral dissertation). University of KwaZulu-Natal, Durban, South Africa.

Ntundu, W., Shillah, S., Marandu, W., & Christiansen, J. L. (2006). Morphological diversity of Bambara groundnut (*Vigna subterranea* [L.] Verdc.) landraces in Tanzania. *Genetic Resources and Crop Evolution, 53*, 367-378. http://dx.doi.org/10.1007/s10722-004-0580-2

Ouedraogo, M., Ouedraogo, J. T., Tignere, J. B., Balma, D., Dabire, C. B., & Konate, G. (2008). Characterisation and evaluation of accessions of bambara groundnuts (*Vigna subterranea* (L.) Verdcourt) from Burkina Faso. *Science and Nature Journal, 5*, 191-197.

Rungnoi, O., Suwanprasert, J., Somta, P., & Srinives, P. (2012). Molecular Genetic Diversity of Bambara Groundnut (*Vigna subterranea* L. Verdc.) Revealed by RAPD and ISSR marker Analysis. *Journal of Breeding and Genetics, 44*, 87-101.

Shegro, A. G., Atilaw A., Pal, U. R., & Geleta N. (2010). Influence of varieties and planting date on productivity of soybean in Metekel Zone, North Western Ethiopia. *Journal of Agronomy, 9*, 146-156. http://dx.doi.org/10.3923/ja.2010.146.156

Shegro, A. G., Jansen Van Rensburg, W. S., & Adebola, P. O. (2013). Assessment of genetic variability in Bambara groundnut (*Vigna subterrenea* [L.] Verdc.) using morphological quantitative traits. *Academia Journal of Agricultural Research, 1*, 45-51.

Shegro, A. G., Van Rensburg, J. S., & Adebola, P. O. (2015). Genetic diversity of *Amaranthus* species in South Africa. *South African Journal of Plant and Soil, 32*, 39-46. http://dx.doi.org/10.1080/02571862.2014.973069

Souza, E., Sorrells, M. E. (1991). Relationships among 70 American oat germplasm. I. Cluster analysis using quantitative characters. *Crop Science, 31*, 599-605. http://dx.doi.org/10.2135/cropsci1991.0011183X003100 030010x

Swanevelder, C. J. (1997). Country reports: South Africa. In J. Heller., F. Begemann., & J. Mushonga (Eds.), *Proceedings of the Workshop on Conservation and Improvement of Bambara groundnut (Vigna subterranea (L.) Verdc.)* (pp. 50-52). Harare, Zimbabwe. International Plant Genetic Resources Institute, Rome, Italy.

Swanevelder, C. J. (1998). *Bambara—food for Africa: Vigna subterranean bambara groundnut.* National Department of Agriculture, South Africa

Vurayai, R., Emongor, V., & Moseki, B. (2011). Physiological Responses of Bambara Groundnut (*Vigna subterranea* L. Verdc) to Short Periods of Water Stress during Different Developmental Stages. *Asian Journal of Agricultural Sciences, 3*, 37-43.

Zongo, J. D., Gouyon, P. H., & Sandmeier, M. (1993). Genetic variability among sorghum accessions from the Sahelian agroecological region of Burkina Faso. *Biodiversity and Conservation, 2*, 627-636. http://dx.doi.org/10.1007/BF00051963

22

Employing Phenology to Delineate Wheat Agro-Climatic Zones in Afghanistan

M. Q. Obaidi[1], Elias Mohmand[2], M. H. Azmatyar[1] & Rajiv Sharma[2]

[1] Agricultural Research Institute of Afghanistan, Kabul, Afghanistan

[2] International Maize & Wheat Improvement Centre (CIMMYT), Kabul, Afghanistan

Correspondence: Rajiv Sharma, International Maize & Wheat Improvement Centre (CIMMYT), Kabul, Afghanistan. E-mail: rk.sharma@cgiar.org

Abstract

Afghanistan grows wheat on about 2.5 million hectares with an average annual production of about five million tonnes. The local research and development efforts make use of country wide research results to recommend varieties and other technologies. Afghanistan has wide ranging climatic variability and its wheat acreage therefore needs to be delineated into wheat climatic zones. A set of 10 different types of wheat varieties were scored for average number of days to 50% flowering (ADF) at 10 locations to delineate Afghan wheat acreage into homogeneous wheat climatic zones based on ADF values. The results obtained hinted at creating eastern, northern, south western and a highland zone for conducting research and recommending wheat technologies.

Keywords: Afghanistan, wheat, agro-climatic, zone, flowering

1. Introduction

Afghanistan is a wheat eating country where per capita wheat consumption stands at over 200 kg/annum. Wheat is a staple food as it provides over 60% of daily caloric requirement of an average Afghan (Persaud, 2012). Decades of war have inhibited growth of both research and development in all fields and more so in agriculture sector as intervention involves moving physically into insecure country side. However, this has not prevented work of CG canters including CIMMYT, and its germplasm introductions led to development and release of several wheat varieties in the country in recent past (Obaidi et al., 2011, 2014, 2015). Afghanistan grows wheat at over 2.5 million hectares with an average annual production of around five million tonnes (Obaidi et al., 2015). Though adequate number of wheat varieties are available, however their dissemination to farmer fields is severely hampered on account of inefficient seed production providing for a less than five per cent seed replacement rate for wheat. The Afghan agricultural research is spearheaded by Agricultural Research Institute of Afghanistan (ARIA) which constitutes Afghanistan's national agricultural research system (NARS). ARIA has over ten functional wheat research stations spread across length and breadth of the country. Current practice of releasing new wheat varieties makes use of yield evaluation trials conducted throughout the country. Afghanistan has wide range of agro-ecologies characterized by cold winters and hot summers. Though eastern Afghanistan does receive monsoon rains, most of the Afghanistan is semi-arid or arid (Saidajan, 2012). It is therefore imperative that country is categorized into all possible wheat climatic zones to not only fine-tune production management but also identify best adapted varieties for each zone. Phenological traits mainly days to flowering is a reflection of agro climatic features of any location and can be aptly employed to characterize different agro climatic zones (Diaz et al., 2012; Pearce et al., 2016).

2. Material & Methods

A total of ten wheat varieties released in Afghanistan and representing different growth habits viz., winter and spring wheat were used to constitute a phenological nursery. The nursery was grown at several locations viz., Balkh, Baghlan, Bamyan, Nangarhar, Kabul, Takhar, Kunduz, Herat, Badakhshan and Kandhar during 2010-11 to 2015-16 crop seasons. Various geological attributes of these locations viz., longitude, latitude and altitude etc., are presented in Table 1 and their geographical locations are shown in Figure 1. The nursery was sown in an unreplicated experiment and each genotype was sown in two rows of two meters each. Standard recommended agronomic practices were adopted to raise a successful crop. Days to 50% flowering was recorded on each genotype at all the locations in each year.

Table 1. Agricultural Research Institute of Afghanistan (ARIA) research stations in Afghanistan

No	Site	Acronym	Institution	Latitude(N)	Longitude(E)	Altitude (m)
1	Badakhshan	BDS	Baharak	36°50′ N	70°9′ E	1733
2	Kunduz	KDZ	Central Farm	36°43′ N	68°51′ E	373
3	Takhar	TKR	Taloqan	36°44′ N	49°30′ E	804
4	Balkh	BLK	Dehdadi	36°65′ N	66°96′ E	387
5	Baghlan	BGL	Posi-e-shan	36°42′ N	67°13′ E	510
6	Bamyan	BMN	Mullah Ghulam	34°43′ N	67°49′ E	2550
7	Kabul	KBL	Darulaman	34°28′ N	69°09′ E	1841
8	Nangarhar	NGR	Shishambagh	34°49′ N	70°74′ E	541
9	Herat	HRT	Urdu Khan	34°18′ N	62°16′ E	927
10	Kandahar	KND	Kokaran	31°35′ N	65°40′ E	630

Figure 1. Locations of various ARIA research stations in Afghanistan

3. Results

Varieties on an average took longest to attain 50% flowering at Badakhshan (161 days) followed by Bamyan (159 days) whereas earliest 50% flowering at an average of 95 days (Table 2) was recorded at Nangarhar in the east of Afghanistan. Such a variation was missing among varieties. The earliest ones took an average of 123 days (PBW 154, Muqawim 09, Ghori 96, Rana 96) to attain 50% flowering across locations whereas Bezostaya and Gul 96 were last to attain this stage after 128 days. When presented graphically (Figure 2), locations clearly differed among themselves more or less maintaining the relative ranking of varieties across locations. Figure 2 also indicates similarities among locations that could come together to form wheat climatic zones. Nangarhar in East is distinct with just 95 average number of days (ADF) to attain 50% of flowering (Table 3) and therefore claims to be a distinct zone (Eastern Zone: EZ) in itself, however it suffers from want of any other location in the zone. Northern region has three locations within the interval of 100 to 120 ADF and therefore along with Balkh (ADF of 124) constitute Northern zone. Southern region has just one location of Kandhar (ADF: 101) and West has one location of Herat (ADF: 111). Owing to similarities in ADF, for the time being a single zone of South West (SW zone) is proposed. Central highlands in Afghanistan are a high altitude region with two regional research stations at Kabul and Bamyan with an average ADF of 153 days. However, Badakhshan in North East is a high altitude region also with an ADF of 161. Though Badakhshan is not geographical contiguous region, however because of similar ADF, a zone comprising of CH and Badakhshan is proposed with the name of Highlands.

Table 2. Average number of days to 50% of flowering of different varieties at several locations over 2010-11 to 2015-16

Locations/Varieties	Balkh	Baghlan	Bamyan	Nangarhar	Kabul	Takhar	Kunduz	Herat	Badakhshan	Kandhar	Average
Bezostaya	130	115	160	97	150	123	114	114	163	103	128
Rana 96	122	107	155	93	144	115	107	111	155	98	123
Mazar 99	125	109	157	94	146	116	110	111	159	100	124
Herat 99	124	109	159	94	146	116	109	110	161	102	124
Solh 02	124	110	159	96	146	114	110	112	164	103	125
Gul 96	127	114	162	97	149	118	115	112	169	104	128
PBW 154	122	108	160	92	147	117	107	110	159	99	123
Ghori 96	122	109	158	95	144	119	108	108	159	98	123
Muqawim 09	122	109	156	95	146	117	109	111	157	99	123
DA 07	126	113	162	96	150	117	111	112	161	103	126
Average	**124**	**110**	**159**	**95**	**147**	**117**	**110**	**111**	**161**	**101**	**124**

Table 3. Regions falling under different categories based on average number of days to 50% flowering (ADF)

ADF Range	E	N	S	W	NE	CH
< 100	NGR (95)	-	-	-	-	
100-120	-	BGL, TKR, KDZ (112)	KND (101)	HRT (111)	-	
120-140		BLK (124)				
> 140					BDK (161)	BMN, KBL (153)
ADF mean for proposed zone	95	118	101	111	161	153

Note. ADF: Average number of days to 50% flowering; E: East; N: North; S: South; W: West; NE: North East; CH: Central Highland.

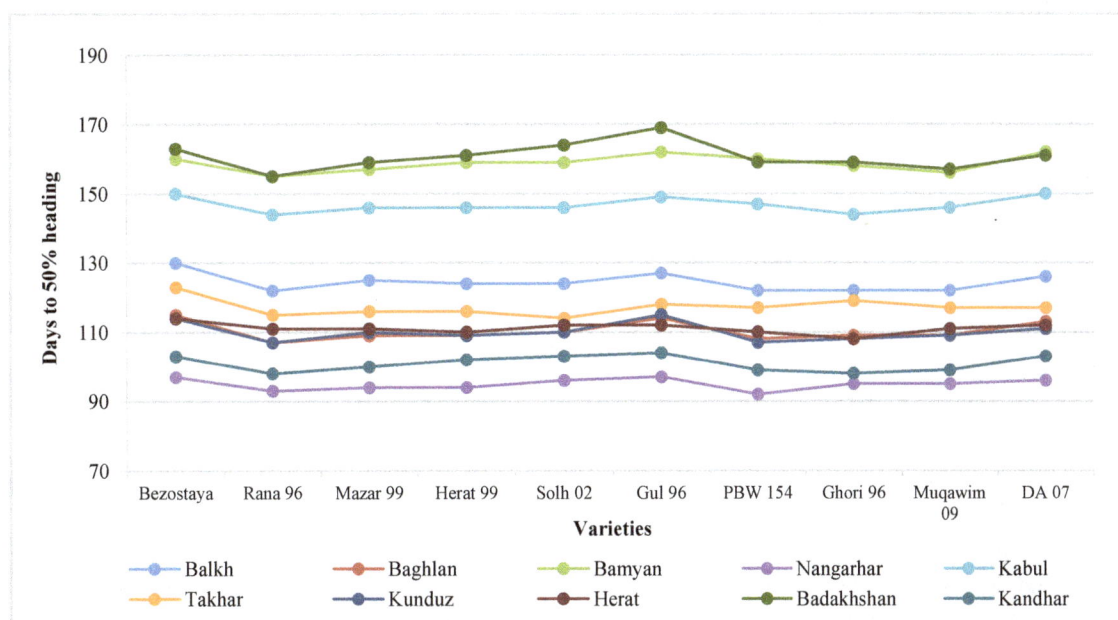

Figure 2. Days to 50% of heading of several wheat varieties across Afghan locations

4. Discussion

Flowering is the development stage signalling transition from vegetative to reproductive stage. Climatic factors like temperature and day length are integrated to regulate flowering time (Nitcher et al., 2014) and are thus critical parameters determining climatic homogeneity in respect of the species in question. Since optimal time of development transition is necessary for reproductive success (Diaz et al., 2012) and to maximize grain yield (Pearce et al., 2016), climatic zonation on the basis of this transition would aid identification of best adapted varieties for a zone. Though, Himani et al. (2013) reported agro climatic zonation of Uttarakhand using remote sensing and GIS, and Jatzold and Kutsch (1982) employed temperature, water supply and length of growing period for zonation of Kenya, Rezaei et al. (2016) distinguished various regions of Iran on the basis of several weather parameters. Additionally, range of weather parameters have been made use of for computing evapotranspiration (Valipour, 2015; Valipour, 2014a, 2014b). However, it will take some time before adequate information on these aspects of Afghanistan is generated to attempt such a zonation. Moreover, we are attempting delineation of Afghan wheat acreage only with respect to a single species. Flowering time (FT) is such an adaptive trait and is the net result of interaction of environmental factors with the species concerned. Therefore, FT can be used to determine climatic homogeneity and thus create zones for the purpose of maximising species performance within the zone. The ADF varied from 95 for Nangarhar in East to a high of 161 in Badakhshan in North East. Interestingly the across location variation is echoed equally by all the genotypes studied. For example, the ADF value ranged between 155 and 169 at Badakhshan, and between 93 and 97 at Nangarhar indicating a much greater role of location in determining ADF. The wheat climatic zones proposed herein *viz.*, East, North, South, West, North-East and Central Highlands are only a tentative working solution based on flowering stage observed among a set of wheat varieties adapted to Afghanistan. However, owing to similar ADF and lack of adequate number of testing sites South-West will be only one zone for the time being. Similarly, North-East is clubbed with Central highlands for same reasons. Wart et al. (2013) suggested an effective balance between zone size and number of zones required to adequately cover the harvested area of major food crop. Afghanistan on the other hand suffers from lack of adequate number of regional research stations to serve as testing sites for new varieties and is therefore constrained to manage only with the available functioning research sites. The regions especially in East, South, West and North East should have more number of research stations to better judge varietal adaptability and to optimize factors of production. The currently proposed wheat climatic zones should be changed to better delineate Afghan wheat area as and when more number of research stations are added in South, West, East and North East.

References

Diaz, A., Zikhali, M., Turner, A. S., Isaac, P., & Laurie, D. A. (2012). Copy number variation affecting the *Photoperiod-b1* and *Vernalization-A1* genes is associated with altered flowering time in wheat (*Triticum aestivum*). *PLoS ONE, 7*, e33234. http://dx.doi.org/10.1371/journal.pone.0033234

Himani, B., Nain, A. S., Shweta, G., & Puranik, H. V. (2013). Agro-climatic zonation of Uttarakhand using remote sensing and GIS. *J. Agrometerology, 15*, 30-35.

Jatzold, R., & Kutsch, H. (1982). Agro-ecological zones of the tropics, with a sample from Kenya. *Tropenlandwirt, 83*, 15-34.

Nitcher, R., Pearce, S., Tranquilli, G., Zhang, X. Q., & Dubcovsky, J. (2014). Effect of the Hope *FT-B1* allele on wheat heading time and yield components. *J. Heredity, 105*, 666-675.

Obaidi, M. Q., Azmatyar, M. H., Elias, M., Habibi, A., Qayum, A., & Sharma, R. (2015). Lalmi 04—A new rainfed wheat variety for Afghanistan. *Triticeae Genomics and Genetics, 6*, 1-3. http://dx.doi.org/10.5376/tgg.2015.06.0001

Obaidi, M. Q., Habibi, A., Ghanizada, A. G., Mashook, M., Azmatyar, M. H., Jan, A., … Sharma, R. (2014). A new high yielding, disease resistant winter wheat variety for Afghanistan. *WIS, 117*, 9-12.

Obaidi, M. Q., Osmanzai, M., Singh, R. P., Pena, J., Braun, H. J., & Sharma, R. (2011). Development of four new Ug99 resistant wheat varieties for Afghanistan. *WIS, 112*, 4-10.

Pearce, S., Kippes, N., Chen, A., Debernardi, J. M., & Dubcovsky, J. (2016). RNA-seq studies using wheat *PHYTOCHROME B* and *PHYTOCHROME C* mutants reveal shared and specific functions in the regulation of flowering and shade-avoidance pathways. *BMC Plant Biology, 16*, 141. http://dx.doi.org/10.1186/s12870-016-0831-3

Persaud, S. (2012). *Long term growth Prospects for Wheat Production in Afghanistan*. Retrieved from http://www.ers.usda.gov/media/193523/whs11101_1_.pdf

Rezaei, M., Valipour, M., & Valipour, M. (2016). Modelling evapotranspiration to increase the accuracy of the estimates based on the climatic parameters. *Water Conserv. Sci. Eng., 1*, 197. http://dx.doi.org/10.1007/s41101-016-0013-z

Saidajan, A. (2012). Effects of war on biodiversity and sustainable agricultural development in Afghanistan. *J. of Developments in Sustainable Agriculture, 7*, 9-13. http://dx.doi.org/10.11178/jdsa.7.9

Valipour, M. (2014a). Analysis of potential evapotranspiration using limited weather data. *Applied Water Sci.* (pp. 1-11). http://dx.doi.org/10.1007/s13201-014-0234-2

Valipour, M. (2014b). Application of new mass transfer formulae for computation of evapotranspiration. *J. Applied Water Engineering and Research, 2*, 33-46. http://dx.doi.org/10.1080123249676.2014.923790

Valipour, M. (2015). Caliberation of mass transfer-based models to predict reference crop evapotranspiration. *Applied Water Sci.* (pp. 1-11). http://dx.doi.org/10.1007/s13201-015-0274-2

Wart, J. van, Bussel, L. G. J. van, Wolf, J., Licker, R., et al. (2013). Use of agro-climatic zones to upscale crop yield potential. *Field Crops Res., 143*, 44-55. http://dx.doi.org/10.1016/j.fcr.2012.11.023

PERMISSIONS

List of Contributors

Zhi-Yuan Meng, Yue-Yi Song, Xiao-Jun Chen, Ya-Jun Ren, Li Ren, Hua-Chen Gen
Jia-Xin Zhu, Quan Yuan and Teng-Fei Li
School of Horticulture and Plant Protection, Yangzhou University, Yangzhou, Jiangsu, P.R. China
Joint International Research Laboratory of Agriculture & Agri-Product Safety, Yangzhou University, Yangzhou, Jiangsu, P.R. China

Chun-Liang Lu
Joint International Research Laboratory of Agriculture & Agri-Product Safety, Yangzhou University, Yangzhou, Jiangsu, P.R. China

Zhi-Ying Xu
Yangzhou Polytechnic College, Yangzhou, Jiangsu, P.R. China

Matthew Caldwell and Manjula Nathan
College of Agriculture, Food and Natural Resources, Division of Plant Sciences, University of Missouri, Columbia, Missouri, USA

Kelly A. Nelson
Greenley Research Center, Division of Plant Sciences, University of Missouri, Novelty, Missouri, USA

Joe L. Parcell
Department of Agricultural and Applied Economics, University of Missouri, Columbia, MO, USA

Glynn T. Tonsor
Department of Agricultural Economics, Kansas State University, Manhattan, KS, USA

Jason V. Franken
School of Agriculture, Western Illinois University, MaComb, IL, USA

Xiaoquan Hua
School of Economics and Management, Huainan Normal University, Huainan, Anhui, China

Moo Jung Kim
Division of Plant and Soil Sciences, West Virginia University, Morgantown, WV, USA
Division of Animal and Nutritional Sciences, West Virginia University, Morgantown, WV, USA

Youyoun Moon, Suejin Park and Nicole L. Waterland
Division of Plant and Soil Sciences, West Virginia University, Morgantown, WV, USA

Dean A. Kopsell
Department of Plant Sciences, The University of Tennessee, Knoxville, TN, USA

Janet C. Tou
Division of Animal and Nutritional Sciences, West Virginia University, Morgantown, WV, USA

T. Thomidis
Alexander Technological Educational Institute of Thessaloniki, Sindos, Macedonia, Greece

S. Pantazis
ANADIAG Hellas, Veria, Macedonia, Greece

K. Konstantinoudis
Technological Educational Institute of Central Macedonia, Serres, Greece

Itzel Galaviz-Villa, Cinthya Sosa-Villalobos, Alicia García-Sánchez, Ma. Refugio Castañeda-Chavez, Fabiola Lango-Reynoso and Isabel Amaro-Espejo
Division of Graduate and Research Studies, Technological Institute of Boca del Río, Veracruz, México

Regina Hlatywayo, Blessing Mhlanga and Upenyu Mazarura
Department of Crop Science, Faculty of Agriculture, University of Zimbabwe, Mount Pleasant, Harare, Zimbabwe

Walter Mupangwa and Christian Thierfelder
International Maize & Wheat Improvement Centre (CIMMYT), Mount Pleasant, Harare, Zimbabwe

N. Balgah Sounders, Tata Emmanuel Sunjo and Mojoko Fiona Mbella
Department of Geography, University of Buea, Cameroon

Neha Rana and Amit J. Jhala
Department of Agronomy and Horticulture, University of Nebraska-Lincoln, Lincoln, NE, USA

Azize Homer and Robin W. Groose
Department of Plant Sciences, University of Wyoming, Laramie, WY, USA

Thomas R. Butts and Bruno C. Vieira
Department of Agronomy & Horticulture, University of Nebraska-Lincoln, North Platte, NE, USA

Joshua J. Miller
Department of Plant Pathology, University of Nebraska-Lincoln, Lincoln, NE, USA

J. Derek Pruitt, Salvador Ramirez II and John L. Lindquist
Department of Agronomy & Horticulture, University of Nebraska-Lincoln, Lincoln, NE, USA

Maxwel C. Oliveira
Department of Agronomy & Horticulture, University of Nebraska-Lincoln, Concord, NE, USA

Amara Evangeline Unigwe and Michael Pillay
Department of Biotechnology, Vaal University of Technology, Vanderbijlpark, South Africa

Abe Shegro Gerrano
Agricultural Research Council-Vegetable and Ornamental Plant Institute (ARC-VOPI), Pretoria, South Africa

Patrick Adebola
Africa Rice Center (AfricaRice), C/o Central Agricultural Research Institute (CARI) Suakoko, Bong County, Monrovia, Liberia

Harbans L. Bhardwaj and Anwar A. Hamama
Agricultural Research Station, Virginia State University, Petersburg, USA

Emerson Borghi
Embrapa Milho e Sorgo, Sete Lagoas, Brazil

Junior C. Avanzi
University of São Paulo, Pirassununga, Brazil

Leandro Bortolon and Elisandra S. O. Bortolon
Embrapa Pesca e Aquicultura, Palmas, Brazil

Ariovaldo Luchiari Junior
Embrapa Informática Agropecuária, Campinas, Brazil

Galdino Xavier de Paula Filho
Education Faculty, Federal University of Amapá, Brazil

Miquéias Freitas Calvi
Forestry Faculty, Federal University of Pará, Brazil

Roberta Rowsy Amorim de Castro
Exact Sciences and Technology Faculty, Federal University of Pará, Brazil

M. Q. Obaidi and M. H. Azmatyar
Agricultural Research Institute of Afghanistan, Kabul, Afghanistan

Elias Mohmand and Rajiv Sharma
International Maize & Wheat Improvement Centre (CIMMYT), Kabul, Afghanistan

Kingsley Osei, Haruna Braimah, Umar Sanda Issa and Yaw Danso
Crops Research Institute, Kumasi, Ghana

Qi Su, Hong Tong, Jiaxu Cheng, Guohui Zhang, Caihua Shi, Chuanren Li and Wenkai Wang
Institute of Insect Sciences, College of Agriculture, Yangtze University, Jingzhou, Hubei, China

Alusaine E. Samura and Osman Nabay
1Sierra Leone Agricultural Research Institute (SLARI), Njala Agricultural Research Centre (NARC), Freetown, Sierra Leone

Kepifri A. Lakoh, Sahr N. Fomba and James P. C. Koroma
Crop Protection Department, School, of Agriculture, Njala University, Freetown, Sierra Leone

Shuai Liu, Dinghua Li, Xiuming Cui, Limei Chen and Hongjuan Nian
Faculty of Life Science and Technology, Kunming University of Science and Technology, Kunming Key Laboratory of Sustainable Development and Utilization of Famous-Region Drug, Key Laboratory of Panax notoginseng Resources Sustainable Development and Utilization of State Administration of Traditional Chinese Medicine, Kunming, China

Alaa Y. Ghidan, Tawfiq M. Al-Antary and Nidá M. Salem
School of Agriculture, the University of Jordan, Amman, Jordan

Akl M. Awwad
Department of Materials Science, Royal Scientific Society, Amman, Jordan

Index

www.ingramcontent.com/pod-product-compliance
Lightning Source LLC
Chambersburg PA
CBHW080537200326
41458CB00012B/4463